低渗透—致密气藏评价方法与应用实例

贾爱林　程立华　冀　光
王国亭　郭　智　孟德伟　编著

石油工业出版社

内 容 提 要

在中国四川、鄂尔多斯、柴达木、塔里木、松辽等盆地先后发现了大量致密气田，资源评估量巨大，致密气开发成为目前的热点，但致密气的有效开发存在诸多问题，迫切需要解决。本书汇总了国内一批天然气开发领域专家的最新研究成果与心得，可以为气藏开发提供理论参考和方法借鉴。

本书可供从事天然气开发的科研人员使用，也可作为高等院校相关专业师生的参考用书。

图书在版编目（CIP）数据

低渗透—致密气藏评价方法与应用实例／贾爱林等编著. — 北京：石油工业出版社，2022.1
ISBN 978-7-5183-5298-2

Ⅰ. ①低… Ⅱ. ①贾… Ⅲ. ①致密砂岩-砂岩油气藏-低渗透油气藏-油藏评价 Ⅳ. ①P618.13

中国版本图书馆 CIP 数据核字（2022）第 062486 号

出版发行：石油工业出版社
（北京安定门外安华里 2 区 1 号　100011）
网　　址：www.petropub.com
编辑部：（010）64523708
图书营销中心：（010）64523633
经　　销：全国新华书店
印　　刷：北京中石油彩色印刷有限责任公司

2022 年 1 月第 1 版　2022 年 1 月第 1 次印刷
787×1092 毫米　开本：1/16　印张：20.25
字数：500 千字

定价：160.00 元
（如出现印装质量问题，我社图书营销中心负责调换）

版权所有，翻印必究

目 录

中国天然气开发技术进展及展望 …………………………………………… 贾爱林(1)

全球不同类型大型气藏的开发特征及经验 ………………… 贾爱林　闫海军　郭建林　等(13)

大型低渗透—致密气田井网加密提高采收率对策——以鄂尔多斯盆地苏里格气田为例
　　………………………………………………………… 贾爱林　王国亭　孟德伟　等(31)

大型致密砂岩气田采收率计算方法 ………………………… 郭建林　郭　智　冀　光　等(47)

鄂尔多斯盆地低渗透—致密气藏储量分类及开发对策 ……… 程立华　郭　智　孟德伟　等(57)

Modeling and analyzing gas supply characteristics and development mode in sweet spots of
　　Sulige tight gas reservoir, Ordos Basin, China ………………………………… Ji Guang(68)

致密砂岩气藏储渗单元研究方法与应用——以鄂尔多斯盆地二叠系下石盒子组为例
　　………………………………………………………… 郭建林　贾成业　闫海军　等(82)

致密砂岩气田储量分类及井网加密调整方法——以苏里格气田为例
　　………………………………………………………… 郭　智　贾爱林　冀　光　等(96)

致密砂岩气藏多段压裂水平井优化部署 …………………… 位云生　贾爱林　郭　智　等(110)

Technical strategies for effective development and gas recovery enhancement of a large
　　tight gas field: A case study of Sulige gas field, Ordos Basin, NW China
　　……………………………………………… Ji Guang　Jia Ailin　Meng Dewei et al.(118)

低渗透—致密气藏开发动态物理模拟实验相似准则 ……… 焦春艳　刘华勋　刘鹏飞　等(140)

神木气田低渗致密储层特征与水平井开发评价 …………… 王国亭　孙建伟　黄锦袖　等(148)

New method in predicting productivity of multi-stage fractured horizontal well in tight gas
　　reservoirs ………………………………… Wei Yunsheng　Jia Ailin　He Dongbo et al.(158)

鄂尔多斯盆地东部奥陶系古岩溶型碳酸盐岩致密储层特征、形成与天然气富集
　　………………………………………………………… 王国亭　贾爱林　孟德伟　等(173)

致密砂岩气藏有效砂体规模及气井开发指标评价——以鄂尔多斯盆地神木气田太原组
　　气藏为例 ……………………………………………… 孟德伟　贾爱林　郭　智　等(188)

3D geological modeling for tight sand gas reservoir of braided river facies
　　…………………………………………………… Guo Zhi　Sun Longde　Jia Ailin et al.(200)

辫状河储层构型规模表征及心滩位置确定新方法——以苏6区块密井网区盒八段为例
　　………………………………………………………… 董　硕　郭建林　李易隆　等(216)

低渗透—致密砂岩气藏开发中—后期精细调整技术 ……… 付宁海　唐海发　刘群明(231)

鄂尔多斯盆地中—东部下石盒子组八段辫状河储层构型
.... 李易隆 贾爱林 冀 光 等(242)

鄂尔多斯盆地东部盒八段致密砂岩储层特征——以子洲气田清涧地区为例
.... 郭 智 冀 光 王国亭 等(259)

Water and gas distribution and its controlling factors of large scale tight sand gas fields:
  A case study of western Sulige gas field, Ordos Basin, NW China
.... Meng Dewei Jia Ailin Ji Guang et al.(272)

砂质辫状河隔夹层成因及分布控制因素分析——以苏里格气田盒八段为例
.... 罗 超 郭建林 李易隆 等(285)

苏里格气田气井压力前缘到达渗流边界时间的判断 ...... 杨炳秀 何东博 王丽娟 等(300)

低渗透致密砂岩储层充注模拟实验及含气性变化规律——以鄂尔多斯盆地苏里格气藏
  为例 .... 徐 轩 胡 勇 邵龙义 等(306)

中国天然气开发技术进展及展望

贾爱林

(中国石油勘探开发研究院,北京 100083)

摘要:在国际油价低位徘徊、国家大力发展绿色能源的背景下,天然气已成为中国油气工业的主营核心业务,其产量及消费量快速攀升,作用更加凸显。为了继续推动中国天然气业务的加速发展,在分析近年来天然气产业发展历程、总结天然气开发技术新进展的基础上,明确了国内天然气开发所面临的挑战,并从产量、需求量、进口量及未来天然气地位等四个方面对中国天然气工业的发展前景进行了展望。研究结果表明:(1)"十二五"以来,国内天然气消费量快速增长、供应多元化、储产量稳定增长、开发效益显著;(2)天然气开发技术在深层天然气开发、大型气田开发调整、致密气提高采收率、页岩气及煤层气开发、工程技术及开发决策体系等六个方面都取得了突破,创新能力显著提升;(3)随着开发程度的深入,受政策、环境的影响及地质条件制约,天然气持续规模效益开发将面临优质储量比例降低、气田开发成本升高、非常规气藏效益开发难度加大、上游效益进一步压缩、主力气田稳产能力减弱和市场竞争愈发激烈等诸多挑战。结论认为,未来国内将进入非常规气与常规气并重的发展阶段,天然气需求旺盛且消费结构呈现多元化趋势,天然气进口量逐年攀升且对外依存度不断加大,天然气将成为国内能源结构调整的主要增长点。

关键词:中国;天然气;开发技术;产量;需求量;进口量;深层气;致密气;页岩气;煤层气;开发决策

1 天然气发展概况

近年来,国内天然气开发取得一系列重大突破[1],已成为石油工业的主营核心业务,产量及消费量快速攀升。"十三五"期间,在国家大力倡导低碳绿色能源、积极进行能源转型的背景下,秉承有质量、有效益、可持续发展的开发理念,积极发展天然气开发业务,对保证中国能源供给、实现能源转型具有重要意义。

1.1 消费量快速增长,竞争力进一步增强,助力国家能源转型

国内天然气消费量快速提高,2010—2016 年由 $1112×10^8 m^3/a$ 上升至 $2100×10^8 m^3/a$,在能源消费中的比例由 4% 提高至 6.2%(图 1)。但与主要能源消费大国相比,国内清洁能源消费比例明显偏低,天然气将在调整和优化能源结构中发挥更大作用。按照国家能源发展战略,2020 年国内天然气占能源消费比重由 2016 年的 6.2% 提高到 10%,预计年消费量将达到 $3000×10^8 m^3$。天然气占能源消费比重的提高,将有助于推动国内能源结构转型,同时也给天然气自身发展带来新的机遇。天然气具有热值高、价格低的特征,国家近期连续推出多项天然气市场改革政策,也将进一步降低终端用户使用成本,从而使天然气对其他能源的竞争力更加凸显。

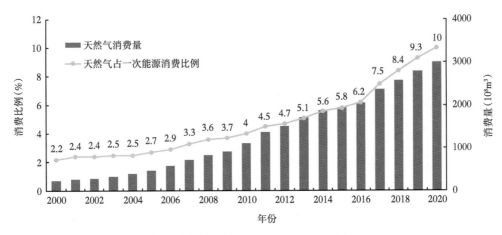

图 1 中国天然气年消费量及能源结构占比

1.2 供应呈现多元化,天然气市场竞争更加激烈

国内天然气供应主要由自产气、进口管道气和LNG组成。目前,大部分天然气产量来自国内自产气,中国石油、中国石化、中国海油三大石油公司均提出加快发展天然气的战略。中国石油处于天然气生产的主体地位,约占国内产量份额的70%,受其他公司天然气业务快速发展的影响,主体地位格局已受到明显冲击。对进口管道气而言,虽然长期贸易合同保障长期稳定供应,但价格和长距离输送降低了其竞争能力。LNG的快速发展使天然气洲际贸易规模化,并进一步缩小了北美、欧洲及亚太国际三大消费市场的价格差距;国内LNG则形成了央企、地方企业、民营企业多元化竞争格局。从价格和灵活性上来看,竞争优势排序为LNG、自产气、进口管道气。

1.3 常规天然气稳定发展,非常规天然气快速上产

天然气产量由常规天然气、非常规天然气(包括致密气、煤层气及页岩气等)组成。总体表现为常规气稳定发展,非常规天然气快速上产的特征。常规天然气在一定的时间内将保持稳定发展,并在较长时间内仍将保持国内主体地位。页岩气产量快速上升,由2015年的44.6×$10^8 m^3$ 增长到2016年的78.9×$10^8 m^3$;煤层气产量稳定增长,由2015年的16×$10^8 m^3$ 增长到2016年的17×$10^8 m^3$。未来天然气产量将保持持续增长的趋势,但产量结构将发生重大变化,非常规天然气占比将逐渐增加(图2)。

1.4 储量与产量稳定增长,供气能力持续加强

中国天然气储量与产量稳定增长,为保证供气能力奠定了坚实基础。从2000年以来,年均保持新增天然气探明储量6600×$10^8 m^3$ 以上,近五年年均新增8800×$10^8 m^3$ 以上,2016年新增探明储量7656×$10^8 m^3$,累计探明储量14.9×$10^{12} m^3$,夯实了发展的资源基础。致密气、深层气、页岩气是近年储量增长的主体。近年来,天然气产量持续增长,2016年生产天然气1371×$10^8 m^3$。国内天然气产量将由快速增长进入稳定增长阶段。

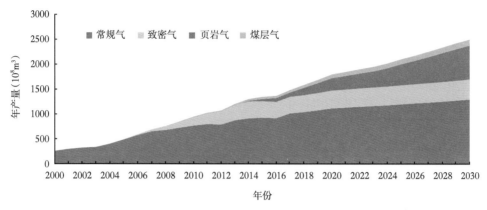

图 2 中国常规天然气与非常规天然气产量规划

1.5 骨干气区开发指标良好,天然气成为盈利主体

中国天然气工业历经近 70 年三个阶段的跨越式发展,已建成鄂尔多斯、塔里木、四川、海域四大天然气生产基地,探明地质储量占全国总量的 89%,产量占全国总量的 87%。天然气开发指标总体良好,气层气动用地质储量 $8.4×10^{12}\ m^3$,已累计产气 $1.45×10^{12}\ m^3$,动用程度 56%,剩余可采储量 $5.4×10^{12}\ m^3$,储采比 39,具备长期稳定发展的基础。

受低油价制约,油气公司上游业务利润大幅度减小,而天然气业务盈利较好,保持了较高的净现金流与净利润,很大程度上减缓了油气公司上游的经营压力。随着天然气产量的不断攀升,国内石油产量与天然气的油当量将会基本持平,这将显著增强油气公司生产的抗风险能力。天然气作为上游核心主营业务的作用更加凸显。

2 天然气开发技术新进展

技术的不断进步,推动了天然气开发从单一气藏到复杂类型气藏,从常规气藏到非常规气藏的转变。气藏描述、产能评价、钻完井、储层改造和采气工艺等技术的综合发展,支撑了苏里格、克拉 2、靖边等不同类型气藏的成功开发[2-5]。"十三五"期间,开发形势发生了较大变化,主要表现为:新增探明地质储量结构发生变化,深层、低渗透—致密、非常规成为主体,开发难度加大;主力气田相继进入稳产期,稳产与提高采收率成为技术攻关的主要方向;非常规天然气开发突破瓶颈技术,开发规模快速增长;提高单井产量和开发效益对工程技术提出了更高的要求;对气田开发规划的指导性和开发指标的科学性提出了更高的要求。在新形势下,天然气开发技术取得以下主要进展。

2.1 深层气藏开发技术显著提升,产建新领域规模发展

中西部盆地深层/超深层气藏开辟了天然气增储上产新领域,以四川盆地龙王庙组、灯影组深层碳酸盐岩气藏、塔里木盆地大北—克深多断块深层致密砂岩气藏为代表。主要形成两项技术系列。

2.1.1 深层碳酸盐岩气藏开发技术

针对四川盆地龙王庙组、灯影组气藏岩溶储层非均质性强、气水分布复杂的特点[6],形成

四项主体开发技术:白云岩岩溶储层描述技术,创新提出颗粒滩、丘滩体岩溶发育模式,建立不同类型储层地震识别方法,形成高产井布井技术;裂缝—孔洞型有水气藏开发优化技术,强化不同类型水侵特征研究,开展水侵监测与调控,降低水侵风险;大斜度井/水平井丛式井组开发技术,增大井筒与储层接触面积;大型气田模块化、橇装化、智能化建设模式,采用全新设计理念,形成气田建设速度、智能化水平、安全环保的新典范。2016年龙王庙组气藏 $110×10^8 m^3$ 年产能全面建成,灯影组气藏 $18×10^8 m^3$ 年产能建设稳步推进,奠定了西南油气田 $300×10^8 m^3$ 战略气区的上产基础。

2.1.2 深层致密砂岩气藏群开发技术

针对塔里木盆地大北—克深多断块气藏储层描述和工程作业难度大的特点,发展了四项主体开发技术:以构造建模为核心的气藏描述技术,通过宽方位三维地震勘探落实构造形态,建立不同构造部位裂缝发育模式,优化井位;以垂直钻井系统国产化为核心的快速钻井技术,自主研发垂直钻井系统、油基钻井液、抗冲击和抗研磨性 PDC 钻头等,使钻井周期和成本大幅下降;以缝网压裂为核心的储层改造技术,重点针对Ⅱ类、Ⅲ类储层,采用缝网酸压和加砂压裂进行增产,单井日产由不足 $30×10^4 m^3$ 提高到 $50×10^4 m^3$ 以上;以超高压压力测试为核心的开发优化技术,突破超高压气井投捞式压力测试技术,滚动评价断块气藏连通性,优化开发井数,实现稀井高产。2016年大北—克深气田群年产量突破 $70×10^8 m^3$,是塔里木气区在克拉2气田、迪那2气田开发调整后,保持气区持续上产的主力气田。

2.2 大型气藏开发调整技术不断完善,进一步提高开发效果

"十三五"期间天然气开发进入上产与稳产并重发展阶段,很多大型气田进入开发调整期,如靖边、克拉2和涩北等气田,针对这些气田形成三种主体稳产模式。

2.2.1 扩边及新层系动用,滚动接替稳产模式

多层系含气、不发育边底水、分布范围广的大型岩性气藏,滚动开发潜力大,代表性气田是位于鄂尔多斯盆地的靖边气田。靖边气田具有多套气层发育的特征,主力产层为奥陶系马五段,保持 $55×10^8 m^3/a$ 规模稳产了13年,是长庆油田稳产的主力气田之一。重点开发技术包括薄层水平井开发技术和富集区优选评价技术。通过毛细沟槽与小幅度构造刻画,实现 2m 薄层水平井开发,推动了外围扩边区每年 $5×10^8$ ~ $6×10^8 m^3$ 弥补递减产能建设,同时深化上古生界气层富集区优选,落实储量 $2441×10^8 m^3$,是"十三五"稳产的主要接替储量。这些技术实现了气田扩边及新层系的动用,实现了滚动接替稳产。

2.2.2 优化指标,调整规模,均衡开采模式

针对边底水活跃的大型整装块状气藏,核心是优化气井指标和生产规模,防止边底水锥进,达到一次井网采收率最大化。若采气速度过高,会造成个别气井水淹、气藏非均匀水侵,给气田稳产带来困难。克拉2气田采取稀井高产开发模式,调峰能力强,高峰年产量达到 $110×10^8 m^3$ 以上,发挥了西气东输主力气田调峰保供作用。主要通过水侵动态分析技术可建立高压气井水侵判别模式,形成千万节点大型数模水侵动态预警机制;通过均衡开发技术可进一步优化采气速度,调整规模。这些技术的应用实现了开发指标优化和气田均衡开采。

2.2.3 治水、控砂、多层系协调动用稳产模式

涩北气田是典型的疏松砂岩气藏,高峰年产量达到 $65×10^8 m^3$,目前稳产规模约 $50×$

$10^8m^3/a$,是青海气区稳产的基石。气藏气层多达上百个,发育多套气水系统,具有气藏出砂出水、储量动用不均、稳产难度大的特点。重点形成了多套井网分层系开采技术及综合治水与防砂技术,划分为五个开发层系,地面井网密度达到5.1口/km^2,减小了多层系干扰,实现了气藏均衡开发,形成以连续油管冲砂为主的工艺技术,优化了压裂充填防砂工艺参数,提高了防砂效果。

2.3 致密气藏提高采收率技术不断升级,有效支撑规模稳产

苏里格气田的成功开发引领了国内致密气规模化发展进程[7-9]。气田目前累计投产9000余口井、产量规模达$220×10^8 \sim 230×10^8m^3$,占全国总产气量的16%。"十三五"期间进入稳产期,针对多井低产、采收率偏低(约30%)的特点,以提高储量动用程度和气田采收率为核心,形成了两项技术系列。

2.3.1 大面积低丰度气藏开发井网优化技术

通过刻画砂体规模尺度、压裂改造范围及气井泄压半径,进行井距优化;在评价砂体几何形态、地应力方位的基础上,明确井网几何形态;论证不同储量丰度区块的经济极限井网密度;形成以直井井网为主,主力层集中型储层采用水平井开发的井型组合。通过密井网区开发先导试验,证实富集区加密的开发效益仍优于非富集区的动用,论证了不同品位储量区的合理动用顺序,明确了加密至4口/km^2后采收率可由600m×800m基础井网的30%提高至45%~50%。

2.3.2 致密气藏提高采收率配套技术

结合地质、气藏工程及改造工艺,形成了致密气藏提高采收率的一系列配套技术:明确了气井工作制度优化可提高采收率1%[10],老井未动用层改造可提高采收率1%~2%,有利目标老井侧钻可提高采收率1%~2%,低产期排水采气可提高采收率2%~3%,即提高采收率综合配套技术可在井网优化的基础上再提升采收率5%~8%。

2.4 页岩气、煤层气开发技术日趋成熟,提产降本效果显著

近年来,3500m以浅海相页岩气开发技术基本成熟[11-15],产量迅速攀升,2016年实现产量$27×10^8m^3$以上;煤层气产量也稳步增长,中—高煤阶开发技术已经成熟,低煤阶开发首次获得突破。两类非常规气藏主体开发技术系列进展分述如下。

2.4.1 页岩气开发技术

2.4.1.1 基于开发尺度的页岩气储层评价技术

形成了主力开发层段小层划分技术,将纵向上研究尺度从几十米精细到几米,优化靶体位置至龙一$_1^1$小层,同时形成了动态储量标定地质储量技术,评价长宁区块动用层段储量丰度约$4.13×10^8m^3/km^2$,五峰组—龙一段储量丰度约$12.3×10^8m^3/km^2$。这些技术为页岩气有效开发提供了地质依据(图3)。

2.4.1.2 3500m以浅钻完井及储层改造技术

四川盆地及周缘3500m以浅页岩气资源量为$2×10^{12}m^3$,经过五年来的攻关和试验,完钻水平井233口,开发技术基本成熟配套,主要包括:以旋转地质导向为核心的优快钻井技术、以低黏滑溜水+低密度支撑剂为核心的体积改造技术及以大井组工厂化作业为核心的工程实施

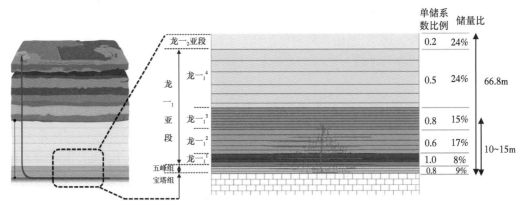

图 3　长宁区块地层结构及小层储量评价技术图

技术,使单井综合成本降低到 5500 万元以内,支撑了页岩气规模有效开发。

2.4.1.3　气井开发指标评价及生产制度优化技术

建立了微裂缝+次裂缝+主裂缝多级次复合裂缝模型,形成概率性产能预测方法,提出了若干关键开发指标,有效指导了页岩气井生产。评价页岩气井单井累计产量平均 $8070×10^4 m^3$,首年日产 $6.5×10^4 \sim 10×10^4 m^3$,初始年递减率 46%~62%,前三年递减率逐步下降到 30% 左右。形成裂缝、储层基质应力敏感定量描述方法,评价放压与控压两种生产方式对单井累计产量的影响。明确了最优生产方式,指出采用控压生产单井累计产量可提高 30% 以上。

2.4.1.4　页岩气开发井距优化技术

提出有效裂缝长度动态预测方法,形成基于产量干扰分析的开发井距优化技术。研究表明,长宁—威远、昭通区块井距可从目前的 400~500m 缩小至 300m,井控储量采出程度可由 25% 提高到 35% 左右。

2.4.2　煤层气开发技术

国内目前已实现中—高阶煤层气开发[16-17],主要开发技术基本成熟,包括以地球物理和储层评价为主的煤储层描述技术、以水平井钻完井和压裂增产改造为主的提高单井产能技术、以排采和防煤粉技术为主的井筒排采技术、以生产剖面测试和动态监测为主的开发调整技术,助推了煤层气产业稳定发展。

同时,低煤阶煤层气开发技术也获得重要进展,二连盆地吉煤 4 井应用填砂分层、低浓度瓜尔胶技术,首次在低煤阶取得重大突破,有助于解放中国石油矿权区内 $6.75×10^{12} m^3$、占总量 51% 的低阶煤层气资源,开辟了煤层气勘探开发新领域。

2.5　工程技术升级发展,有力支撑降本增效

天然气开发的快速推进,很大程度上受益于国内天然气工程技术的进步和发展。核心技术国产化、成本的大幅降低助推天然气的效益开发,主要形成三个方面技术系列。

2.5.1　大井组—多井型—工厂化钻井规模化应用

近年来,天然气开发井型从直井、丛式井,发展到平台水平井,目前天然气钻完井实现了大井组—工厂化的根本性变革,钻井周期大幅缩短,降低了成本、提高了效率。在鄂尔多斯盆地

东部,形成多井型大井组立体开发的典型代表,研发三维绕障、三维水平井轨迹控制、低摩阻钻井液等配套技术,最大单平台混合井组达15口井。至2016年底,多井型大井组累计应用1200个井丛,节约土地2万余亩。

2.5.2 储层改造工艺、工具装备不断取得新突破

以往储层改造工艺以直井多层和水平井多段常规改造为主,体积改造是近年来兴起的新性储层改造技术[18-19]。经过攻关实践,国内自主研发的体积改造技术成熟配套,已实现规模化应用,与工厂化作业模式结合,成为非常规低成本开发的关键技术。其主体技术为大通径桥塞分段压裂技术和低黏滑溜水液体体系,配套技术包括桥塞泵送与分簇射孔、连续混配与连续输砂、压裂液回收利用等。在工具装备方面,国内自行研制的可溶桥塞压裂技术已达到国际领先水平,实现了从技术模仿到技术引领的转变。2016年累计现场试验10井次,压裂141段,成功率100%。

2.5.3 形成适应多气藏类型的采气工艺技术系列

针对不同类型气藏的开发特征形成了相应的排水采气技术系列[20-22]。如低压低丰度低渗透气藏形成泡沫排水、速度管柱、柱塞气举等系列技术,疏松砂岩气藏形成泡排、井间互联气举技术,火山岩气藏形成泡沫排水、速度管柱技术,四川盆地石炭系老气田采用电潜泵排水采气技术。撬装式、移动式排水采气设备的试验与应用,增加了排水采气的灵活性,节约了成本。

2.6 建立开发决策体系,有效支撑天然气科学发展

随着气田开发程度的不断深入,逐步建立了较科学、完善的天然气开发决策体系:标定了主要类型气藏在不同开发阶段的关键指标体系,包括单位压降采气量、产量、采出程度等,为气藏开发设计提供了依据(表1);形成大型气田群长期稳产技术对策,以气田群整体开发效益最大化为原则,优化设计,实现了主力气田与卫星气田间的协同开发;建立了一套能够快速评价、突出效益、风险可控的开发战略规划决策支撑系统,包括多情景供气规模分析模型及全生命周期气藏技术经济评价方法,支撑了公司战略规划决策的有效制定。

表1 三类气藏不同开发阶段关键开发指标变化规律量化标定表

气藏类型	开发阶段	产量	压降	可采储量采出程度(%)	单位压降产量
高压气藏	评价阶段	小产气量	无		稳定
	建产阶段	快速上升	基本无	<10	稳定
	稳产阶段	保持稳定	70%~80%	70~80	平稳下降
	递减阶段	20%	80%~90%	80~90	快速下降
	低产阶段	20%以下	>95%	>95	下降趋稳
低渗透—致密气藏	评价阶段	小产气量	无	<5	上升
	建产阶段	阶段上升	基本无	<10	上升
	稳产阶段	保持稳定	40%~60%	50~60	上升趋稳
	递减阶段	20%	60%~75%	60~70	稳定
	低产阶段	20%以下	>80%	>70	稳定

续表

气藏类型	开发阶段	产量	压降	可采储量采出程度(%)	单位压降产量
带边底水的孔隙—裂缝型气藏	开发评价	小产气量			稳定
	建产阶段	快速上升	<10%	5~10	稳定
	稳产阶段	保持稳定	60%	55~65	稳定
	递减阶段	20%	80%	75~80	上升趋稳
	低产阶段	20%以下	85%~90%	>80	趋稳

3 天然气开发面临的挑战

随着开发的深入、油价的下跌和政策、环境的影响,受地质条件制约,国内气层优质储量比例降低,气田开发成本升高,非常规气藏效益开发难度加大,上游效益进一步压缩,主力气田稳产能力减弱,市场竞争越发激烈。结合技术、效益及管理等几方面,总结了中国天然气开发面临的六项挑战。

3.1 储量产量比例发生变化,保持增速面临挑战

新增探明储量与当年产量比例由2005年的8降低到目前的5左右,持续上产的资源保障能力有所下降。新增储量中,低渗透—致密等非常规储量占70%以上,优质储量比例偏低,平均标定采收率呈下降趋势。储量替代率是年新增探明可采储量与当年开采消耗储量的比值,反映储量的接替能力,已由4.5以上逐年降低到2.5左右,制约了产量的高速增长,未来国内将进入产量稳定增长期。

3.2 主力气田稳产能力较弱,稳产形势面临挑战

根据国内天然气资源特点,除少数气田具有自然稳产能力外(如榆林南气田、克拉2气田),大部分气田稳产都需要新建产能来弥补,主要包括苏里格、靖边、涩北等气田。苏里格气田已进入稳产期,气田综合递减率平均为20%左右,保持$250\times10^8 m^3/a$稳产,每年需弥补递减产能$50\times10^8 m^3$,效益稳产面临挑战。涩北气田为典型的疏松砂岩气田,开发的主要挑战为出水和压力下降,目前近40%的层组产量递减率大于10%,整体综合递减率近五年为8%左右,长期稳产面临挑战。

3.3 气田开发成本逐年上升,开发效益面临挑战

随着资源品位的降低和开发难度的增大,公司天然气完全成本由2011年的631元/$10^3 m^3$上升到2015年的883元/$10^3 m^3$,成本上涨252元/$10^3 m^3$,而气价同期上涨300元/$10^3 m^3$,即气价上升所带来的利润增益几乎被成本的上升所抵消。从全球资源型企业发展历程来看,依靠价格上升拉动的利润增长不可持续。随着技术发展与管理提升,综合成本降低是必然趋势,例如美国致密油综合成本由2013年的70美元/bbl降低到目前的30美元/bbl左右。

3.4 非常规气开发取得突破,技术效益面临挑战

国内非常规气开发取得了长足的进步,但与北美地区对比来看,在钻完井、压裂改造技术

及单井开发效果等方面依然存在着较大差距(表2),在技术瓶颈突破以前,进一步降低钻井与压裂周期面临挑战。目前公司页岩气开发综合投资5500万元左右,考虑实际补贴,仍处于边际效益,随着补贴的降低,效益开发面临挑战。

表2 中美页岩气开发参数对比

区块	井身全长(m)	水平段长(m)	压裂段数	压裂进度(段/d)	钻井周期(d)	压裂周期(d)	单井EUR(10^8m^3)
美国最长水平井PurpleHayes 1H	8244	5652	124	5.3	17.6	23.5	6.8~9.0
中国石油区块	4747	1520	18	2~3	81.0	10.0	0.5~1.5

3.5 天然气效益链分配不均,上游效益面临挑战

天然气效益链分配不均,上游利润偏低。以长庆气区为例,将天然气从平均埋深3500m的复杂致密储层中采出,平均生产利润为0.4元/m^3,而陕京线下游管道运输、北京燃气终端销售平均利润分别为0.35元/m^3和0.6元/m^3。下游工业用户用气价格中,省网与城市管网配气费占40%~50%,终端销售企业利润高于天然气生产与输送部分。公司近期仍以进一步控制成本、增加效益为主;中长期要强化产、运、销国家政策中利润切割的争取与引导,力争上游利润在总产业链中的比例。

3.6 天然气供给气源多元化,市场竞争面临挑战

天然气供给气源呈现多元化。在国际气源供给方面,进口管道气、LNG均对自产气形成了竞争。多份长贸合同的签订保障了进口管道气的长期稳定供应,进口LNG的价格优势对自产气的供应形成了挑战,波罗的海散运指数(航运价格指数)由高峰期10000以上降到1000左右,大幅降低了LNG的运输成本。在国内气源供给方面,形成了央企、地方企业、民营企业多元化竞争的天然气产业格局,地方政府积极参与,建设的LNG接收站主导了购销一体的布局结构,民企经营具有更强的灵活性。全国各大气区中,四川气区是受LNG冲击最小的高端市场,但国内多元的生产主体和充足的产量形成竞争(规划总产量620×10^8m^3/a,市场规模500×10^8m^3/a左右)。

4 天然气发展展望

立足天然气开发历程和国内外开发形势,从产量、需求、进口及未来天然气地位等四个方面对国内天然气工业发展进行展望。

4.1 天然气产量迅速攀升,常规气与非常规气并举

2016年国内天然气总产量1371×10^8m^3,2020年预计总产量1800×10^8m^3,2030年预计总产量2520×10^8m^3,将进入常规与非常规并重的发展阶段,表现为常规气稳定发展、致密气长期稳产、页岩气快速上产、煤层气稳步推进的发展态势。常规气2016年产量918×10^8m^3,2020年预计产量1115×10^8m^3,2030年预计产量1300×10^8m^3。非常规气2016年产量453×10^8m^3,2020年预计产量685×10^8m^3,2030年预计产量1220×10^8m^3,其中页岩气是产量增长主体。

4.1.1 常规气稳定发展

常规气2017—2020年发展要充分结合新区突破、在建气田上产与已开发气田稳产,其中塔里木气区克深—大北气田群、库车山前勘探新区预计2020年新增产量$50×10^8m^3$,四川气区川中震旦系、川东北高含硫、川西海相预计2020年新增产量$120×10^8m^3$,深海海域预计2020年新增产量$30×10^8m^3$。2020—2030年,目前已开发的常规气田大部分已进入递减期,新发现气田品质差,新建产能主要弥补递减,2030年保持产量缓慢增长至$1300×10^8m^3$。

4.1.2 致密气长期稳产

依靠鄂尔多斯盆地苏里格气田外围、神木气田及盆地东部的新区上产,预计致密气产量由2016年$330×10^8m^3$增加到2020年$360×10^8m^3$,2020—2030年依靠勘探新区缓慢上产至$400×10^8m^3$。苏里格外围主要包括苏里格东部及南部滚动扩边,预计2020年新增产量$8×10^8m^3$。神木气田及盆地东部一期、二期方案设计规模$18×10^8m^3/a$,预计2020年产量$35×10^8m^3$,新增产量$24×10^8m^3$。

4.1.3 页岩气快速上产

依靠深层突破,国内页岩气将迎来跨越式发展。四川盆地及周缘3500m以浅的海相页岩气资源量为$2×10^{12}m^3$,可工作面积$3500km^2$,2020年可上产$220×10^8$~$260×10^8m^3$;3500~4500m海相页岩气资源量$10×10^{12}m^3$,可工作面积达$2×10^4km^2$,技术和效益突破后,预计2030年产量可上升到$600×10^8$~$800×10^8m^3$。

4.1.4 煤层气稳步推进

煤层气以中—高煤阶为主,预计产量可由2016年$44×10^8m^3$增加至2020年$75×10^8m^3$,其中中国石油沁南、鄂东、蜀南筠连三个区域2020年可上产$40×10^8m^3$,其他公司区块可由2016年$28×10^8m^3$上产至2020年$35×10^8m^3$。二连盆地、鸡西、白家海等低煤阶区块有望获得规模突破,助推2030年煤层气产量上产$120×10^8m^3$。

4.2 天然气需求旺盛,消费结构呈现多元化

壳牌公司预测,至2030年,全球天然气年需求量将由2010年的$3.1×10^{12}m^3$增长到$5×10^{12}m^3$,跨区域贸易量将增长到约$1.3×10^{12}m^3$,天然气需求增长最快的地区是亚洲。在国内经济增速换挡、资源环境约束趋紧的新常态下,能源转型与消费革命进一步激发天然气需求的上涨。预计2020年国内天然气需求量为$3000×10^8m^3$,2030年需求量为$5220×10^8m^3$。与发达国家相比,国内工业、发电及居民用气比例处于较低水平,未来天然气消费结构将呈现多元化发展:在节能减排政策的促进下,发电和工业燃料气代煤将加速,有望成为消费主体;城镇化持续推进,城市燃气消费量将随之稳定增长;在城镇化和价格优势两个因素的驱动下,天然气交通仍有发展潜力。预计到2020年,国内天然气消费结构中,发电、工业燃料、城市燃气及交通运输分别占24.5%、32%、26.4%及16.3%,到2030年分别占30%、25.6%、20.2%及15%。

4.3 天然气进口量上升,对外依存度加大

随着国民经济的发展,国内天然气的产量与需求的缺口越来越大。在全球能源市场供需宽松的背景下,进口管道气、进口LNG迎来新的机遇。长期贸易合同保障了进口管道气的长

期稳定供应,进口 LNG 的快速发展使天然气洲际贸易规模化,并进一步缩小了国际三大消费市场的价格差距。进口气包括中亚的管道气及广东大鹏和福建进口的 LNG。国内天然气进口量预计 2020 年达到 $1200 \times 10^8 m^3$,对外依存度达到 40%,2030 年进口 $2700 \times 10^8 m^3$ 左右,对外依存度将超过 50%。

4.4 天然气将成为能源结构调整的主要增长点

2016 年世界能源消费结构中,煤炭、石油、天然气占比相对均衡,分别为 28.1%、33.3% 和 24.1%。而在中国,煤炭仍是能源消费的主体,在能源结构中的比例为 61.8%。天然气作为最具潜力的清洁能源,是能源转型的主力军。近年来天然气需求完善,发展迅猛,在能源结构中的比例上升到 6.2%,但远低于世界平均水平。国务院办公厅《能源发展战略行动计划(2014—2020 年)》、国家发展和改革委员会《天然气发展"十三五"规划》均明确提出,到 2020 年天然气在国内一次能源消费比重将达到 10%。

5 结论

在国家大力倡导低碳绿色能源的背景下,秉承有质量、有效益、可持续发展的理念,积极发展天然气开发业务。进入"十三五"时期,结合开发对象与气田开发阶段的新变化,在深层天然气气藏开发技术、大型气田开发调整技术、致密气提高采收率技术、页岩气及煤层气开发技术、工程技术及开发决策体系等方面取得了显著技术进步,创新能力不断提升,天然气产量稳定增长。

随着开发的深入、油价的下跌和政策、环境的影响,受地质条件制约,国内气田优质储量比例降低,气田开发成本升高,非常规气藏效益开发难度加大,上游效益进一步压缩,主力气田稳产能力薄弱,市场竞争越发激烈,天然气持续规模效益开发面临诸多挑战。

未来国内将进入常规气与非常规气并重的发展阶段。预计 2030 年全国天然气产量 $2520 \times 10^8 m^3$,其中常规气占 51.6%、非常规气占 48.4%。天然气需求持续旺盛,将呈现出以发电气、工业燃料气、城市燃气及交通运输气为主体的消费结构多元化。国内天然气产量与需求的缺口不断扩大,进口气量不断攀升,天然气对外依存度不断加大。蓬勃发展的天然气产业是推动国家能源结构转型的主力军。

参 考 文 献

[1] 李海平,贾爱林,何东博,等. 中国石油的天然气开发技术进展及展望[J]. 天然气工业,2010,30(1):5-7.
[2] 贾爱林,唐俊伟,何东博,等. 苏里格气田强非均质致密砂岩储层的地质建模[J]. 中国石油勘探,2007,12(1):12-16.
[3] 李保柱,朱忠谦,夏静,等. 克拉 2 煤成大气田开发模式与开发关键技术[J]. 石油勘探与开发,2009,36(3):392-397.
[4] 何江,方少仙,侯方浩,等. 风化壳古岩溶垂向分带与储集层评价预测——以鄂尔多斯盆地中部气田区马家沟组马$五_5$—马$五_1$亚段为例[J]. 石油勘探与开发,2013,40(5):534-539.
[5] 吴永平,王允诚. 鄂尔多斯盆地靖边气田高产富集因素[J]. 石油与天然气地质,2007,28(4):473-478.
[6] 金民东,曾伟,谭秀成,等. 四川磨溪—高石梯地区龙王庙组滩控岩溶型储集层特征及控制因素[J]. 石

油勘探与开发,2014,41(6):650-661.

[7] 马新华,贾爱林,谭健,等. 中国致密砂岩气开发工程技术与实践[J]. 石油勘探与开发,2012,39(5):572-579.

[8] 谭中国,卢涛,刘艳侠,等. 苏里格气田"十三五"期间提高采收率技术思路[J]. 天然气工业,2016,36(3):30-40.

[9] 卢涛,刘艳侠,武力超,等. 鄂尔多斯盆地苏里格气田致密砂岩气藏稳产难点与对策[J]. 天然气工业,2015,35(6):43-52.

[10] 陆家亮. 中国天然气工业发展形势及发展建议[J]. 天然气工业,2009,29(1):8-12.

[11] 聂海宽,金之钧,马鑫,等. 四川盆地及邻区上奥陶统五峰组—下志留统龙马溪组底部笔石带及沉积特征[J]. 石油学报,2017,38(2):160-174.

[12] 贾爱林,位云生,金亦秋. 中国海相页岩气开发评价关键技术进展[J]. 石油勘探与开发,2016,43(6):949-955.

[13] 郭彤楼,张汉荣. 四川盆地焦石坝页岩气田形成与富集高产模式[J]. 石油勘探与开发,2014,41(1):28-36.

[14] 陈作,薛承瑾,蒋廷学,等. 页岩气井体积压裂技术在我国的应用建议[J]. 天然气工业,2010,30(10):30-32.

[15] 聂海宽,张金川,张培先,等. 福特沃斯盆地Barnett页岩气藏特征及启示[J]. 地质科技情报,2009,28(2):87-93.

[16] 田炜,王会涛. 沁水盆地高阶煤煤层气开发再认识[J]. 天然气工业,2015,35(6):117-123.

[17] 罗平亚. 关于大幅度提高我国煤层气井单井产量的探讨[J]. 天然气工业,2013,33(6):1-6.

[18] 何明舫,马旭,张燕明,等. 苏里格气田"工厂化"压裂作业方法[J]. 石油勘探与开发,2014,41(3):349-353.

[19] 凌云,李宪文,慕立俊,等. 苏里格气田致密砂岩气藏压裂技术新进展[J]. 天然气工业,2014,34(11):66-72.

[20] 张书平,白晓弘,樊莲莲,等. 低压低产气井排水采气工艺技术[J]. 天然气工业,2005,25(4):106-109.

[21] 杨涛,余淑明,杨桦,等. 气井涡流排水采气新技术及其应用[J]. 天然气工业,2012,32(8):63-66.

原文刊于《天然气工业》,2018,38(4):77-86.

全球不同类型大型气藏的开发特征及经验

贾爱林 闫海军 郭建林 何东博 魏铁军

(中国石油勘探开发研究院,北京 100083)

摘要:据 AAPG 资料,全球已发现的 370 个大气田,集中分布在西西伯利亚盆地、波斯湾盆地、扎格罗斯盆地、卡拉库姆盆地、墨西哥湾盆地、卡那封盆地以及东西伯利亚盆地,分布在这些盆地的大型气藏占全球总数量的 45%,其储量占到大气田储量的 68.3%。在分析所有大型气藏资料的基础上,研究了大型气藏的数量、储量、地区、地层、圈闭类型、深度、发现时间等特征。依据"实用性、针对性和科学性"分类原则,主要从岩性、规模形态、物性、压力、流体等特征将全球大型气藏分为厚层整装高渗砂岩气藏、低渗透砂岩气藏、边底水裂缝型碳酸盐岩气藏、"三高"气藏及凝析气藏五种类型,并研究了不同类型大型气藏的开发特征。

关键词:大型气藏;分布特征;类型划分;划分原则;开发特征;高效开发

1 概述

大气田在世界石油工业的发展中具有举足轻重的地位,现有研究资料表明世界上绝大多数天然气赋存在少数大气田中。近年来,中国天然气勘探开发获得了长足的发展,截至 2012 年,全年国内生产天然气 $1077×10^8m^3$,同比增长 6.5%,中国已经迈入天然气生产大国。随着中国天然气工业的发展,天然气开发对象越来越复杂,中国天然气行业的健康发展迫切需要吸收国内外大型气藏开发的成功经验,中国天然气工业的快速发展依赖于大气田的持续发现和高效开发。因此,研究全球大气田的分布特征及开发特征对于中国天然气事业的持续、高效、安全、快速发展具有重要的指导意义。

根据国际惯例,大型气藏指最终天然气可采储量超过 3tcf($850×10^8m^3$)的气藏;特大型气藏是指最终可采储量超过 300tcf($8500×10^8m^3$)的气藏;巨型气藏指最终可采储量超过 300tcf($85000×10^8m^3$)的气藏[1]。世界大油气田分布一直受到国内外学者的广泛关注:AAPG 出版过四部有关全球大油气田的专著:AAPG Memoir 14[2],AAPG Memoir 30[3],AAPG Memoir 54[4],AAPG Memoir 78[5];李国玉、金之钧等也出版了《世界含油气盆地图集》[6];李国玉、唐养吾也出版了《世界气田图集》[7];宋芊和金之钧对全球油气田的基本特征做过一个统计分析,但该研究利用的数据资料只到 1993 年底,而且没有对大气田的分布特征做出阐述[8];白国平和郑磊依据全球发现的 355 个大气田对其分布特征进行分析,对世界大气田首次进行了系统研究,同时指出了大气田分布主控因素,但是没有阐述大型气藏的开发特征[9]。本文依据 AAPG 统计气田资料,对全球范围内发现的大气田分布特征进行系统分析研究,全面梳理大气田的分布特征。同时,为了便于研究大气田的开发特征,对全球大型天然气藏进行类型划分,分析不同类型气藏的开发特征,从中找出大型气藏开发规律和开发模式,以期为中国天然气的开发提供借鉴和指导。

2 大型气藏分布特征

2.1 大型气藏概述

据 AAPG 资料,全球已发现 370 个大型天然气藏,这些气藏是相当长一段时间内全球天然气开采的主力军,对于世界天然气行业的市场稳定起着至关重要的作用(表1)。本文主要从数量和储量分布、地区及沉积盆地分布、储层分布、圈闭类型及深度分布、发现时间等特征对全球大气田进行系统梳理,总结全球大型气藏的分布特征。

表 1 世界前十大型天然气藏基本情况表

序号	气田名称	国家	盆地	级别	最终可采储量(tcf)	发现年份	圈闭类型	岩性	深度(m)
1	北方(North Field)	卡塔尔	波斯湾(阿拉伯湾)	巨型	272816	1971	构造	白云岩	2765
2	南帕斯(Pars South)	伊朗	波斯湾(阿拉伯湾)	巨型	131479	1991	构造	白云岩	2849
3	乌连戈伊(Urengoy)	俄罗斯	西西伯利亚	巨型	99512	1966	构造	砂岩	2286
4	扬堡(Yamburg)	俄罗斯	西西伯利亚	特大型	44448	1969	构造	砂岩	1098
5	扎波利亚尔(Zapolyarnoye)	俄罗斯	西西伯利亚	特大型	35984	1965	构造	砂岩	1119
6	哈西鲁迈勒(Hassi R'Mel)	阿尔及利亚	古达米斯	特大型	35176	1957	构造	砂岩	2125
7	阿斯特拉罕(Astrakhan')	俄罗斯	滨里海	特大型	33329	1976	礁	有孔虫灰岩	3850
8	西北穹隆(Northwest Dome)	卡塔尔	波斯湾(阿拉伯湾)	特大型	22639	1976	构造	碳酸盐岩	—
9	卡拉恰加纳克(Karachaganak)	哈萨克斯坦	滨里海	特大型	22412	1979	礁	珊瑚灰岩	4480
10	拉格萨费德(Rag-E-Safid)	伊朗	扎格罗斯	特大型	22190	1964	构造	碳酸盐岩	1341

注:据 AAPG 资料整理而成,北方气田和南帕斯气田构造上属于同一个气田,但在统计时往往作为两个气田来处理。

2.2 大型气藏分布特征

2.2.1 大型气田数量和储量分布特征

根据国际上划分大型气田标准,全球范围内可采储量超过 3tcf($850×10^8 m^3$)的 370 个大型气田中,巨型气田、超大型气田、大型气田之比为 3:25:342。而从可采储量上来看,巨型气田、超大型气田、大型气田之比为 31:28:41。可以看出,大型气田个数按级别的分布有"绝对集中"的特点,而大型气田储量按级别的分布有"相对分散"的特点(表2)。

表2 世界大型气田可采个数以及可采储量按级别分布表

级别	个数	个数所占百分比（%）	可采储量（$10^8 m^3$）	储量所占百分比（%）
巨型气田(megagiant)	3	0.81	503807	30.94
超大型气田(supergiant)	25	6.76	457300	28.09
大型气(giant)	342	92.43	666978	40.97
合计	370	100.00	1628085	100.00

2.2.2 大型气田地区分布特征

按地区来说,中东和东欧、中亚、俄罗斯大型气田个数占总数的46%,而可采储量占总储量的75%;西欧、非洲、亚洲大洋洲、北美洲和中南美洲气田个数占总数的64%,但是可采储量仅占总可采储量的25%(如图1、表3)。

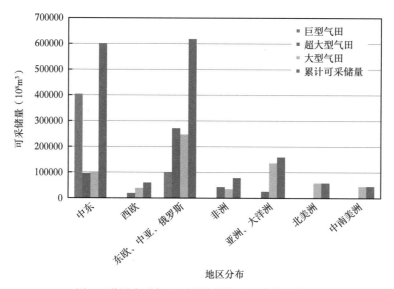

图1 世界大型气田可采储量按地区分布柱状图

表3 世界大型气田可采储量按地区分布表

地区	巨型气田		超大型气田		大型气田		合计		
	个数	可采储量（$10^8 m^3$）	个数	可采储量（$10^8 m^3$）	个数	可采储量（$10^8 m^3$）	个数	可采储量（$10^8 m^3$）	所占比例（%）
中东	2	404295	6	94973	47	101651	55	600919	36.91
西欧	0	0	2	20879	30	40585	32	61464	3.78
东欧、中亚、俄罗斯	1	99512	13	271186	101	248221	115	618919	38.02
非洲	0	0	2	43870	25	36027	27	79897	4.91
亚洲、大洋洲	0	0	2	26392	76	134085	78	160477	9.86
北美洲	0	0	0	0	37	59196	37	59196	3.64
中南美洲	0	0	0	0	26	47213	26	47213	2.90
总计	3	503807	25	457300	342	666978	370	1628085	100.00

按沉积盆地来说,370 个大型气藏分布在 94 个沉积盆地,其中 55%以上的可采储量分布在波斯湾盆地和西西伯利亚盆地(图 2)。世界上分布大型气藏最多的是西西伯利亚盆地(57个)、波斯湾盆地(26 个)、扎格罗斯盆地(25 个)、卡拉库姆盆地(20 个)、墨西哥湾盆地(15个)、卡那封盆地(12 个)以及东西伯利亚盆地(11 个),45%的大型气藏分布在这七个盆地中(图 3)。世界上大型气田储量的分布按地区有"高度集中"的特点。

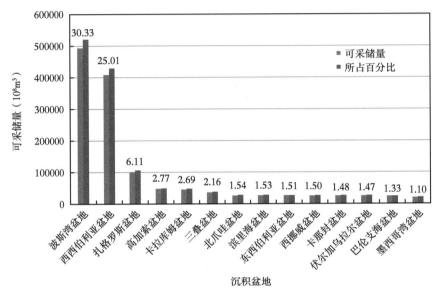

图 2 世界大型气田可采储量按盆地分布柱状图

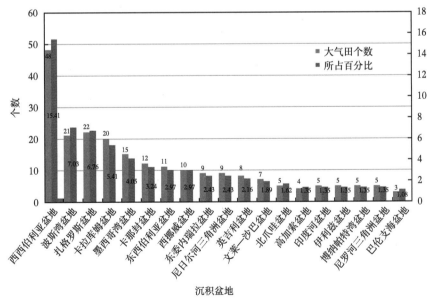

图 3 世界大型气田个数按沉积盆地分布柱状图

2.2.3 大型气田按储层分布特征

从层系来看,大气田的分布层系相当广泛,除了志留纪之外从元古宙至第四纪均有分布;随着储层时代变老,大气田的个数降低(图4);大气田主要分在石炭系—新近系,这些层系内发现大气田个数为338个,占大气田总数的91%;大气田储量主要存在于白垩系、三叠系和二叠系,可采储量分别占到大气田总可采储量的23%、22%和17%(图5)。

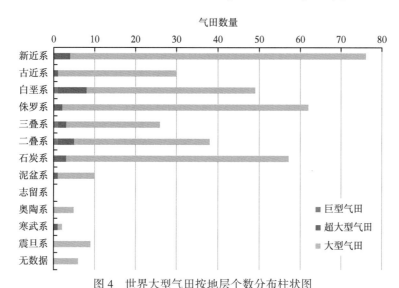

图 4 世界大型气田按地层个数分布柱状图

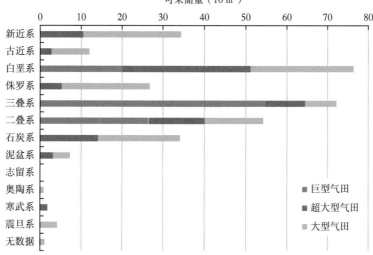

图 5 世界大型气田按地层可采储量分布柱状图

从储层岩性来说,大气田储集岩类型集中分布在砂岩和碳酸盐岩,其中砂岩气藏的个数和可采储量均占整个大型气田的一半以上,是大型气藏储层类型的主体。但是,碳酸盐岩气藏储量规模明显高于砂岩气藏,碳酸盐岩气藏个数仅占总个数的26%,而其可采储量却占到46%,砂岩个数占总个数的71%,而其可采储量占总储量的54%(表4)。

表 4 世界大型气田按储集岩类型统计表

	巨型气田		超大型气田		大型气田		合计		所占百分比(%)	
	个数	可采储量($10^8 m^3$)	个数	可采储量($10^8 m^3$)	个数	可采储量($10^8 m^3$)	个数	可采储量($10^8 m^3$)	个数	可采储量
砂岩	1	99512	16	289043	245	485301	262	873856	70.810	53.674
碳酸盐岩	2	404295	9	168257	87	169527	98	742079	26.487	45.58
浊积岩	0	0	0	0	3	4624	3	4624	0.008	0.003
无数据	0	0	0	0	7	7256	7	7256	0.019	0.005

2.2.4 大型气田按圈闭类型及深度分布特征

从圈闭类型来看,大型气田中,构造圈闭无论是个数还是可采储量均在整个大型气藏圈闭类型中占有绝对优势,其个数和可采储量占整个大型气田的百分比分别为80%和86%(表5)。

表 5 世界大型气田按圈闭类型统计表

圈闭类型	巨型气田		超大型气田		大型气田		合计		所占百分比(%)	
	个数	可采储量($10^8 m^3$)	个数	可采储量($10^8 m^3$)	个数	可采储量($10^8 m^3$)	个数	可采储量($10^8 m^3$)	个数	可采储量
地层	0	0.0	1	17149.0	49	123278.0	50	140427.0	13.5	8.6
生物礁	0	0.0	2	55741.0	9	12774.0	11	68515.0	3.0	4.2
构造	3	503807.0	22	384410.0	272	515036.0	297	1403253.0	80.3	86.2
无数据	0	0.0	0	0.0	12	15890.0	12	15890.0	3.2	1.0

从深度特征来看,由于44个气田没有深度数据,仅对326个气田的深度数据进行统计,大型气田深度相对集中分布在1500~3000m的深度段内,气田个数占总统计个数的52%,其储量占整个大型气田储量的64%。小于1500m的气田个数占21%,大于2500m的气田个数占26%(图6)。

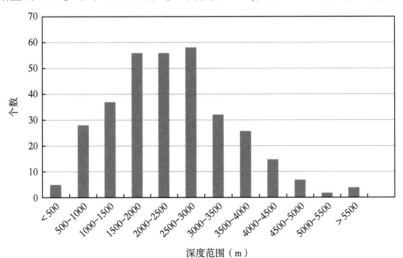

图 6 大型气田深度概率分布柱状图

2.2.5 大型气田发现时间分布特征

纵观全球整个天然气的勘探发现历史可以发现全球天然气整体可采储量的增长统理论技术进步、政府优惠政策以及天然气气价密切相关。整体上来说,全球大气田发现经历四个高峰期:1954—1959 年、1963—1979 年、1988—1992 年、1997—2001 年,四次高峰期内所发现气田从构造气田向构造岩性气藏过渡,其构造气藏比例依次为 100%、87.6%、71.4%、65%,构造岩性气藏比例依次为 0、9%、18%、26%(图 7、图 8)。

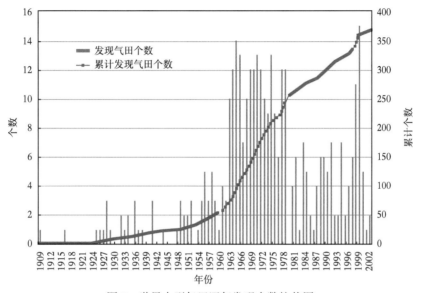

图 7 世界大型气田逐年发现个数柱状图

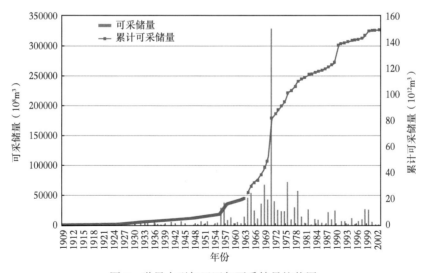

图 8 世界大型气田逐年可采储量柱状图

针对某一具体盆地来说,天然气可采储量的增长除了同非自然因素(包括公司的投入以及政策支持)和盆地具体特征(包括储层特征、沉积特征、流体特征等)有关外,更重要的是与

— 19 —

理论技术的进步密切相关。如利用层系地层预测油气发现模式在挪威上侏罗区块中，油气藏的发现遵循由高位构造油气藏(Troll 油气藏)到海侵地层油气藏(Draugen 油气藏)，最后是低位深水油气藏(Fram 油气藏)(图9)。理论技术的进步为单一特征盆地天然气发现潜力指明道路，为关键技术发展趋势指出方向。总之，对一特定盆地，大气田的发现可以持续几十年，盆地的地质条件越复杂，发现的持续时间越长。这些特征决定了中国的天然气勘探开发是一个长期过程，随着理论技术的进步及对盆地认识程度的增加，相信在相当长时间内我国的天然气探明储量将保持持续增长。

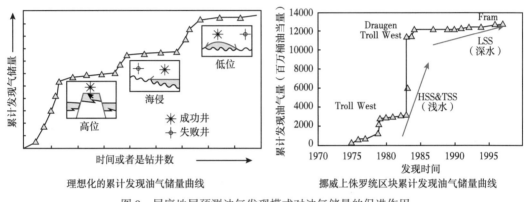

图9 层序地层预测油气发现模式对油气储量的促进作用

3 大型气藏开发特征

大型气藏的储层地质、圈闭类型、流体分布等特征各具特色，为了研究大型气藏的开发特征有必要对已发现大型气藏进行梳理，划分其主要类型，研究其开发特征。

3.1 大型气藏类型划分

3.1.1 大型气藏类型划分原则

在大气藏类型划分的过程中，遵循"实用性""针对性""科学性"三大原则，对全球大气藏进行类型划分[10]。需要说明的是，三者重要性不是等同的，实用性是首要原则，针对性和科学性是次要原则。

3.1.2 大型气藏类型划分结果

根据三条划分原则，围绕岩性、厚度、规模、物性、压力、流体等因素将全球大型气藏划分为以下几种类型：(1)厚层整装高渗砂岩气藏；(2)低渗砂岩气藏；(3)边底水裂缝型碳酸盐岩气藏；(4)"三高"气藏；(5)凝析气藏。大型气藏类型划分为其开发特征的研究奠定基础。

3.2 大型气藏开发特征

3.2.1 厚层整装高渗砂岩气藏开发特征

厚层整装高渗砂岩气藏的主要特征是"规模大、储层厚、物性好、或存在边底水"，该类气藏开发以格罗宁根气藏最典型。

3.2.1.1 地质概况

该气藏为西荷兰盆地南部二叠系气藏,气藏产层为斯特洛奇特伦段(河流砾岩、砂岩和风成砂岩)和顿布厄段(粉砂质细砂质黏土岩);主力层厚度158m,含气面积800km^2,气藏可采储量$2.8×10^{12}m^3$;孔隙度15%~20%,渗透率0.1~3000mD。

3.2.1.2 开发概况

该气藏发现于1959年,1963年投产,壳牌公司和埃克森美孚公司合资(NAM)开发。气藏有边底水,气、水界面为-2970m,气层压力35.5MPa,井深3000m,为弹性水驱;同时天然气组成中甲烷占81.3%,乙烷以上重烃含量2.84%,N_2占14.32%,CO_2为0.87%,属干气气藏。气藏最高日产气$3.5×10^8m^3$,2009年$9300×10^4m^3/d$,年产气$350×10^8m^3$;截至2009年,累计产气$1.7×10^{12}m^3$,60%的原始可采储量已经被采出。

3.2.1.3 气藏开发特征

(1)承担"调峰生产"作用,保证地区安全平稳供气。

由于该类气藏产气量高,因此气田生产在整个国家甚至地区的天然气供应中担当调峰的作用。荷兰政府自从石油危机之后采取保护格罗宁根的政策,支持发现和开采尽量多的小气田,有将近格罗宁根气田一半储量的气田被发现。在供气紧张时,格罗宁根气田大规模生产,保证安全供气;在供气不紧张时,把它作为储气库用。这样既稳定了格罗宁根气田压力,提高了格罗宁根气田的采收率,又解决了安全平稳供气的问题。截至目前,小气田产量占每年产量的30%,格罗宁根气田占70%,大气田的"调峰生产"和小气田的"持续生产"有力地保证了地区安全平稳供气。

(2)采用"井组布井"方式生产,实现气田高效开发。

由于气田厚度大,储层物性好,较好的气藏储层地质条件为高效井组的布井奠定了基础。同时,格罗宁根气田位于人口稠密地区,为了少占耕地、安全和环境保护,气田开发采用井组式布井方式。气田开发设计为25个井组,每个井组8~10口井,实际建成28个井组,生产井285口,日生产能力达到$5×10^8m^3$。井组间距2.4km,每个井组占地8公顷,包括8~10口3050m深的生产井,分为两排,每排4~5口,地面井距70m,钻开125m厚的储层。井组式布井方式开发降低了气田开发的各项成本,实现了气田的高效开发。

(3)加强水侵及压力动态监测,保持气藏压力均衡生产。

由于气藏规模较大,同时含有边、底水,因此在整个开发过程中,要坚强动态监测,保持气藏压力均衡,避免由于压降漏斗造成边、底水的突进。针对格罗宁根气藏,为了随时观察气、水界面的变化情况,防止边底水的不均匀推进造成气井产量递减,采取了三条措施:①打定向斜井和控制打开程度,为避免在储层局部地区集中采气,过早形成水锥,用钻定向井的方法布置地下井位,尽量加大地下井距,同时把射孔下限限定在距气、水界面50m;②建立观察点,在气田北部布置了一批钻穿气、水界面的含水层观察井,采用斯伦贝谢公司脉冲中子测井仪器实测气、水界面移动的位置监测水侵变化。③开展水锥试验,在含水层附近钻了一批水锥试验井,这些井全部钻穿气、水界面,并在允许条件下以最高速度进行生产,以便在早期发现水锥。

同时,在气藏开发初期,为避免边底水过早地窜入气层,采取首先开采构造顶部的气,即先开发气田南部。同时,为了保持气藏压力均衡,避免南部气层压力下降过快,造成气田过早上压缩机,从1970年开始在气田中部和北部投产新的井组,并提高开采速度。当气田北部产量提高约50%时,北部和南部的压力逐渐趋于一致,1983年以前气田在最大压差不超过2MPa条件下进行配产(图10),从而保证整个气藏压力的均衡生产。

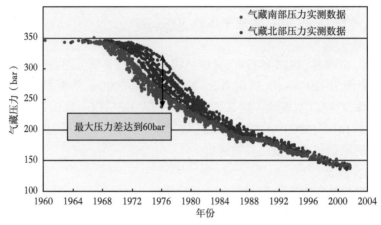

图10　格罗宁根气藏压力逐年变化曲线图

3.2.2　低渗透砂岩气藏开发特征

低渗透砂岩气藏的主要特征是"规模大、物性差、一般不存在边底水",该类型气藏以我国的苏里格气藏为代表。

3.2.2.1　地质概况

苏里格气田构造形态为由北东向南西方向倾斜的单斜。该气田含气层位主要为二叠系下石盒子组盒八段及山西组山一段。气层埋深3200~3600m,储层孔隙度5%~12%,渗透率0.06~2.00mD,压力系数为0.87,储量丰度$1×10^4~2×10^4m^3/km^2$,是典型的低渗透率、低压力、低丰度的"三低"气田。截至2012年,累计探明天然气地质储量$3.49×10^{12}m^3$。

3.2.2.2　开发概况

苏里格气田产能建设始于2006年,截至2012年底,累计投产气井5862(水平井388)口,日均开井4693口,日均产气$5412.44×10^4m^3/d$,平均单井产量$1.15×10^4m^3/d$,日产量压降速率控制在0.013MPa/d。

3.2.2.3　气藏开发特征

(1)储层物性差,气井基本无自然产能,储层改造是实现气井经济有效开发的基础。

苏里格气田的"三低"特征以及气藏的强非均质性,导致有效砂体连续性和连通性差。气井试气成果表明,苏里格气田除少数气井无阻流量大于$10×10^4m^3/d$,超过90%的气井无阻流量小于$10×10^4m^3/d$,且其中一半的气井无阻流量小于$4×10^4m^3/d$[11]。苏里格气田1口直井一般可钻遇2~4个气层,最多可钻遇6~7个气层,只有通过对个气层的充分改造,提高剖面上储层的动用程度,才能提高单井产量,实现效益开发。自2000年以来,苏里格气田持续进行改造技术的试验攻关,不断取得阶段性突破,经历了大规模合层压裂、适度规模压裂、直井水平井分段多层压裂、体积压裂等几个阶段。2012年苏里格气田进行了10口井的体积压裂现场试验,平均无阻流量达到$68.07×10^4m^3/d$,取得了较好的增产效果。因此,低渗砂岩气藏通过储层改造技术的不断进步可以大幅度提高单井产量,从而实现该类型气藏经济有效开发。

（2）单井产量递减较快,区块接替+井间接替方式是气田稳产的主要方式。

通过分层分段压裂、大规模体积压裂,苏里格气田初期单井产量较高,实现了气藏储量的有效动用。但是,由于气藏基质物性较差,单井控制储量较小,有限的气不能无限制高速的供向井筒,造成气井表现为产量低、稳产期短、地层压力下降快、关井压力恢复缓慢的特点[10]。苏里格气田气井稳产能力差,一般稳产三年,有的气井几乎没有稳产期,一直呈现递减趋势,因此气田稳产面临巨大挑战。对于该类气田的稳产,一方面,由于气井控范围有限,大量剩余气无法有效动用,因此可通过多种手段进行井网评价,对气井不断进行加密调整保证气田稳产。另一方面,该类气藏的开发往往是富集区优先开发,随着技术的进步以及气藏特征认识程度的增加,早期认为不可动用的区块利用现有技术可以实现有效动用。因此,整个气田的稳产可以通过滚动扩边从而实现整个气田的稳产、甚至上产。因此,对低渗透砂岩气藏来说,井间接替和区块接替是实现气田稳产的主要方式[12-13]。

（3）气井控制范围有限,井控储量小,气井加密调整是提高气藏采收率的有效手段。

苏里格气田早期井网1200m×600m,气井单井控制储量低,一般在$1000×10^4 \sim 3500×10^4 m^3$之间,平均$2100×10^4 m^3$,有大量的剩余气依靠目前井网无法动用[14]。因此,气井加密是苏里格气田提高采收率的有效手段。针对苏里格气田单井控制储量低的特点,形成了一套针对低渗砂岩气藏开发井距优化系列评价方法[12]。综合利用地质、测井及生产动态等资料,以储层沉积学和测井地质学的理论为指导,对实施加密井进行砂体解剖;结合井组干扰井试井成果,进一步验证砂体规模与连通性;利用相控建模对储层砂体井间分布和储层物性的变化规律进行预测,建立高精度的储层三维地质模型;利用丛式井、水平井等多种井型井网提高储层平面及纵向动用程度。在有效储层规模及空间展布规律研究的基础上,利用动储量评价、经济极限法、数值模拟法等对气田井网井距进行优化。优化结果表明,平均储量丰度$1.2×10^8 m^3/km^2$,合理单井控制面积$0.48 km^2$,井距800m×600m,该井网较前期开发井网(1200m×600m)更合理,可以提高苏里格气田最终采收率约15%。随着天然气价格的提高以及开发成本的进一步降低,相信在800m×600m井网基础上还有不断加密的空间,可进一步提高气藏采收率[15]。

（4）气藏构造相对简单,有效储层普遍发育,建立规模化丛式井组、采用工厂化作业是实现气藏高效开发的关键。

苏里格气田构造相对简单,有效储层大面积分布,为工厂化作业奠定了基础。同时,井型井网的优化促进了苏里格气田开发方式的转变,也为工厂化作业带来了契机。气田从2007—2008年开展丛式井试验,通过两年的技术攻关,完善了"富集区块整体部署,评价区随钻部署"的丛式井部署流程,形成了丛式井开发配套技术。2009年开始大力推广丛式井开发,在优化井场布置、节约用地面积、减少采气管线、优化生产管理、降低综合成本、科技绿色环保等方面起到了举足轻重的作用。2009年全年完钻丛式井占总井数的56.1%,平均钻井周期缩短至20天左右,Ⅰ类+Ⅱ类钻井比例高达87.5%,丛式井开发取得了良好效果。因此,采用规模化丛式井组开发模式和精细化管理,将钻井、压裂、试气等作业程序"流程化、批量化、标准化",从组织模式、资源配置、流程设计、技术支撑、作业管理等多方面进行革新,集中现有资源和技术优势,专业化施工、模块化组织、程序化控制、流程化作业。苏里格气田形成了具有特色的工厂化钻完井作业模式,实现了"三低"气田的规模效益开发,为同类型气田高效开发树立了样板[16-17]。

3.2.3 边底水裂缝型碳酸盐岩气藏开发特征

边底水裂缝型碳酸盐岩气藏的主要特征是"规模大、裂缝发育、存在边底水",该类气藏以奥伦堡气藏为代表。

3.2.3.1 主要地质特征

奥伦堡气田处于伏尔加—乌拉尔盆地乌拉尔山前坳陷带南端的西侧,于1966年被发现,气田受奥伦堡长垣构造控制。气藏为裂缝—孔隙型大气田,主力气层为二叠纪碳酸盐岩,含气面积 $1500km^2$,气藏中部含气层厚度为525m,西部275m,气藏平均埋深1700m,产层有效厚度 $89.4 \sim 253.6m$,孔隙度11.3%,渗透率 $0.098 \sim 30.6mD$,气藏原始地层压力20.33MPa,天然气储量 $1.9 \times 10^{12} m^3$。

3.2.3.2 主要开发概况

奥伦堡气田含气面积广,储量大,储层非均质性强,气井产能差别大,气井见水早。该气田1968年开发,1974年开始工业化开采,根据气田不同部位的地质特征、天然气和地层水化学组分、地层水活跃程度和开采特征的差异,将气田分为15个开采区。气田最高年产量 $450 \times 10^8 m^3$,1991年到现在气田进入产量递减期,现在每年生产 $180 \times 10^8 m^3$。气藏为边底水混合驱动,气藏选择性水侵严重,气井过早水淹。底水主要沿中部裂缝发育带上窜侵入,1971年投产8个月后即有1口井产水,1981年有49口井产水,单井日产水 $10 \sim 125m^3$,距水侵方向近的井产水量高达 $100m^3/d$ 以上,另外有125口井见出水显示[18],为了弥补大量出水而递减的气产量,每年要新投50口新井。

3.2.3.3 气藏开发特征

俄罗斯奥伦堡气藏为裂缝—孔隙型碳酸盐岩储层,与中国威远气田极为相似。作为受水侵影响的边底水型碳酸盐岩气藏,气田开发的整个生命周期主要围绕防水治水开展工作。

(1)水侵是该类气藏开发最主要特征,严重影响气藏的采出程度。

在气藏开发过程中,边底水侵入含气区必须具备的条件,一是气区压力低于含水区压力,二者压差越大,水侵速度越快;二是含气区至含水区存在高渗透裂缝渗透通道[18-19]。裂缝性有水气藏水侵有两种形式,一是边底水大面积侵入含气区;二是生产压差使底水很快沿裂缝窜至局部气井,生产压差越大水窜越快,很多气井投产短时间就见地层水而气水同产。气井见水后,使得近井地带储层的含水饱和度急剧增加,储层孔隙通道有效空间减小甚至堵塞,阻碍气的通过,最终导致气相渗透率的降低和产能的下降。由于水锁效应以及地层水的非均质水窜,易行成"封闭气"和"死气区",致使大量的气被地层水分隔包围,不能采出。奥伦堡气田18块不同岩心的渗吸水驱气实验结果表明,驱替系数(含束缚水)为 $0.42 \sim 0.723$(平均0.504)平均残余气饱和度为 $27.7\% \sim 58\%$(平均36.6%),平均封闭气量达到49.1%。水侵区大量的剩余气没有被采出从而大大降低气藏的采出程度。

(2)加强动态监测,避免气藏大幅度水淹。

对于裂缝型油水气藏,水侵是其主要开发特征,因此动态监测是贯穿气藏开发整个生命周期的一项工作。奥伦堡气田在开发过程中,建立了完善的气藏监测系统,包括气体动力学监测井,主要用于监测气藏各部位的压力和温度、研究开采井的产能特征、观察气藏不同区块压降漏斗分布特点和排流程度、研究井间干扰等。水文地质学监测井,主要任务是观察底水推进特点和预报采气井水侵的可能性,在水文观察井中进行地层水压力、化学组分、含气饱和度、溶解气组分等的观察。矿场地球物理监测井,用于研究储层的非均质性、裂缝对水侵过程的影响、评价在开采过程中不同区块含气饱和度的变化等。由于奥伦堡气田设置了完整的观察和监测系统,及时掌握了气藏的动态和水侵特点,因而可适时调整开发系统,避免气藏大幅度水淹[21]。

(3)制定科学治水对策,提高气藏最终采出程度。

奥伦堡气田于1974年投产不久,发生了严重的水侵。针对此问题,原苏联国家科技人员在室内进行了岩心一维和三维毛细管渗吸、径向水驱气以及高压水驱气采收率研究,发现水驱气的主要特征是水淹区内封闭气量较大,在大岩心实验中,微裂缝能促使水选择性地从裂缝面向岩心中心运动,增加了封闭气量,此时封闭气须在发生膨胀且占据50%以上孔隙空间时才能流动。因此,对于奥伦堡气藏,首先采用早期整体治水,避免或减弱水侵程度,延长无水采气期。第二,采用阻水工艺,阻水工艺是指在气水界面含水一侧打开排水井以减缓边、底水的侵入,然后在地层水活跃的高渗透断裂带、裂缝发育带,用高分子聚合物黏稠液建立阻水屏障,阻止边、底水进入气藏。该方法在1982年在奥伦堡气田进行现场试验。在地层水侵入裂缝发育带,与水侵通道方向垂直方位,布3口井为1组井排,射开水动力相连水侵层位;井组两边的井做排水井,中间注黏稠液。经过数值模拟计算,如果不建立屏障,稳产期仅6年,采收率仅40%,而建立阻水屏障可稳定开采22年,采收率高达93%。第三,在水淹气藏中,可采用人工举升助排工艺、结合自喷井带水采气排出侵入储气空间的水和井筒积液,使部分"水封气"解堵,变为可动气而被采出,称"二次采气技术",约可提高采收率10%~20%。奥伦堡的治水措施可以总结为:早期合理布井,控制采气速度;气水边界含水一侧打排水井,拖住边、底水推进;高渗透带用高分子聚合物黏稠液建立阻水屏障,减少气水接触。通过综合治水对策的有效实施,可提高该类气藏的最终采出程度。

3.2.4 "三高"气藏开发特征

"三高"气藏的主要特征是"高温或高压或高含硫"。法国拉克气田Inferieure气藏是一个典型的深层"三高"气藏。

3.2.4.1 主要地质概况

拉克气田位于阿奎坦盆地南部,该气藏东西长16km,南北宽10km,闭合面积120km^2,闭合高度1400m,北缓南陡,气田地质储量为3226×10^{8}m^3,气藏构造为一向东南方向倾斜的背斜构造,气藏埋深超过3000m。气藏储层是一组巨厚的碳酸盐岩储层,分为上下两层,上部层位以下白垩统尼欧克姆亚石灰岩为主,下部层位为上侏罗统马诺白云岩为主。储集空间以孔隙为主,裂缝为主要的渗流通道。储层厚度大,上部层位有效厚度200~300m,下部层位有效厚度150~200m,全气藏有效厚度350~500m,上下两层物性相差不大。

3.2.4.2 主要开发概况

拉克气藏是典型的深层三高无边底水的封闭气藏,平均井深3800m,最深井5000m。原始气层压力达66.15MPa,气层温度140℃。天然气组分中甲烷占69%,乙烷占3%,硫化氢占15.6%,CO_2占9.3%,其他组分占1.9%。拉克气田开发经历了四个阶段:第一阶段(1952—1957年)为试采阶段,主要对三口井进行试采,检验井底及井口设备的抗硫防腐性能,同时获取气藏动态参数;第二阶段(1957—1964年)为产能建设阶段,共有26口生产井,气田日产量由82×10^{4}m^{3}上升至2156×10^{4}m^3,平均单井产量为80×10^{4}m^3/d,采气速度2.4%;第三阶段(1964—1983年)为稳产阶段,通过在构造高点打 10 口加密井,气田日产量为 $1906\times10^4 \sim 2361\times10^4 m^3$,平均单井产量 $50\times10^4 \sim 60\times10^4 m^3/d$,采气速度 2.6%,稳产期长达 19 年,稳产期可采储量采出程度为 65% 左右;第四阶段(1983 年至今)为产量递减阶段,1994 年气田日产量递减为 $405\times10^4 m^3$,气田累计产气 $2258\times10^8 m^3$,地质储量采出程度为 70%。

3.2.4.3 气藏开发特征

(1)注重前期评价,弄清气藏基本特征后再开始大规模开发。

针对拉克气田"三高"特征,在准备开发和开发过程中对气井钻井完井、钻采系统和地面工程防腐、增产、净化和回收硫黄等技术进行了系统研究,从发现到投产,历时七年,做了大量研究工作,对该类气藏开发积累经验,为后续大规模开发奠定基础。主要表现在开发初期采用双层油管完井,因为管径小、限制了气井产能问题,随着气藏压力的降低,采用大直径(5in)单层抗硫油管,解决了地层能量的合理利用问题。另外,由于抗硫油管、套管的成本高,开发初期只在气藏连通性好的构造顶部打生产井。最后,开发初期地面天然气采输系统和净化处理系统规模不宜太大,为后续大规模开发积累经验。

(2)井下和集气系统整体防腐,实现气田开发有效防腐。

拉克气田在开发过程中,研制了不同型号的钢材,同时根据气藏开发不同阶段采用不同的油管序列。油管的失重腐蚀在五年后才开始变得明显,且常发生在井的下部,腐蚀形成硫化物,堵塞油管,维修困难,于是决定每月往井内灌 5% 缓蚀剂的柴油 $8 m^3$,同时为了防止硫化物沉淀产生堵塞,要注入轻质循环油不停循环。而对于集输管网防腐,对管材采用 B 级钢的无缝钢管,同时气体进入集气系统前脱水,每 $10^6 m^3$ 气中加 30L 防腐剂。通过井下和集气系统的整体防腐,实现气田开发的安全生产。

(3)实时监测评价气田生产过程,提高精细化科学管理水平。

气藏开发过程中定期进行必要的静态和动态监测,深入了解气井、气藏的开采特征和开采规律,严格控制溶有硫化氢的地下水侵入气井,减少设备管材的腐蚀,为气田的稳定生产提供保证。首先,估算边、底水的体积,拉克气田边底水体积约为气藏体积的三倍左右。其次,评价边底水的活跃度。根据测井解释,最下层水层渗透率低,水层的产能很低,垂向的夹层分布较多,底水活跃性差。但是气藏南北两翼较陡,储层翼部的高渗透段出露在气水界面以下,因此有可能造成边水活跃的局面。第三,在编制开发方案时分析了打开程度对底水锥进的影响,巨厚储层内部垂直连通性分析,Ⅲ段顶部夹层对开发效果的影响,水体大小对开发效果的影响等。在此基础上,采用了一套井网开发古近系砂砾岩和白垩系巴什基奇克组砂岩,总体上布三排井,以轴部布井为主,井距 900m 左右,总生产井 28 口(含观察井),单井产量 $100\times10^4 \sim 210\times10^4 m^3/d$ 间,采气速度 4%,年产气 $100\times10^8 m^3$,稳产 14 年,稳产期末采出程度 56.5%。

3.2.5 凝析气藏开发特征

大型凝析气藏主要特征是"压力温度异常、相态复杂、常常发生反凝析现象",俄罗斯乌克蒂尔气藏是典型的凝析气藏。

3.2.5.1 主要地质概况

乌克蒂尔气田位于俄罗斯地台东北部与乌拉尔地槽的过渡带上,属蒂曼—伯朝拉油气区的一部分。构造主要是一个长轴背斜,产层是中石炭纪 Moscow 层和 Bashkir 层。构造高度为 1440m,由石灰岩层和白云岩层交替组成,夹层平均厚度为 1.5m。气层有效厚度 170m,含气面积 356km², 最终可采储量 4300×10⁸m³, 凝析油储量 1.42×10⁸t。整个气藏中储层特性差异性大, 产层孔隙空间的特征复杂, 孔、洞、缝分布不均, 地层明显特点是裂缝发育, 存在着大量方向不同的张开裂缝[21]。

3.2.5.2 主要开发概况

乌克蒂尔气田于1963年钻井, 1964年发现下二叠统气层, 该气田发现时储层原始饱和度为77.5%, 压力为36MPa, 温度61℃, 小部分边缘含有轻质原油。该气田下有个含水区, 但水驱作用不明显且横向分布不均匀。考虑气田裂缝发育及存在小断层等, 采气速度和压差过大将引起气井过早水淹和降低采收率, 方案年产气 $150×10^8m^3$, 设计稳产七年, 平均单井产量 $5.28×10^4m^3/d$。为防止底水锥进, 气井井底在气水界面以上 100~150m, 保持稳定采气压差 5.8~7.8MPa。

3.2.5.3 气藏开发特征

(1) 取全取准PVT相态数据是凝析气藏开发的基础。

对凝析气藏来说, 在高温状态下, 当地层压力低于凝析气的露点压力时, 从地层中析出凝析油残留和吸附在岩石颗粒的表面, 地下形成油气两相, 称为"反凝析"现象。在凝析气藏开发过程中, 储层油气体系在地下和地面都会发生反凝析现象, 气井既产气又产凝析油。因此, 凝析气藏开发比一般气藏开发具有其特殊性和复杂性。同时, 在开发过程中, 随着压力的下降和温度的变化, 油气体系相态和组成随时随地都会发生变化, 所以一定要十分重视获得气藏原始压力、温度条件下的准确的、有代表性的凝析油气样品, 有高质量的PVT和相态分析实验数据, 很好地拟合状态方程参数, 建立凝析气相态模型, 为组分模型和数值模拟技术的准确应用打下扎实的基础。乌克蒂尔气藏在开发初期就对地层流体等数据进行取样, 同时相应地发展一套先进适用的油气取样和实验分析技术, 为气藏的开发提供坚实的资料基础[22]。

(2) 优选凝析气田开发方式是该类气藏高效开发的关键。

凝析气藏由于其含有凝析油的特殊性, 决定其在开发方式的选择上与常规气藏有较大区别。为了尽可能地提高干气、凝析油和原油的采收率, 凝析气藏的开发方式显得尤为重要。凝析气田的开发方式可分为衰竭开采和保持压力开采。衰竭开采主要用于原始地层压力高, 气藏面积小, 凝析油含量少, 地质条件差, 边水比较活跃的气藏。由于乌克蒂尔气田地质情况比较复杂, 因此采用衰竭气驱作为一次生产机理来开发, 而没有实施循环注气方法。到开发后期, 乌克蒂尔气田储层压力为 3.5~5MPa, 天然气采收率在83%左右, 而凝析油的采收率仅仅32%, 因此大约还有 $1×10^8t$ 的凝析油滞留在储层中。针对这种情况, 气田实施了一系列不同的先导性试验来开采滞留的凝析油。先导性试验结果表明, 使用丙烷和丁烷溶剂带对增加凝析油采收率效果不太明显; 注干气的方法可以明显提高采收率, 但是需要注入大量的干气; 先注入溶剂, 然后注入干气, 注入量达到一定体积后, 气井恢复生产, 在六个月到一年半的时间内, 气井产液量提高了20%~40%, 随后有下降到原来的产量水平。实践证明, 凝析气藏的开发要不失时机地选择相适应的开发方式可最大限度地提高凝析油的采收率[23]。

(3) 建立全过程的凝析气田监测系统。

由于凝析气藏的独特特征, 建立完善的开发监测系统有利于凝析气田开发和开采动态分析、开发方案的及时调整和修改, 是凝析气田开发方案设计的配套系统工程之一。开发监测的主要任务是保证综合观察气藏开采全过程, 目的是评估采纳的开发系统的有效性, 以及地质技术措施的可行性。同时要决策调整和完善开发过程,以达到高的最终采收率和最佳的开发效益。乌克蒂尔气田开发方案研究中,一开始就重视气田监测系统的建立。气田开发计算钻探和探井改建六口井用于监测气藏气水界面、地层压力等,在气田开发动态分析和开发规模调整中起到重要作用。

4 大型气田开发值得借鉴的经验

（1）大气田的开发宜采取"整体部署,分步实施"的技术思路。

俄罗斯乌连戈伊气田、亚姆布尔气田、梅德维日等大型气田的开发均采用主力气藏先行开发,主力气藏的主力区块先行开发,然后逐步加深,实现区块和层间的接替。同时,"三高"气田的开发宜开辟试验区先行试采,再大规模开发,积累经验,逐步推进[24]。

（2）取全取准每口井的静动态资料,早期识别储层类型、驱动类型是大型气田有效开发的基础。

天然气开发具有"较稀井网开发"的特点,这增加了认识气藏的难度。气藏基本特征和驱动类型的认识错误,会导致气田开发方案和建设的严重失误。只有在少数探井的基础上,珍惜每次窥探地下的机会,取全取准有限的静动态资料,开展早期评价,分析气藏类型以及驱动类型,才能为气藏的大规模开发奠定基础。

（3）井型和井网井距优化是大型气田有效开发的核心。

气藏不是"铁板一块",要努力寻找高产发育区,避免打无效或者是低效井。要综合地质、地震、气藏工程等学科技术,进行储层横向预测,气藏描述要优选超前进行,寻找富集区、裂缝发育带布井。只要对气田的地质特征有比较客观、实际的认识,开发的技术和方法总会有效;反之,若在地质认识上发生了偏差,那么即使很强的技术实力也不会取得好的效果[25]。对于大型气田开发井型、井网部署,要因地制宜,根据不同类型气藏的静态和动态特征采用不同的井型、井网部署原则;要努力寻找高产富集区,尽量减少无效和低效井;同时要努力保持全气藏均衡开采,动用尽可能多的储量,提高气藏的最终采出程度。

（4）优化气藏开发规模和气井产量是大型气田高效开发的核心。

优化气井产量及气藏开发规模是任何类型气藏方案设计和方案调整的重要任务,要做到开发规模、稳产供气年限和市场需求的有机结合。气井配产主要考虑无阻流量、排泄区内的压降储量和有无地层水干扰等三个因素限制。一般大气田的采气速度要低一些,有水气藏采气速度比气驱气藏要低。

（5）动态监测是大型气田有效开发的保证。

全球大型天然气藏的开发,均设立了一大批观察井、水层测压井,以观察底水上升、气藏压力变化等。贯穿整个生命周期的动态监测让气藏流体、压力以及腐蚀等变化了然于胸,可以提前预知气藏的动态变化,为制定有针对性的对策留足回旋余地,实现气藏的安全高效开发。

（6）发展有针对性的设备、工艺技术流程是大型气藏高效开发的关键。

大型气藏的开发涉及的产量高,时间长,影响范围广,因此对大型气藏的开发要形成有针对性的配套技术和材料装备。例如丛式井组中大直径井、定向井、水平井和复杂结构井的钻井、完井、开采工艺、固井、防水治沙工艺、排水采气工艺、安全防腐工艺等方面形成系列配套技术。

5 结论

（1）大型气藏主要分布在中东和东欧、中亚、俄罗斯的西西伯利亚盆地、波斯湾盆地、扎格罗斯盆地、卡拉库姆盆地等几个大型盆地中；大型气藏主要分布在石炭纪到新近纪的地层中，深度介于 1500~3000m；构造气藏仍是大型气藏的主要类型；大型气藏的可采储量增长同理论技术的进入、优惠政策以及天然气价格密切相关。

（2）依据"实用性、针对性、科学性"划分原则，围绕岩性、厚度、规模、物性、压力、流体等因素将大型气藏划分为五种类型：厚层整装高渗砂岩气藏；低渗砂岩气藏；边底水裂缝型碳酸盐岩气藏；"三高"气藏以及凝析气藏。厚层整装砂岩气藏主要特征是："规模大、储层厚、物性好、或存在边底水"；低渗透砂岩气藏的主要特征是规模大、物性差、一般不存在边底水；边底水裂缝型碳酸盐岩气藏的主要特征是规模大、裂缝发育、存在边底水；"三高"气藏的主要特征是高温或高压或高含硫；凝析气藏的主要特征是压力温度异常、相态复杂、常常发生反凝析现象。

（3）全球大型气藏的分布特征为我国天然气藏的勘探指明了方向，全球大型气藏开发过程和开发特征为我国相同类型气藏的经济、高效、安全开发提供弥足珍贵的经验。

参 考 文 献

[1] Halbouty M T. Giant oil and gas fields of the 1990s: an introduction [M]. Halbouty M T. Giant Oil and Gas Fields of the Decade 1990-1999, AAPG Menoir 78 Tulsa: AAPG, 2003:1-13.

[2] Halbouty M T. Geology of Giant Petroleum Fields, AAPG Memoir 14[M]. Tulsa: AAPG, 1970:1-575.

[3] Halbouty M T. Giant Oil and Gas Fields of the Decade 1968-1978, AAPG Memoir 30[M]. Tulsa: AAPG, 1980:1-596.

[4] Halbouty M T. Giant Oil and Gas Fields of the Decade 1978-1988, AAPG Memoir 54[M]. Tulsa: AAPG, 1992:1-526.

[5] Halbouty M T. Giant Oil and Gas Fields of the Decade 1990-1999, AAPG Memoir 78[M]. Tulsa: AAPG, 2003:1-340.

[6] 李国玉,金之钧. 世界含油气盆地图集(上下册)[M]. 北京:石油工业出版社,2005.

[7] 李国玉,唐养吾. 世界气田图集[M]. 北京:石油工业出版社,1991:1-67.

[8] 宋芊,金之钧. 大油气田统计特征[J]. 石油大学学报:自然科学版,2000,24(4):11-14.

[9] 白国平,郑磊. 世界大气田分布特征[J]. 天然气地球科学:2007,18(2):161-167.

[10] 贾爱林,闫海军,郭建林,等. 不同类型碳酸盐岩气藏开发特征[J]. 石油学报,2013,34(5):914-923.

[11] 何光怀,李进步,王继平,张吉. 苏里格气田开发技术新进展及展望[J]. 天然气工业,2011,31(2):12-16.

[12] 何东博,贾爱林,冀光,等. 苏里格大型致密砂岩气田开发井型井网技术[J]. 石油勘探与开发,2013,40(1):79-89.

[13] 何东博,王丽娟,冀光,等. 苏里格致密砂岩气田开发井距优化[J]. 石油勘探与开发:2012,39(4):458-464.

[14] 王丽娟,何东博,冀光,等. 低渗透砂岩气藏产能递减规律[J]. 大庆石油地质与开发,2013,32(1):82-86.

[15] 凌宗法,王丽娟,胡永乐,等. 水平井注采井网合理井距及注入量优化[J]. 石油勘探与开发:2008,35(1):85-91

[16] 马新华,贾爱林,谭健,等. 中国致密砂岩气开发工程技术与实践[J]. 石油勘探与开发,2012,39(5):572-579.
[17] 李海平,贾爱林,何东博,等. 中国石油的天然气开发进展及展望[J]. 天然气工业,2010,30(1):5-7.
[18] 孙志道. 裂缝性有水气藏开采特征和开发方式优选[J]. 石油勘探与开发,2002,29(4):69-71.
[19] 闫海军,贾爱林,郭建林,等. 龙岗礁滩型碳酸盐岩气藏气水控制因素及分布模式[J]. 天然气工业,2012,32(1):67-70.
[20] 李士伦. 气田开发方案设计[M]. 北京:石油工业出版社,2004.
[21] 胡文瑞,马新华,李景明,等. 俄罗斯气田开发经验对我们的启示[J]. 天然气工业,2008,28(2):1-6.
[22] 李士伦,潘毅,孙雷. 对提高复杂气田开发效益和水平的思考与建议[J]. 天然气工业,2011,31(12):76-80.
[23] 胡永乐,李保柱,孙志道. 凝析气藏开采方式的选择[J]. 天然气地球科学,2003,14(5):398-401.
[24] 方义生,徐树宝,李士伦. 乌连戈伊气田开发实践和经验[J]. 天然气工业,2005,25(6):90-93.
[25] 李士伦,汪艳,刘延元,等. 总结国内外经验,开发好大气田[J]. 天然气工业,2008,28(2):7-11.

原文刊于《天然气工业》,2014,34(10):33-46.

大型低渗透—致密气田井网加密提高采收率对策
——以鄂尔多斯盆地苏里格气田为例

贾爱林　王国亭　孟德伟　郭　智　冀　光　程立华

(中国石油勘探开发研究院,北京 100083)

摘要:提高采收率是中国低渗透—致密气田稳产期间面临的核心问题,确定合理的加密井网是提高储量动用程度的关键。明确了有效储层规模尺度与四种空间组合类型,评价指出气田动静储量比仅 15.3%,储量动用程度低,剩余储量规模大,划分出直井未动用、水平井遗留、井间剩余三种剩余储量类型。提出井网加密是提高井间剩余储量动用程度的有效措施,构建了采收率、采收率增量、平均气井产量、加密井增产气量、产量干扰率等井网加密评价指标体系,确立了合理加密井网需满足的标准。结合地质模型、数值模拟、密井网试验数据验证等手段,综合评价认为合理加密井网应与有效储层组合类型相匹配、与气价及成本条件密切相关。在目前气价波动范围及经济技术条件下,苏里格气田采用 4 口/km² 的加密井网是合理的。

关键词:低渗透致密气田;储量动用程度;剩余储量;提高采收率;井网加密;合理井网

中国低渗透—致密砂岩气藏主要分布于鄂尔多斯、四川、塔里木等沉积盆地,具有巨大的资源潜力和储量规模[1-5]。苏里格气田是中国低渗透—致密砂岩气藏的典型代表,位于鄂尔多斯盆地北部中段,由中区、西区、东区及南区四个区块构成(图1)。气田叠合含气面积约 $4×10^4 km^2$,探明储量(含基本探明)规模达 $4.77×10^{12} m^3$,储层孔隙度介于 2%~14%,渗透率介于 0.01~2.0mD,属大型低渗透—致密砂岩气藏[6]。近三年来气田产量持续超过 $220×10^8 m^3/a$,是中国目前探明储量、年产量和累计产气量最高的气田。

苏里格气田目的层为上古生界下石盒子组盒八段和山西组山一段,属陆上辫状河沉积,有效储层主要为辫状河心滩和辫状河道底部粗粒沉积[7-8]。经过多年持续攻关,形成了甜点区筛选、快速钻井、储层改造、井型井网优化等一系列关键开发技术,实现了气田规模效益开发[9-11]。随着开发深入,气田各

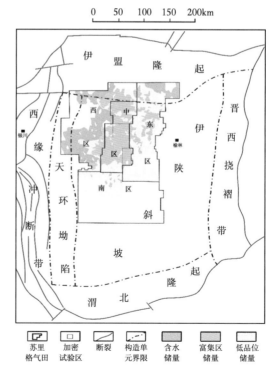

图 1　鄂尔多斯盆地苏里格气田区块构成图

区储量品质差异逐渐显现,中区储量品质佳且开发效果最好,因受地层水严重影响西区储量有效动用面临挑战,东区、南区储量品位较低开发风险较高[12-16]。气田开发形势由持续上产转变为实现长期稳产,受储量品质不均衡的制约,采用区块接替的方式维持气田稳产终将不可持续,因此开展储量动用程度分析、明确剩余储量类型及分布,并确定合理加密技术对策对实现气田长期稳产具有重大意义。

1 有效储层规模与空间组合

苏里格气田储层非均质性强,表现出"二元"结构特征,有效储层呈"透镜状"包裹于厚层基质砂岩中[17-18]。辫状河沉积体系空间发育具有明显的分带性,不同区带有效储层的规模和空间组合结构都具有较大差异[19]。上述差异是造成有效储层强非均质性的主要原因,也是造成气井生产动态差异的根本原因。明确有效储层规模尺度和空间组合类型是苏里格气田提高采收率研究的地质基础。

1.1 有效储层规模

有效储层规模分析包括厚度、宽度、长度及展布面积等参数的评价。在开发早期阶段受动静态资料有限的制约,主要依靠有限钻井、野外露头及室内模拟等开展储层规模评价,因此具有较大不确定性。随着气田开发深入,苏里格气田钻井数量快速增加、已达万口规模,加密试验区、井间干扰试验的数量也不断增加,为有效储层精细解剖提供了良好条件。目前气田共有加密井网试验区八个,面积 $2.5 \sim 9.9 km^2$,井网密度 $2.4 \sim 5.0$ 口$/km^2$,并开展了42个井组的干扰试验(表1)。

表1 苏里格气田加密试验区综合数据

参数	苏6	苏36-11	苏14南部	苏14北部				
				A	B	C	D	E
面积(km^2)	9.90	2.60	4.80	3.30	2.80	2.30	2.80	2.50
储层丰度($10^8 m^3/km^2$)	1.46	2.17	2.16	1.25	1.56	1.28	1.80	2.62
区内井数(口)	31.00	13.00	18.00	11.00	8.00	7.00	7.00	7.00
井网密度(口$/km^2$)	3.10	5.00	3.80	3.30	2.90	2.50	2.50	2.80
干扰试验(组)	12.00	5.00	8.00	0	3.00	5.00	4.00	5.00

解剖分析表明:苏里格气田盒八段及山一段有效单砂体的厚度范围为 $1.0 \sim 8.0m$,主要分布在 $1.5 \sim 4.0m$ 之间;有效单砂体宽度范围 $100 \sim 600m$,主要分布在 $300 \sim 500m$ 之间,有效单砂体长度范围为 $200 \sim 800m$,主要分布于 $400 \sim 600m$;有效单砂体面积范围为 $0.02 \sim 0.50km^2$,主要范围为 $0.12 \sim 0.30km^2$(图2)。基于上述分析可知,苏里格气田有效储层规模差异较大,随着气田开发精细程度的提高,基于平均规模有效砂体开展粗线条式评价研究以宏观指导气田开发已经难以适应精细开发需求,因此有效储层空间组合精细描述刻画将是未来开发地质研究的关键。

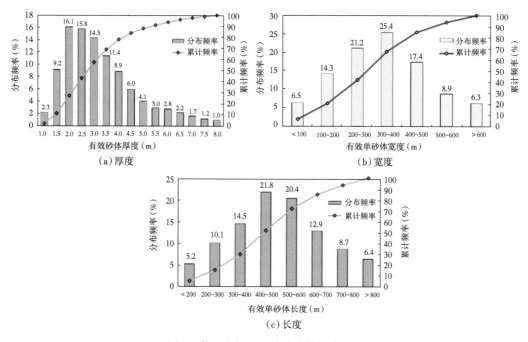

图 2 苏里格气田有效单砂体规模统计

1.2 有效储层空间组合类型

苏里格辫状河沉积体系是地质历史时期物源、水动力、古地形、可容空间以及沉积物供给等地质因素共同作用的结果,空间分布具有区域性差异,可分为辫状河体系叠置带、过渡带和体系间三个宏观相带,沉积水动力由强到弱、可容空间由大到小、砂体叠置期次由多到少、砂体连通性及连续性由好到差[19]。结合辫状河体系带划分理论,基于重点区块解剖,综合动静态多参数指标将有效储层空间组合划分为高丰度厚层连通型、中丰度多期叠置半连通型、低丰度多期分散叠置型及特低丰度薄层孤立型四种主要类型(表2)。

高丰度厚层连通型发育于水动力条件最强的辫状河体系叠置带主体部位,有效储层总厚度大、储量丰度高、气井累计产量高,由规模尺度最大、连通性最好的有效储层多层分布而成。中丰度多期叠置半连通型分布于水动力条件较强的叠置带边翼部位,有效储层总厚度较大、丰度中等、气井累计产量较高,由规模较大的有效储层多层切割叠置而成,连通性相对较好。低丰度多期分散叠置型发育于水动力条件中等的辫状河体系过渡带,有效储层总厚度较小、储量丰度偏低、气井累计产量也较低,以中等规模有效砂体多层分散叠置而成,连通性差。特低丰度薄层孤立型发育于水体条件最弱的辫状河体系带间,有效层总厚度最小、储量丰度最低、气井累计产量最小,由规模尺度最小的有效储层孤立发育而成(图3)。结合上述有效储层空间组合类型划分,可进行平面展布评价,从而为气田提高采收率分析奠定地质基础。前三种类型构成苏里格气田优质储量,是目前效益开发动用的主体,后一种类型储量品位低,目前尚难开发动用。

表2 苏里格气田有效储层空间组合划分

类型	丰度 ($10^8 m^3/km^2$)	有效储层总厚(m)	气井平均EUR ($10^4 m^3$)	组合特征 层数	组合特征 主力砂厚(m)	组合特征 规模尺度	沉积部位
高丰度厚层连通型	≥2.0	≥18.0	≥3600	≥3	≥6.0	以宽大于600m、长大于800m厚层块状、规模尺度较大的有效砂体多层空间组合为主	辫状河体系叠置带主体部位,水动力条件最强,有效砂体规模最大、连通性最好
中丰度多期叠置半连通型	1.5~2.0	12.0~18.0	2400~3600	≥3	4.0~6.0	主要为宽400~600m、长600~800m的中等规模有效砂体,空间多层切割叠置组合为主	辫状河体系叠置带侧边翼部位,水动力条件相对较强,有效砂体规模较大、连通性较好
低丰度多期分散叠置型	1.0~1.5	6.0~12.0	1200~2400	2~3	2.0~4.0	以宽200~400m、长300~600m中等规模有效砂体多层、分散叠置空间组合为主	辫状河体系过渡带,水体条件中等,有效砂体规模中等、连通性差
特低丰度薄层孤立型	<1.0	<6.0	<1200	<2	<2.0	主要为宽小于200m、长小于300m薄层孤立、小规模有效砂体空间组合为主	辫状河体系带间,水体条件最弱,有效砂体规模最小、连通性最差

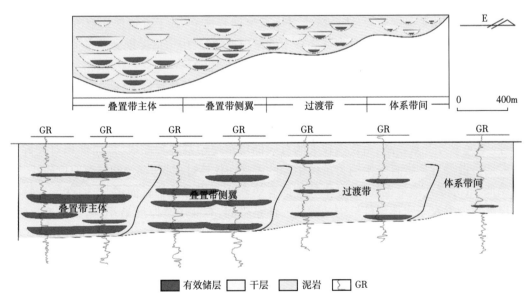

图3 苏里格气田有效储层空间发育组合特征

2 气田剩余储量评价

2.1 储量动用程度

苏里格气田目前存在多种井网井型,包括600m×800m、600m×1200m、500m×500m及试验区密井网及不规则井网等直井开发井网,还包括直井/水平井混合井网及水平井井网等,总体以600m×800m直井基础开发井网为主。中区是气田重要组成部分,开发相对较早、储量品质较优,在其内选取重点区块开展储量动用程度分析具有较强代表性。气田或区块内所有气井动态控制储量累加和是目前已经动用的储量,其与地质储量的比值即动静储量比,能够有效反映地质储量的动用程度。气田中区四个典型区块1500余口井的评价结果表明,动静储量比为11.4%~18.8%,平均仅为15.3%(表3)。评价表明,气田当前储量动用程度偏低,剩余储量规模较大,采收率具备大幅提升的空间。

表3 苏里格气田重点开发区块储量动用程度评价

典型区块	直井		水平井		区块已动储量（$10^8 m^3$）	区块地质储量（$10^8 m^3$）	动静储量比（%）
	单井动态控制储量（$10^8 m^3$）	井数（口）	单井动态控制储量（$10^8 m^3$）	井数（口）			
苏36-11	0.3210	277	0.7829	67	141.4	752.8	18.8
苏6	0.2768	266	0.6831	29	93.4	819.4	11.4
苏14	0.3007	569	0.7283	70	222.1	1288.6	17.2
桃2	0.3154	291	0.7877	53	133.5	1010.1	13.2
平均或合计	0.3035	1503	0.7455	219	590.4	3870.9	15.3

2.2 剩余储量分类

剩余储量是气田未来开发动用的重点,也是气田长期稳产的基础。结合气田目前开发井网井型及储层改造措施,将剩余储量划分为直井未改造型、水平井遗留型、井间剩余型三种主要类型。

苏里格气田气井钻遇多个有效砂体,储层改造措施主要针对物性好、含气饱和度高、厚度大的层段,少量物性及含气性较差的层段未进行压裂改造,形成直井未压裂改造型剩余储量(图4)。结合未改造气层厚度、气井数量、储层物性及含气性资料,开展此类剩余储量规模分析。以提高单井产量为目标,2010年后苏里格气田大力实施水平井开发,开发的主要地质目标是针对垂向储量集中度大于60%的层段[20]。由于水平井对开发地质目标的选择性,导致水平段剖面方向上遗留了大量储量,即水平井遗留型剩余储量(图5)。结合水平井储量集中度、水平井数量及泄气范围等,进行此类剩余储量规模评价。井间剩余储量是指分布于气井之间的、泄气范围之外的未动用储量,其规模为地质储量与开发已动用储量、直井未压裂改造型剩余储量、水平井遗留型剩余储量的差值。

剩余储量评价结果表明,直井未压裂改造型占总储量规模的比例为1.37%,水平井遗留型占比为3.03%,而井间型占比为80.3%(图6)。因此,井间剩余型是苏里格气田剩余储量主体,是未来提高气田采收率的重要目标。在当前经济技术条件下,井网加密是提高此类剩余储量的最有效措施。

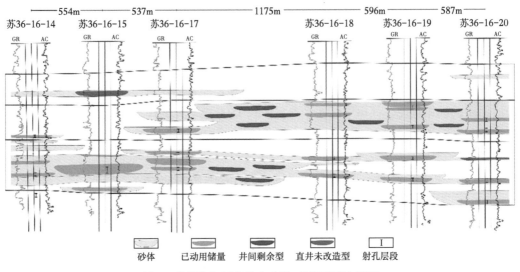

图4 苏里格气田直井未改造、井间型剩余储量

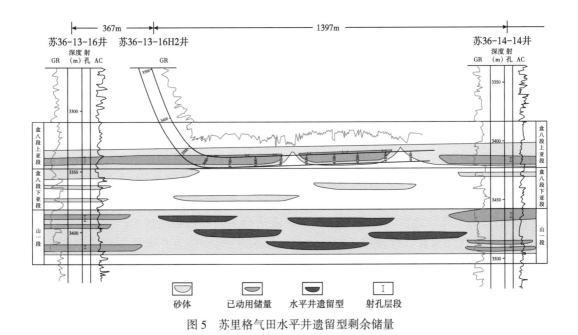

图5 苏里格气田水平井遗留型剩余储量

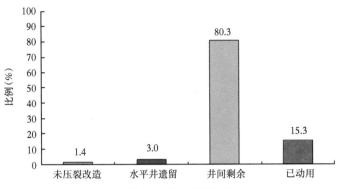

图 6 苏里格气田剩余储量类型及比例

3 井网加密技术对策

国内外低渗透—致密气的开发实践表明,井网加密是提高井间剩余储量动用程度最有效的技术措施,评价方法主要有地质统计法、移动窗口法、动态评价法等[21-25]。地质统计法在典型气井选取的基础上,明确气井泄气面积频率分布曲线,并获取泄气面积与单井最终可采储量对应关系,计算当前井网密度与气井平均泄气面积的差值,将此差值对应的单井最终可采储量作为加密气井最终产气量,将其与经济下限产量比较从而评价加密的可行性。移动窗口法首先选定合适大小的"窗口",利用生产动态数据评价"窗口"内是否存在井间干扰,不断移动"窗口"直至覆盖整个研究区域,筛选能够进行井网加密的潜力区,再针对每个潜力区进行分析,找出未动用区域,确立加密井井位,并评价加密后的新增可采储量。动态评价法通过分析井间干扰率与井网密度之间的关系,建立开发井网优化数学模型,得到气田采收率和井网密度之间的定量描述,并以井间不产生大量干扰作为合理加密井网确定的重要前提条件。

前两种方法虽可用于加密井网快速评价,但未能充分考虑有效储层空间组合的差异型和强非均质性,也缺乏系统评价指标体系,难以适应气田精细开发需求;后一种方法以不接受井间大量干扰为前提条件,在目前中国天然气供需失衡、气荒时常发生,且未来中国天然气需求巨大的背景下,以上述前提作为合理加密井网确定的依据将难以适应天然气深入开发的需求。基于上述不足,构建了井网加密开发评价系统指标,并充分考虑有效储层组合结构类型,开展了系列井网密度下指标评价,明确了不同气价情景下合理加密技术对策,并落实井间干扰率程度与不同加密井网密度的定量关系,以适应低渗透致密气田的精细化开发的需求。

3.1 井网加密开发评价指标的构建

苏里格气田开发实践表明,早期制定的系列开发评价指标有效指导了气田开发,包括井型井网、采气速度、气井配产、采收率等。随着气田开发形势的转变,提高采收率成为未来气田开发的核心目标,早期指标难以适用于开发中后期井网调整加密评价的需求,因此构建了采收率、采收率增量、平均气井产量、加密气井增产气量、产量干扰率等系列开发指标组合。

3.1.1 采收率

采收率是衡量气藏整体开采效果的重要指标,指从气藏中最终采出天然气量占探明地质储量的比值。600m×800m 基础开发井网下苏里格气田采收率约为30%，相对于国外同类气藏50%以上的采收率[17]，还有较大差距。计算方法为

$$c_n = q_n/r \times 100\% \tag{1}$$

3.1.2 采收率增量

对于低渗透—致密砂岩气藏而言，随着开发井网密度增加，采收率逐渐提高。为了定量评价随着井网密度增加采收率的提升幅度，提出了采收率增量指标，是指井网密度每增加 1 口/km² 的情况下采收率的提高幅度。计算方法为

$$\Delta c = c_n - c_{n-1} \tag{2}$$

3.1.3 平均气井产量

平均气井产量是衡量气井产能的关键指标，是指总产气量之和与气井数量的比值，即气井最终累计产气量 EUR。在明确气井单井综合投资的条件下，可基于此进行整体开发效益评价。计算方法为

$$w_n = q_n/(n \times s) \tag{3}$$

3.1.4 加密气井增产气量

随着开发井网变密，气井间连通有效储层增多，早期部署气井产量会受到加密气井的影响。加密井采出的天然气由两部分组成：一是与早期气井连通的有效储层产出的天然气，二是钻遇到的新有效储层产出的天然气（图7）。对于前者，即使不部署加密井，早期井网气井也能够将最终其采出，此部分天然气对采收率提高无贡献作用，加密气井仅提高了采气速度。此外，加密时机影响此部分气量大小，部署越早越大、越晚则越小。

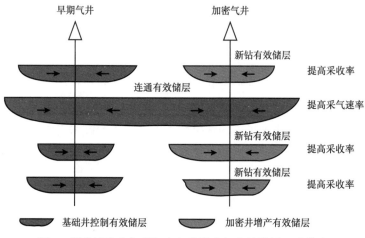

图 7 低渗透致密气藏加密气井增产气有效储层地质模式

对于钻遇到的新有效储层产出的天然气，是加密井真正多增产的天然气量，这部分天然气是对提高采收率起真正的贡献作用。将加密井增产气量定义为井网密度每增加 1 口/km² 的

情况下每口新增加密井从地层中多采出的天然气量。计算方法为

$$\Delta w = (q_n - q_{n-1})/(1 \times s) \quad (4)$$

加密井增产气量是很重要的评价指标,对评价合理井网密度具有关键作用。若加密井增产气的收益能够回收加密井综合投资,是可进行加密井部署的必要条件之一。

3.1.5 产量干扰率

井网加密后井间连通有效砂体增加,导致早期 600m×800m 基础井网气井产量的降低。为了定量评价加密井对早期气井产量的影响,提出产量干扰率的评价指标,是指加密前、后基础井网骨架气井产量降低的百分比。计算方法为

$$g = (w_o - w_n)/w_o \times 100\% \quad (5)$$

上述五个指标总体可归纳为气田总体指标和气井指标两类,总体指标包括 c_n、Δc,气井指标包括 w_n、Δw、g。低渗透—致密气藏井网调整加密后,既能明显提高采收率,又能满足一定的经济效益才是合理的。基于苏里格气田开发实际,确立了上述指标需同时满足的标准:气田整体达到较高的开采程度,采收率 c_n 在 45%以上;采收率增量提升明显,井网密度每增加 1 口/km² 采收率增量 Δc 在 5%以上;加密后气井平均产量 w_n 满足开发方案设计要求的收益率标准;加密井增产气量 Δw 能够回收其综合投资成本,即加密井能够自保;产量干扰率有助于定量评价基础井网气井产量的降低幅度(表4)。

表4 苏里格低渗透—致密砂岩气田合理加密井网评价指标标准

指标体系	区块/气田指标		气井指标		
	采收率 c_n(%)	采收率增量 Δc(%)	平均气井产量 w_n($10^4 m^3$)	加密井增产气量 Δw($10^4 m^3$)	产量干扰率 g(%)
需同时满足的标准	井网加密后整体达到较高开采程度,采收率≥45%	井网密度每增加1口/km²,采收率提升效果明显,采收率增量≥5%	井网加密后,总体能实现较好经济效益,满足方案设计要求的内部收益率标准,一般≥12%	加密气井增产天然气量能够回收其单井综合投资成本,即加密井自身能够自保,内部收益率≥0	为辅助指标,在其他指标能够满足的条件下,可接受不同程度的产量干扰

3.2 井网加密论证

以苏里格气田有效储层空间组合类型为基础,优选典型区块进行精细地质建模,并设计 2~8 口/km² 的井网密度系列,采用数值模拟手段系统论证加密指标体系,以密井网试验区实际资料对模拟结果进行验证,并结合气价及气井综合投资情景,最终确定合理加密井网。苏里格气田目前气井固定成本为 800 万元,以银行贷款 45%、利率 6%、操作成本 120 万元、折旧 10 年为模型,并综合考虑销售税金、城市建设、教育附加、资源税等,确定了系列气价下达到不同收益率标准所需最低气井产量(表5)。在固定单井综合投资成本的情景下,合理开发井网与气价条件密切相关,气价越高达到要求的收益率标准所需气井产量则越低,合理开发井网密度则越大,气价越低则反之。以高丰度厚层连通型、中丰度多期叠置半连通型、低丰度多期叠置

分散型等有效储层空间组合类型为基础,分别开展不同气价下合理井网密度的确定(图8)。

表5 苏里格气田不同气价情境下气井经济产量数据

气价 (元/10³m³)	达到收益率 0所需最低气井 产量(10⁴m³)	达到收益率 12%所需最低气井 产量(10⁴m³)	气价 (元/m³)	收益率 0所需最低气井 产量(10⁴m³)	收益率 12%所需最低气井 产量(10⁴m³)
1.00	1278	1786	1.55	756	1057
1.05	1203	1681	1.60	728	1019
1.10	1135	1587	1.65	703	984
1.15	1075	1504	1.70	680	951
1.20	1021	1428	1.75	658	920
1.25	972	1360	1.80	637	892
1.30	928	1298	1.85	617	865
1.35	887	1242	1.90	600	839
1.40	850	1190	1.95	582	815
1.45	816	1142	2.00	566	793
1.50	785	1098	2.05	551	773

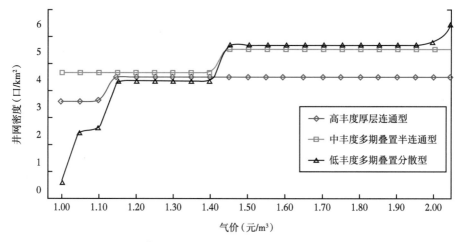

图8 苏里格气田合理开发井网密度与气价关系

3.2.1 高丰度厚层连通型

为能准确获取系列指标,建模区的选取应该具有代表性,面积以不小于10km²为宜,动静态资料丰富。针对高丰度厚层连通型有效储层,在气田中区选取建模区10.9 km²,丰度为2.2×10⁸m³/km²,储量规模为14.45×10⁸m³。模拟结果表明:由于储层连通性好,2口/km²基础井网下即可达到45.4%的采收率,随着井网密度增加采收率呈先快、后慢的增加趋势,采收率增量先大、后小,3~4口/km²井网密度下采收率增量在5%以上,采收效果提升较为明显,高于4口/km²时采收率提升效果降低;由于储量丰度大、连通性好,平均气井产气量高,基础井网气井产量高达 4997.1×10⁴m³,随井网密度增加呈降低趋势,受储层连通性好的影响,加密井增产气量呈快速降低的趋势;基础井网气井产量受干扰严重,3 口/km² 井网密度下干扰程度达 20.49%,随井网密度增加干扰率逐渐增加(图9,表6)。

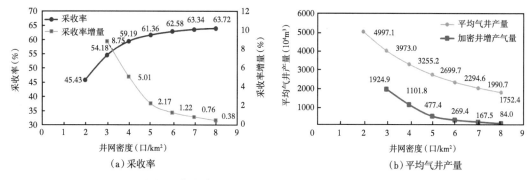

图 9 高丰度厚层连通型有效储层井网加密指标

表 6 高丰度厚层连通型有效储层井网加密指标数值模拟及现场加密试验成果数据

井网密度 (口/km²)	数值模拟结果					现场加密试验结果				
	c_n (%)	Δc (%)	w_n (10⁴m³)	Δw (10⁴m³)	g (%)	c_n (%)	Δc (%)	w_n (10⁴m³)	Δw (10⁴m³)	g (%)
2	45.43	—	4997.1	—	—	45.27	—	4912.1	—	—
3	54.18	8.75	3973.0	1924.9	20.49	54.06	8.79	3910.2	1906.4	20.40
4	59.19	5.01	3255.2	1101.8	34.86	59.08	5.02	3205.2	1090.2	34.75
5	61.36	2.17	2699.7	477.4	45.98	61.15	2.07	2654.1	449.8	45.97
6	62.58	1.22	2294.6	269.4	54.08	—	—	—	—	—
7	63.34	0.76	1990.7	167.5	60.16	—	—	—	—	—
8	63.72	0.38	1752.4	84.0	64.93	—	—	—	—	—

为验证数值模拟结果的准确性,采用实际现场加密试验数据进行验证。苏 36-11 加密试验区是此类储层代表,试验面积为 2.6km²,储量丰度为 2.17×10⁸/km²,井网密度达 5 口/km²,并进行了五组井间干扰试验测试,且区内气井生产时间都相对较长,总之试验区动静态资料较多,为验证分析提供了必要的数据资料(图10,表1)。系统评价加密试验区井网密度在 2～5 口/km² 范围内的各项指标,并将其与模拟结果进行对比(表6)。结果表明,二者各项指标基本接近,验证了模拟结果的可靠性。

基于表 4 确定的井网加密指标,针对高丰度厚层连通型有效砂体开展 1.0～2.0 元/m³ 气价范围内合理开发井网密度评价。

分析结果表明,当气价低于 1.15 元/m³ 时,3 口/km² 为合理开发井网密度;当气价为 1.15～2.00 元/m³ 时,4 口/km² 加密井网更为合理,由于受储层品质好、井间连通程度高的影响,井网密度大于 4 口/km² 时加密井增产气量低,难以回收单井综合投资成本(图8)。

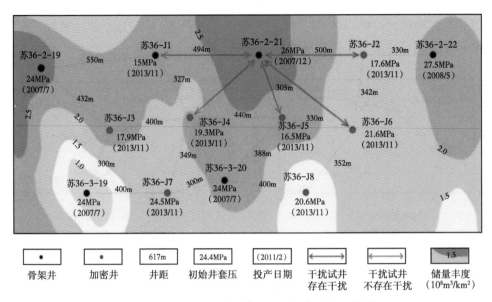

图 10　苏里格气田中区苏 36-11 加密试验区参数

3.2.2　中丰度多期叠置半连通型

针对中丰度多期叠置半连通型有效储层选取建模区 11.41km²，丰度为 1.67×10⁸m³/km²，储量规模为 19.05×10⁸m³。模拟分析结果表明，受储层连通程度降低的影响，2 口/km² 基础井网下采收率为 33.64%，随着井网密度增加，采收率呈先快、后慢的增加趋势，采收率增量先大、后小，3~5 口/km² 井网密度范围下采收率增量在 5% 以上，采收效果提升明显，高于 6 口/km² 时，采收率提升效果降低；储量丰度相对较高，平均气井产量也较高，基础井网下气井产量为 2892.8×10⁴m³，随井网密度增加呈逐渐减低趋势，由于储层连通性较好、产量干扰率较高，加密井增产气量呈逐渐降低趋势；基础井网气井产量受干扰明显，3 口/km² 井网密度下产量干扰率为 13.53%，随井网密度增加干扰率呈逐渐增加趋势（图 11，表 7）。

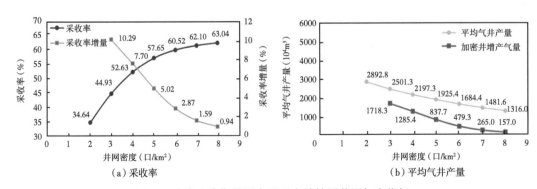

图 11　中丰度多期叠置半通型有效储层井网加密指标

表7 中丰度多期叠置半连通型有效储层井网加密指标数值模拟及现场加密试验成果数据

井网密度 (口/km²)	数值模拟结果					现场加密试验结果				
	c_n (%)	Δc (%)	w_n ($10^4 m^3$)	Δw ($10^4 m^3$)	g (%)	c_n (%)	Δc (%)	w_n ($10^4 m^3$)	Δw ($10^4 m^3$)	g (%)
2	34.64	—	2892.8			34.61		2803.4		
3	44.93	10.29	2501.3	1718.27	13.53	44.86	10.25	2422.7	1661.2	13.58
4	52.63	7.70	2197.3	1285.37	24.04	52.49	7.62	2125.7	1234.8	24.17
5	57.65	5.02	1925.4	837.66	33.44	—				
6	60.52	2.87	1684.4	479.32	41.77					
7	62.10	1.59	1481.6	264.98	48.78					
8	63.04	0.94	1316.0	156.96	54.51					

选取气田中区苏14区南部加密试验数据对上述模拟结果进行检验,南部加密试验区面积为6.3km²,丰度为1.62×10⁸/km²,加密后平均井网密度为3口/km²,局部可达4口/km²,并开展了八组干扰试验分析,气井生产时间较长,动静态资料比较丰富(图12,表1)。系统评价加密试验区井网密度在2～4口/km²时各项指标(表7),并将其与模拟结果进行对比,各项指标基本接近,证实模拟结果可信度较高。基于表4确定的井网加密可行性评价标准,当气价低于1.45元/m³时,4口/km²为合理开发井网密度;当气价为1.45～2.00元/m³时,5口/km²开发井网密度更为合理(图8)。

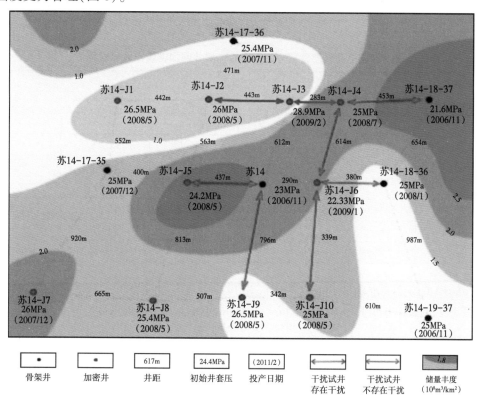

图12 苏里格气田中区苏14区块南部加密试验区参数

3.2.3 低丰度多期叠置分散型

针对低丰度多期叠置分散型有效储层,选取建模区面积为 12.83km², 丰度为 $1.27×10^8 m^3/km^2$, 储量规模为 $16.29×10^8 m^3$。模拟分析表明:受储层分散孤立的影响,2 口/km² 基础井网下采收率仅为 26.55%, 随着井网密度增加采收率呈明显增加趋势,3~5 口/km² 井网密度范围下采收率增量在 5% 以上,采收效果提升明显,高于 6 口/km² 时,采收率提升效果降低;由于储量丰度较低,此类储层平均气井产量低,基础井网下气井产量为 $1686.23×10^4 m^3$,随井网密度增加呈近直线降低趋势,受储层连通性差、产量干扰率低的影响,加密井增产气量呈缓慢降低趋势;基础井网气井产量受干扰较低,3 口/km² 井网密度下产量干扰率仅 3.04%,随井网密度增加干扰率也逐渐程度增加(图 13,表 8)。

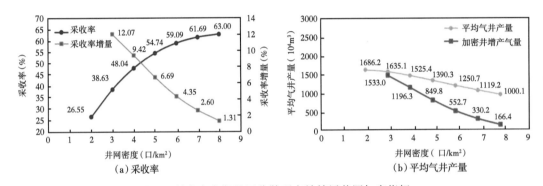

图 13 低丰度多期叠置分散型有效储层井网加密指标

表 8 低丰度多期叠置分散型有效储层井网加密模拟成果数据

井网密度 (口/km²)	c_n (%)	Δc (%)	w_n ($10^4 m^3$)	Δw ($10^4 m^3$)	g (%)
2	26.55	—	1686.2	—	—
3	38.63	12.07	1635.1	1532.9	3.04
4	48.04	9.42	1525.4	1196.3	9.55
5	54.74	6.69	1390.3	849.8	17.58
6	59.09	4.35	1250.7	552.7	25.87
7	61.69	2.60	1119.2	330.2	33.69
8	63.00	1.31	1000.1	166.4	40.76

为检验模拟结果的准确性选取气田中区苏 14 区块北部加密试验 C 区试验数据进行验证,试验面积为 2.3km², 丰度 $1.28×10^8 m^3/km^2$, 平均井网密度为 3.0 口/km², 为一次性成型井网,完成干扰试验五组(表 1),由于试验区井网密度有限,仅可对上述模拟部分数据进行验证。验证结果表明,当井网密度为 3 口/km² 时采收率为 38.96%、采收率增量为 12.12%、平均气井产量为 $1662.1×10^4 m^3$、加密井增产气量为 $1551.9×10^4 m^3$、产量干扰率为 3.21%, 与模拟结果基本一致(表 10)。

评价结果表明,当气价低于 1.05 元/m³ 时,平均气井产量低于内部收益率 12% 要求的最低产气量,不具备开发条件;当气价为 1.05~1.10 元/m³ 时,2 口/km² 的基础井网为合理开发井网;当气价为 1.15~1.40 元/m³ 时,4 口/km² 为合理开发井网;当气价为 1.45~2.00 元/m³ 时,5 口/km² 为合理开发井网密度,受储量品位低的影响,加密井增产气量有限,更高井网密度下加密井增产气量所获收益难以回收单井综合投资成本(图8)。

综合上述分析可知,低渗透—致密砂岩气田合理加密井网与有效储层类型、气价条件密切相关。随着有效储层品质的提升,即丰度增加和连通性变好,合理开发井网密度降低,而随着天然气价格的升高,合理开发井网密度增加。目前苏里格气田出厂气价波动区间为 1.15~1.40 元/m³,上述三类有效储层总体采用 4 口/km² 的调整加密井网是合理的,若气价进一步提升,后两类储层具备加密到 5 口/km² 的潜力。此外,加密井的存在会对早期基础井网气井产量产生明显影响,若能满足提高采收率、确保目标收益、加密井自保的要求,不同程度的井间干扰是可以接受的。

4 结论

(1)紧密结合苏里格气田加密试验区的动静态资料开展有效储层规模分析,明确了有效储层规模参数的分布范围,将有效储层空间组合类型划分为高丰度厚层连通型、中丰度多期叠置半连通型、低丰度多期叠置分散型及薄层孤立型四种。

(2)选取典型区块开展储量动用程度评价,气田目前动静储量比平均仅为 15.3%,动用程度较低;结合目前井网井型及储层改造措施,将剩余储量划分为直井未改造、水平井遗留、井间剩余三种类型,并明确井间剩余型是主要的剩余储量类型。

(3)提出井网加密是提高井间剩余储层动用程度的有效措施,构建了加密井网评价指标组合,制定了合理加密井网需满足的标准。确定了系列气价条件下不同类型有效储层的合理井网密度,指出合理井网应与有效储层类型相匹配、与气价条件密切相关。目前气价波动区间内,苏里格气田总体采用 4 口/km² 的调整加密井网是可行的。

符号说明

n—某开发井网密度,口/km²;c_n—井网密度为 n 时区块采收率,%;q_n—开发井网密度为 n 时气田或区块总产气量,10^4 m³;r—探明地质储量,10^4 m³;w_n—井网密度为 n 时平均气井产量,10^4 m³;s—开发面积,km²;$n×s$—总开发井数,口;Δw—井网密度为 n 时加密气井增产气量,10^4 m³;q_{n-1}—井网为 $n-1$ 时总产气量,10^4 m³;$1×s$—新增加密井数量,口;g—产量干扰率,%;w_0—基础井网气井最终产气量,10^4 m³。

参考文献

[1] 戴金星,倪云燕,吴小奇. 中国致密砂岩气及在勘探开发上的重要意义[J]. 石油勘探与开发,2012,39(3):257-264.

[2] 杨华,刘新社. 鄂尔多斯盆地古生界煤层气勘探进展[J]. 石油勘探与开发,2014,41(2):129-138.

[3] 杨华,付金华,刘新社,等. 苏里格大型致密砂岩气藏形成条件及勘探技术[J]. 石油学报,2012,31(S1):27-36.

[4] 王珂,张惠良,张荣虎,等. 超深层致密砂岩储层构造裂缝特征及影响因素——以塔里木盆地克深2气田为例[J]. 石油学报,2016,37(6):715-727,742.

[5] 王香增. 延长石油集团非常规天然气勘探开发进展[J]. 石油学报,2016,37(1):137-144.

[6] 贾爱林,闫海军,郭建林,等. 全球不同类型大型气藏的开发特征及经验[J]. 天然气工业,2014,34(10):33-46.

[7] 贾爱林,唐俊伟,何东博,等. 苏里格气田强非均质致密砂岩储层的地质建模[J]. 中国石油勘探,2007,12(1):12-16.

[8] 何东博,贾爱林,田昌炳,等. 苏里格气田储集层成岩作用及有效储集层成因[J]. 石油勘探与开发,2004,31(3):69-71.

[9] 马新华,贾爱林,谭健,等. 中国致密砂岩气开发工程技术与实践[J]. 石油勘探与开发,2012,39(5):572-579.

[10] 周大伟,张广清,刘志斌,等. 致密砂岩多段分簇压裂中孔隙压力场对多裂缝扩展的影响[J]. 石油学报,2017,38(7):830-839.

[11] 谭中国,卢涛,刘艳侠,等. 苏里格气田"十三五"期间提高采收率技术思路[J]. 天然气工业,2016,36(3):30-40.

[12] 窦伟坦,刘新社,王涛. 鄂尔多斯盆地苏里格气田地层水成因及气水分布规律[J]. 石油学报,2010,31(5):767-773.

[13] 王泽明,鲁保菊,段传丽,等. 苏里格气田苏20区块气水分布规律[J]. 天然气工业,2010,30(12):37-40.

[14] 代金友,李建霆,王宝刚,等. 苏里格气田西区气水分布规律及其形成机理[J]. 石油勘探与开发,2012,35(5):524-529.

[15] 郭智,贾爱林,冀光,等. 致密砂岩气田储量分类及井网加密调整方法——以苏里格气田为例[J]. 石油学报,2017,38(11):1299-1309.

[16] 卢涛,刘艳侠,武力超,等. 鄂尔多斯盆地苏里格气田致密砂岩气藏稳产难点与对策[J]. 天然气工业,2015,35(6):43-52.

[17] 何东博,贾爱林,冀光,等. 苏里格大型致密砂岩气田开发井型井网技术[J]. 石油勘探与开发,2013,40(1):79-89.

[18] 何东博,王丽娟,冀光,等. 苏里格致密砂岩气田开发井距优化[J]. 石油勘探与开发,2012,39(4):458-464.

[19] 郭智,孙龙德,贾爱林,等. 辫状河相致密砂岩气藏三维地质建模[J]. 石油勘探与开发,2015,42(1):76-83.

[20] 刘群明,唐海发,冀光,等. 苏里格致密砂岩气田水平井开发地质目标优选[J]. 天然气地球科学,2016,27(7):1360-1366.

[21] 万玉金,罗瑞兰,韩永新. 透镜状致密砂岩气藏井网加密技术与应用[J]. 科技创新导报,2014,28:41-44.

[22] 严谨,史云清,郑荣臣,等. 致密砂岩气藏井网加密潜力快速评价方法[J]. 石油与天然气地质,2016,37(1):125-128.

[23] McCain W D,Voniff G W,Hunt E R,et al. A Tight gas Field Study:Carthage(CottonValley)Field[C]. SPE 26141,1993.

[24] Voniff G W,Craig Cipolia. A New Approach to Large-Scale Infill Evaluation Applied to the Ozona(Canyan)Gas Sands[C]. SPE 35203,1996.

[25] 李跃刚,徐文,肖峰,等. 基于动态特征的开发井网优化——以苏里格致密强非均质砂岩气田为例[J]. 天然气工业,2014,34(11):56-61.

原文刊于《石油学报》,2018,39(7):802-813.

大型致密砂岩气田采收率计算方法

郭建林 郭 智 冀 光 贾成业 程立华 王国亭 孟德伟

(中国石油勘探开发研究院,北京 100083)

摘要:大型致密砂岩气田含气面积大,储层物性差,非均质性强,气田储量与产量规模大,气田采收率与最终采气量是指导气田长期稳定生产、制定开发技术对策以及衡量气田开发效果的关键指标。致密砂岩气田孔喉小,渗流机理复杂,常规的实验室模拟方法难以得到准确的采收率数据。以苏里格大型致密砂岩气田为研究对象,优选中区、东区、西区、南区等典型区块进行精细解剖,根据地质特征及开发效果将投产井分成三类;以沉积相带为约束,确定各类井区的面积比例;选取生产时间较长、基本达到拟稳态的大量井为分析样本,利用产能不稳定分析及生产曲线积分等方法,评价各类井的井均动态储量及最终累计产量;结合储层规模、结构与生产动态特征,论证单井控制范围;对各类井区以面积比例加权,模拟预测井网足够完善时区块的技术极限采气量、动用储量及采收率。研究表明,气田各大区技术极限采收率为 71.2%~78.5%,平均 75.3%,远低于常规气藏的 80%~90%。气田技术极限采气量为 $1.96\times10^{12}\,\mathrm{m^3}$,目前经济极限采气量 $1.17\times10^{12}\,\mathrm{m^3}$,未来通过技术进步,降低开发成本,气田可增产 $7900\times10^8\,\mathrm{m^3}$。

关键词:致密砂岩气;苏里格气田;技术极限采收率;动用储量;动态储量

致密砂岩气是一种非常重要的非常规天然气资源,其储量和产量分别占中国天然气总储量的 30% 和总产量的 26%[1]。气田采收率与最终采气量是指导气田长期稳定生产、制定开发技术对策以及衡量气田开发效果的关键指标。传统的采收率定义为气田报废时累计采气量与探明储量的比值。探明储量是指在气田评价钻探阶段完成后计算的地质储量,是开发方案编制、产能建设的依据,一般根据容积法获得。对大型致密砂岩气田而言,以探明储量做分母计算采收率主要存在两方面的问题:一是储层地质条件复杂,勘探阶段井距大、资料少、精度低,获得的储层参数不够准确,根据容积法得到的地质储量可靠性不强[2]。尽管随着开发过程的深入,能够获得更加准确的气田储量参数,但在目前的储量管理模式下,探明储量一旦提交,很少更改或核销[3];二是未结合生产动态,计算的采收率往往无法反映出气藏的开发规律,不利于不同类型气藏的横向对比[4]。致密气藏渗流能力弱,井网未控制的区域储量难以动用。因此,建议采用动用储量代替采收率计算公式中的探明储量。动用储量是指按照开发方案进行产能建设、技术井网覆盖区域内生产利用的那部分储量。较规则的井网与不规则的储层分布难以完全匹配,会造成一定程度的面积损耗,使部分探明储量难以动用;且在开发过程中,对储层非均质性的认识不断深化,获得的储层参数准确性不断提高,使得利用开发井网计算的动用储量往往会小于探明储量。需要注意,动态储量与动用储量有所区别,动态储量是以投产井为计算单元,利用动态方法得到的气井泄流面积内控制的储量之和,是设想气藏地层压力降为零时,能够渗流、流动的那部分地质储量。因此,就储量规模而言,探明储量>动用储量>动态储量。

致密砂岩气田孔喉半径小,渗流机理复杂,采收率的实验室标定方法难以模拟出地下多地质要素及其组合关系;储层非均质性强,各区块的地质及动态特征差异明显,采收率也不尽相同。本研究以苏里格大型致密砂岩气田为例,充分利用气田开发过程中的海量地质和生产动态资料,将采收率的计算方法分为几个步骤:(1)从各大区挑选典型区块,开展测井资料复查,落实孔隙度、含气饱和度、有效厚度等地质参数,对储层进行精细解剖,明确储层规模、分布及控制因素;(2)挑选生产时间长、数据可靠的开发井为样本,结合地质与开发特征,对投产井进行分类评价,以沉积相带分布为约束,圈定各类井区所控制的面积;(3)优选产能评价方法,确定各类气井的井均动态储量,结合废弃条件,预测井均最终累计产量;(4)综合储层规模、结构样式和泄气面积,分析各类井的井均控制范围,明确各类井不产生干扰时的最大井网密度;(5)以各类井的井均储层参数、最终累计产量、不产生干扰时的井网密度,井区所占面积等数据为依据,预测区块最终累计产气量及最终动用储量,将两者相除,得到采收率。这里的采收率是技术极限采收率,是指不考虑经济成本,在井网完善的情况下最大限度地动用储量所能达到的极限采收率,是气田开发的终极技术目标。

1 气田基本情况

苏里格气田是中国致密砂岩气田的典型代表,位于鄂尔多斯盆地伊陕斜坡的西北侧(图1),

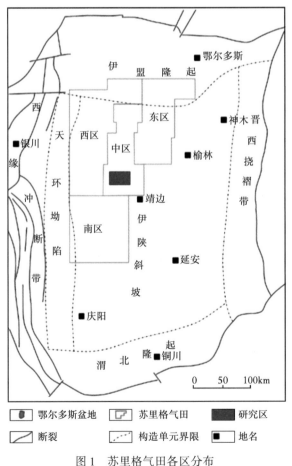

图1 苏里格气田各区分布

具有储层物性差、非均质性强、储量丰度小、优质储量比例低、单井产量低、产能递减快等特点。苏里格气田也是中国已开发的最大气田[5]，2016年产量226.5×10⁸m³，约占全国天然气年产量的1/5。截至2017年6月，气田累计投产9723口井（水平井1242口），累计产气1580×10⁸m³。气田于2014年进入稳产阶段[6]，部分区块开发进入中后期，随着开发的深入，将要投入开发的储量品质日益降低，气田稳产难度大，需要明确采收率及最终可采气量，为开发调整、剩余气挖潜提供依据。

1.1 河道多期叠置，形成大规模辫状河体系

气田主要产层为二叠系盒八段和山一段，主体沉积环境为陆相辫状河，发育心滩、河道充填、泛滥平原等沉积微相。地质历史时期在宽缓的构造背景下，河道多期改道、叠置，形成几千至上万平方千米的大规模致密砂岩区，呈片状连续分布，称为"辫状河体系"[7]。辫状河体系对应比单期河道更高的沉积层级，平面上为千米级规模，垂向上为砂组级地层（30~40m），包含2~3个开发小层[8]。根据物源、水动力、可容纳空间、古地貌的演化特征，辫状河体系可划分为叠置带、过渡带和体系间三个区带（图2）。辫状河体系带对沉积微相的发育与储层的结构、规模及分布有较强的控制作用。

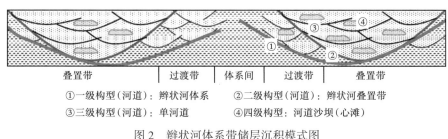

① 一级构型（河道）：辫状河体系　② 二级构型（河道）：辫状河叠置带
③ 三级构型（河道）：单河道　　　④ 四级构型：河道沙坝（心滩）

图2　辫状河体系带储层沉积模式图

1.2 基质砂体及有效砂体呈"砂包砂"二元结构

在沉积、成岩的双重作用下[9]，有效砂体不同于基质砂体，是在普遍致密背景下，孔渗值相对高（孔隙度大于5%，渗透率大于0.1mD），含气性相对好（含气饱和度大于45%）的"甜点"。有效砂体发育规模小，频率低，在空间上呈分散分布，累计厚度仅占基质砂体厚度的1/3~1/4，空间预测难度大，与连片的基质砂体呈"砂包砂"二元结构。气田储层物性差，气体流动性差，经过压裂改造才可获工业气流。气藏边界突破传统的概念，储层改造到哪，气藏的边界就到哪，因此致密气藏也被称为"人工气藏"。有效砂体是压裂改造的主要对象，有效砂体+近井带裂缝对气井早期产气起主要贡献，基质砂体对气井长期产气起一定作用。

1.3 单井产量低，递减快，气田开发依靠井间加密

气田平均储量丰度低（1.2×10⁸m³/km²），地层压力低（压力系数为0.8~0.9），单井产量低（平均日产气0.95×10⁴m³），泄气范围小，经济有效开发难度大。致密气藏能量衰减快，气井没有严格意义上的稳产期，投产之后即递减。初期有效砂体供气，产气量大但递减快，后期基质砂体供气，产气量小却递减慢，单井生命周期平均递减率20%~24%。开发实践表明，井间加密是气田提高储量动用程度和采收率的最主要途径。

1.4 储层非均质性强,各区带差异明显

受物源、水动力、成岩作用改造和后期构造运动等多因素影响,储层非均质性强。气田勘探面积近 $5×10^4 km^2$,主要分为中区、西区、东区和南区等几个大区(图1),各大区地质特征与开发效果差异明显(表1)。中区储层质量相对好,投产时间长,开发效果最好;西区储层物性与中区接近,但大面积含水[10],造成超过 $7000×10^8 m^3$ 的储量难以动用;东区岩屑含量高,储层较致密,北部部分层段见水;南区距离沉积物源远,岩石粒度细,储层厚度薄,仅有部分区块建产,基本还处在开发评价阶段。

表1 苏里格气田各大区储层基本参数

区块	孔隙度 (%)	含气饱和度 (%)	有效厚度 (m)	储量丰度 ($10^8 m^3/km^2$)	三年期日产 ($10^4 m^3$)
中区	8.5	60.0	13.2	1.5	1.16
东区	6.6	55.0	11.5	1.2	0.97
西区	7.8	45.0	8.0	1.0	0.87
南区	7.3	50.0	8.5	1.1	0.55

2 采收率影响因素

相比于油藏采用注水开发的方式补充地层能量,气藏开发主要依靠地层天然能量,采用衰竭式开采。气藏采收率是泄流系数和压降效率的综合函数,压力场的分布及压降变化是衡量气藏开发的主要指标。苏里格致密砂岩气田主要受到储层连续性和连通性、水锁、封闭孔隙、废弃压力等因素的制约,导致其采收率偏低[11]。

2.1 有效储层连续性和连通性差

苏里格气田有效储层连续性和连通性差,对采收率提高主要提出了两项挑战:一是有效砂体预测难度大,确定合适的井网井距与有效砂体的规模及分布相匹配难度大;二是有效储层非均质性强,水平井实钻剖面表明复合砂体内是不连通的,发育多个细粒泥质夹层——"阻流带",系心滩垂积过程中形成的落淤夹层[12],宽度 10~30m,间隔 50~150m。水平井通过多段压裂改造可以克服阻流带的影响[13-14],而直井压裂工艺难以对储层充分改造,致使直井钻遇的部分储层在开发过程中压降效率低。

2.2 成岩作用强烈,部分含气孔隙封闭

气藏形成过程中遭受了强烈的成岩作用,充注气体后的部分孔隙及其连通喉道在压实、胶结等破坏性成岩作用下,变成封闭的死孔隙,这部分天然气难以开采。

2.3 气水两相渗流区小

致密气藏储层孔喉结构复杂,气体充注程度低,含水饱和度相对较大。气水两相共渗区小,水相封闭易产生滞留剩余气,形成水锁,使得部分天然气难以流动。

2.4 废弃压力高

气井开发需要一定的井底流压。致密气藏地层能力小,单井产量低,携液能力弱,废弃压力高,导致部分天然气残留在地下地层中,难以开采。

3 气田中区典型区块采收率计算

3.1 采收率计算方法

气田投产井数近1万口,资料繁杂,各区块差异明显,全部应用困难较大。针对各大区分别挑选典型区块开展研究。苏中大区是苏里格气田开发的重要组成部分,面积6300km², 平均储量丰度$1.5×10^8m^3/km^2$,投产井4207口,产能$100×10^8m^3/a$,投产井数和产能均占气田总投产井数和总产能的40%以上。苏里格中区包括10个开发区块,从中优选苏14区块作为研究区(图1):(1)研究区面积大(850km²),储层条件较好,对气田中区有代表性;(2)区块2006年投产,是苏里格气田最早投产的几个区块之一,气井开发时间长,动态资料较可靠;(3)区内动、静态资料完备,共投产井数646口,开展加密试验6井区,适合开展综合研究。

根据"分类讨论、重点解剖"的研究思路,基于开发过程中的海量数据,确定了采收率的计算方法:对投产井开展综合分类,确定各类井的井数比例;以沉积相带为约束,圈定各类井区所控制的面积;优选产能评价方法,计算各类井平均动态储量、预测最终累计产量及控制面积,确定各类井不产生干扰时的最大井网密度;预测区块技术极限产气量及最终动用储量,得到最终采收率。明确储层规模与叠置样式、分析生产动态特征、研究储层与井网的匹配关系是采收率计算的前提和保障。

3.2 储层规模及控制因素

受沉积和成岩控制,研究区有效砂体规模小,在空间分布零星,70%以上的有效砂体为单期孤立型。根据研究区646口井2520个单砂体测井解释结果,井均钻遇有效砂体3~5个,单砂体厚度主要分布在1~5m,在此范围内的有效砂体占有效砂体总数的86%,平均厚度3.2m。野外露头观测和沉积物理模拟表明[15-16],鄂尔多斯盆地二叠系盒八段、山一段心滩、河道充填宽厚比为50~120,长宽比为1.2~4,平均1.5。根据密井网解剖、干扰试井分析,结合有效单砂体厚度与宽厚比、长宽比数据,认识到有效单砂体宽度主要分布在100~500m之间,平均310m(图3);有效单砂体长度主要分布在300~700m之间,平均520m。单砂体平均面积0.16km²,单井钻遇的多个有效砂体在平面叠合后面积在0.18~0.23km²之间。充分考虑人工裂缝半长、储层物性等参数,根据大量的生产井动态评价,认识到气井泄气范围有限,63%的井泄流面积小于0.24km²,24%的井在0.24~0.48km²之间,仅13%的井大于0.48km²,平均0.20km²(图4),证实了对有效砂体规模的分析是基本可靠的。

有效砂体与心滩底部、河道充填底部等粗砂岩相有较好的对应关系,然而沉积微相在空间具有很强的不均一性,其分布特征难以预测。研究表明,辫状河体系带是控制沉积微相和有效砂体分布的关键地质因素。叠置带水动力强,砂地比大于0.5,心滩发育比例高(平均58%),富集了气田70%以上的有效砂体,有效砂体规模相对较大;过渡带处在叠置带边部,水体能量减弱,砂地比0.3~0.5,主要发育河道充填微相(比例高达72%),心滩发育比例仅为28%,包含约25%的有效砂体,有效砂体规模有所减小,连续性较差。体系间水动力弱,砂地比小于0.3,以粉砂质、泥质等细粒沉积为主,有效砂体基本不发育。

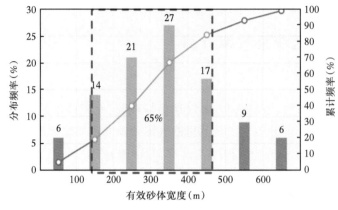

图3　单砂体宽度与长度

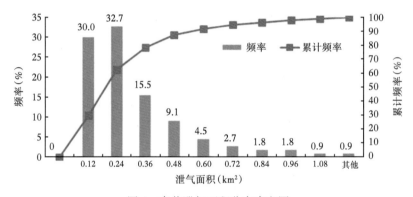

图4　直井泄气面积分布直方图

3.3　生产井分类评价

气井按钻遇的储层特征及开发效果,可明显分为Ⅰ类、Ⅱ类、Ⅲ类井。提出了三类井的划分评价标准(表2),包括单层厚度、累计有效厚度两个地质参数和无阻流量、初期产量两个动态参数。Ⅰ类井单层厚度大于5m,累计厚度大于10m,开发效果最好;Ⅱ类井单层厚度为3~5m,多层叠置后形成一定的储层规模,累计厚度6~10m,开发效果较好;Ⅲ类井单层厚度小于3m,有效砂体个数少,累计厚度小于6m,开发效果差。在开发早期迅速判断井类型,是合理配产、生产制度优化的重要依据,有助于将气井产能发挥到最大。

在研究区选取生产时间长(大于500天),基本达到拟稳态的531口生产井进行分析。根据三类井划分评价标准,Ⅰ类、Ⅱ类、Ⅲ类井分别为97口、305口、129口,井数比例分别为18.3%、57.4%、24.3%(表3)。结合产量不稳定分析和生产曲线积分两种方法判断各类井的井均动态储量。流体从储层流向井筒经历两个阶段,即开井初期的不稳定流动段和后期的边界流动段。通过不稳定流动段的拟合可以计算气井的表皮系数、储层渗透率、裂缝长度等,通过边界流动段拟合可以计算气井动态储量。为保证计算结果的准确可靠,对每口参与计算的气井进行了数据质量控制,对异常点进行排查处理,同时在参数选取上针对每口井提取射孔层沟通的有效储层厚度,并以该有效储层厚度为基础,通过加权平均的方式获取孔隙度和渗透率物性数据。经过计算,Ⅰ类、Ⅱ类、Ⅲ类井的井均动态储量分别为$5672×10^4m^3$、$2916×10^4m^3$、$1450×10^4m^3$,结合单井开发废弃条件(井口压力小于3MPa,日产气小于$1000m^3$),得到三类井的井均最终累计产量分别为$5167×10^4m^3$、$2482×10^4m^3$、$1096×10^4m^3$。从储层地质规模和动态泄气范围两方面考虑,得到三类井井均控制面积分别$0.29km^2$、$0.22km^2$及$0.14km^2$,即三类井在不发生干扰时对应的最大井网密度分别为3.4口、4.5口、7.1口。

表 2 苏里格气田三类井划分评价标准

气井	静态分类		动态分类	
	单层有效厚度 (m)	累计有效厚度 (m)	无阻流量 ($10^4m^3/d$)	初期产量 ($10^4m^3/d$)
Ⅰ类	>5	>10	>10	>1.5
Ⅱ类	3~5	6~10	4~10	0.8~1.5
Ⅲ类	<3	<6	<4	<0.8

表 3 苏14区块三类井综合评价

井类型	井数 (口)	井数比例 (%)	井均动态储量 (10^4m^3)	井均预测最终累计产量 (10^4m^3)	井均控制范围 (km^2)
Ⅰ类	97	18.27	5672	5167	0.29
Ⅱ类	305	57.44	2926	2482	0.22
Ⅲ类	129	24.29	1450	1096	0.14
总计/平均	531	100.00	2691	2266	0.20

3.4 苏14区块采收率计算

前文已述,辫状河体系带对沉积微相和有效砂体分布有较强的控制作用,对不同类型生产井的分布也有一定的预测作用。通过分析对比,认识到Ⅰ类井主要分布在辫状河体系叠置带(砂地比大于0.5),Ⅱ类井主要分布在过渡带(砂地比0.3~0.5)和体系间(砂地比小于0.3),Ⅲ类井大部分分布在体系间,少部分分布在过渡带(图5)。以辫状河体系带为约束,圈定三类井面积,其所占比例分别为16.5%、53.6%及29.9%。当三类井区井网加密到足够密时,预计最终动用储量分别为$286.1×10^8m^3$、$653.6×10^8m^3$、$286.7×10^8m^3$,累计$1226.4×10^8m^3$,占探明储量的95.2%。三类井区最终累计产量分别为$249.4×10^8m^3$、$514.4×10^8m^3$、$198.9×10^8m^3$,累计$962.7×10^8m^3$。Ⅰ类、Ⅱ类、Ⅲ类井区采收率分别为87.2%、78.7%、69.4%,区块采收率为78.5%(表4)。因苏14区块储层规模、储层物性、储量丰度与苏中较为接近,对苏中具有较好的代表性,将苏14区块的采收率作为苏里格中区的采收率。

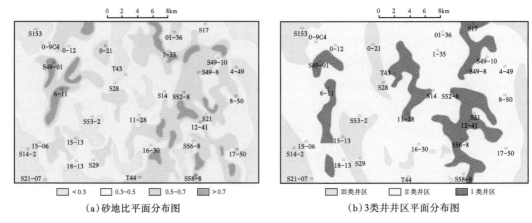

图 5 苏 14 区块砂地比与三类井井区平面分布图

4 气田采收率计算

按照同样的方法,从苏里格气田东区、西区、南区分别优选了典型区块,开展地质和动态综合分析,计算了各大区的采收率。各区三类井的比例、开发效果、控制范围有所差异,导致它们的采收率各有不同。中区储层条件最好,探明储量可靠性最强,可动用程度高,开发效果好,Ⅰ类+Ⅱ类井比例大于 70%,采收率 78.5%。西区大面积含水,导致大规模的储量难以动用,但本身储层物性并不差,故探明储量采收率偏低,为 26.8%,动用储量采收率仅次于中区,为 76.6%。东区相对致密,物性较差,Ⅰ类+Ⅱ类井比例 62.1%,可动储量采收率为 73.8%。南区最致密,开发效果较差,Ⅰ类+Ⅱ类井比例不足 40%,可动储量采收率为 71.2%(表 5)。

表 4 苏 14 区块采收率计算表

井类型	控制面积 (km^2)	控制面积比例 (%)	平均储量丰度 ($10^8 m^3/km^2$)	探明储量 ($10^8 m^3$)	动用储量 ($10^8 m^3$)	最终累计产量 ($10^8 m^3$)	采收率 (%)
Ⅰ	140	16.52	2.10	294.9	286.1	249.4	87.2
Ⅱ	456	53.60	1.51	688.0	653.6	514.4	78.7
Ⅲ	254	29.88	1.20	305.7	286.7	198.9	69.4
总计/平均	850	100.00	1.52	1288.6	1226.4	962.7	78.5

表 5 苏里格各区块采收率

区块	投产井数 (口)	Ⅰ类+Ⅱ类井数比例 (%)	Ⅰ类+Ⅱ类井区面积比例(%)	探明储量 ($10^8 m^3$)	预计最终动用储量 ($10^8 m^3$)	技术极限采气量 ($10^8 m^3$)	探明储量采收率 (%)	动用储量采收率 (%)
苏中	3774	72.7	68.1	9377	8908	6996	74.6	78.5
苏东	2413	62.1	59.8	9071	7256	5355	59.0	73.8
苏西	1360	34.8	32.3	11524	4026	3084	26.8	76.6
苏南	934	41.2	39.5	8285	5800	4129	49.8	71.2
总计/平均	8481	60.1	57.3	38257	25990	19564	51.1	75.3

在综合分析各大区的基础上,计算整个苏里格气田的采收率。截至2017年6月,气田探明地质储量(含基本探明)$4.77×10^{12}m^3$,其中气田开发区范围内为$3.83×10^{12}m^3$(靖边气田上古生界$0.94×10^{12}m^3$)。在现有技术条件下,预测气田最终动用储量为$2.60×10^{12}m^3$,技术极限采气量为$1.96×10^{12}m^3$,技术极限采收率为75.3%。

气田目前主体开发井网为600m×800m,井网密度2口/km^2,预计最终采气量为动用储量的30%左右,井网对储量控制不足。考虑经济因素,在单井钻完井及储层改造成本800万/口、气价1.15元/m^3、所有井平均具有12%收益率的条件下,气田可整体加密至4口/km^2,动用储量采收率可由目前的30%提升至45%,经济极限采气量$1.17×10^{12}m^3$。未来气田通过技术进步[17]、降低开发成本,一方面使原来不可动用的储量得以动用,另一方面,大幅度提高动用储量的产气量,预计可累计增产达$0.79×10^{12}m^3$。

5 结论

大型致密砂岩气藏含气面积大,不同区块储量品质不一,勘探提交探明储量时计算参数标准没有体现储层差异,或者部分区块没有提交探明储量,导致气田采收率计算不准或者无法求取。选取开发程度较高的区块作为研究对象,采用分类统计方法,在单井分类评价的基础上,建立不同储层条件下系列开发指标,推广到其他区块,能够较准确的预测出不同区块累计采气量和最终动用储量两个指标,用最终动用储量代替探明储量计算气田采收率,该计算方法科学可靠,对现场开发具有很好的指导意义。

受复合砂体内阻流带、废弃压力高、水锁气、封闭气影响,苏里格气田技术极限采气量为$1.96×10^{12}m^3$,各大区动用储量采收率为71.2%~78.5%,平均75.3%,远小于常规气藏的80%~90%。目前条件下,通过井网加密的经济极限采收率为45%左右,经济极限采气量为$1.17×10^{12}m^3$。未来通过技术进步,降低开发成本,可累计增产$0.79×10^{12}m^3$,相当于一个巨型气田的储量,开发潜力大。

参 考 文 献

[1] 何东博. 致密气藏有效开发与提高采收率技术[R]. 中国石油勘探开发研究院,2016.
[2] 李忠兴,郝玉鸿. 对容积法计算气藏采收率和可采储量的修正[J]. 天然气工业,2001,21(2):71-74.
[3] 王永祥,张君峰,段晓文. 中国油气资源/储量分类与管理体系[J]. 石油学报,2011,32(4):645-651.
[4] 李安琪,郝玉鸿. 对油气田储量和采收率的新认识[J]. 天然气工业,2005,25(8):81-84.
[5] 马新华,贾爱林,谭健,等. 中国致密砂岩气开发工程技术与实践[J]. 石油勘探与开发,2012,39(5):572-579.
[6] 卢涛,刘艳侠,武力超,等. 鄂尔多斯盆地苏里格气田致密砂岩气藏稳产难点与对策[J]. 天然气工业,2015,35(6):43-52.
[7] 文华国,郑荣才,高红灿,等. 苏里格气田苏6井区下石盒子组盒8段沉积相特征[J]. 沉积学报,2007,25(1):90-98.
[8] 郭智,贾爱林,何东博,等. 鄂尔多斯盆地苏里格气田辫状河体系带特征[J]. 石油与天然气地质,2016,37(2):197-204.
[9] 何东博,贾爱林,田昌炳,等. 苏里格气田储集层成岩作用及有效储集层成因[J]. 石油勘探与开发,2004,31(3):69-71.

[10] 孟德伟,贾爱林,冀光,等. 大型致密砂岩气田气水分布规律及控制因素——以鄂尔多斯盆地苏里格气田西区为例[J]. 石油勘探与开发,2016,43(4):607-614.

[11] 何东博,贾爱林,冀光,等. 苏里格大型致密砂岩气田开发井型井网技术[J]. 石油勘探与开发,2013,40(1):79-89.

[12] 王丽娟,何东博,冀光,等. 阻流带对子洲气田低渗透砂岩气藏开发的影响[J]. 天然气工业,2013(5):56-60.

[13] 李建奇,杨志伦,陈启文,等. 苏里格气田水平井开发技术[J]. 天然气工业,2011,31(8):60-64.

[14] 卢涛,张吉,李跃刚,等. 苏里格气田致密砂岩气藏水平井开发技术及展望[J]. 天然气工业,2013,33(8):38-43.

[15] 贾爱林,何东博,何文祥,等. 应用露头知识库进行油田井间储层预测[J]. 石油学报,2003,24(6):51-53.

[16] 郭建林,贾爱林,何东博,等. 滦平上侏罗统—下白垩统扇三角洲露头层序地层学研究[J]. 中国地质,2007,34(4):628-635.

[17] 贾爱林,郭建林. 智能化油气田建设关键技术与认识[J]. 石油勘探与开发,2012,39(2):118-122.

原文刊于《石油学报》,2018,39(12):1389-1396.

鄂尔多斯盆地低渗透—致密气藏储量分类及开发对策

程立华 郭 智 孟德伟 冀 光 王国亭 程敏华 赵 昕

(中国石油勘探开发研究院,北京 100083)

摘要:鄂尔多斯盆地低渗透—致密气藏储量规模大、储层物性差、非均质性强,储量动用程度低且差异大,实现气藏的长期稳产及效益开发难度大。为此,以鄂尔多斯盆地五个主力气田为研究对象,以效益开发为导向,以内部收益率为核心评价指标,结合动、静态特征对低渗透—致密气藏进行储量评价单元划分、储量分类评价和储量接替序列的建立,并针对不同类型的储量提出相适应的开发技术对策。研究结果表明:(1)该盆地单井动态储量小,产气量低,产气类型可以划分为多层协同供气和单层主力供气两种;(2)基于地质条件和单井动态特征相近的原则,结合开发管理区块分布情况,将该盆地内五个主力气田划分为 11 个储量评价单元,以内部收益率 30%、8%和 5%为界限,把储量评价单元划分为高效、效益、低效和难动用四种储量类型;(3)以内部收益率 8%为有效开发的基准,将其对应的井均估算最终开采量(EUR)与各个储量评价单元实际的井均 EUR 对比,按照效益由高到低的顺序,建立了储量评价单元经济有效动用序列;(4)高效储量适宜采取增压开采和局部井网调整对策,效益储量需通过井网加密提高储量动用程度,低效储量采取富集区优选、滚动开发对策,难动用储量需加大技术攻关以实现效益开发。结论认为,该研究成果有利于提高鄂尔多斯盆地储量动用程度,可以为该盆地天然气长远开发战略的制度提供技术支撑。

关键词:鄂尔多斯盆地;低渗透—致密气;储量评价单元;储量分类;储量接替序列;内部收益率;开发对策

低渗透—致密气是一种非常重要的天然气资源[1],在中国主要分布在鄂尔多斯、四川、松辽及塔里木盆地[2],其中以鄂尔多斯盆地储量规模最大。鄂尔多斯盆地低渗透—致密气的开发已成为中国该类型天然气开发的典范,历经多年的技术攻关和生产实践,开发理念和技术不断创新,实现了气藏的规模开发和持续稳产,年产气量持续保持国内领先地位[3-4]。在国家大力发展天然气的战略背景下,鄂尔多斯盆地作为中国最大的天然气生产基地,实现低渗透—致密气的长期稳产与效益开发意义重大。由于该盆地天然气储量基数大、储层物性差、非均质性强,储量动用程度差异大,在气藏开发过程中储量动用面临以下两个方面的问题:(1)根据国内外开发经验,巨型气田采气速度介于 1%~2%是较为合理的[5-6],而按照鄂尔多斯盆地总的年产量和储量规模测算,目前的采气速度仅为 0.6%,理论上还有较大的提升空间,但实际生产情况表明进一步扩大生产规模的难度非常大;(2)由于相对优质储量已逐步动用,剩余储量的品位不断降低,找到适合建产的"甜点区"难度越来越大,长期稳产及效益开发面临极大的挑战。关于储量分类,前人多是从地质或生产动态评价的角度切入[7-9],忽略了经济性评价,而只有通过经济性指标才能衡量气藏开发效益的高低。为此,笔者以鄂尔多斯盆地五个主力气田(GF1、GF2、GF3、GF4 和 GF5,其合计天然气储量占该盆地天然气总储量的 90%)为研究对象,以效益开发为导向,以内部收益率为核心评价指标,结合动态、静态特征对该盆地低渗透—致密气藏进行储量评价单元划分、储量分类评价和储量接替序列的建立,并针对不同类型的储量提出相适应的开发技术对策,以期为国内大型低渗透—致密气藏的长期稳产和效益开发提供借鉴。

1 主要地质特征

上述五个主力气田分布在鄂尔多斯盆地伊陕斜坡北部,主要发育下古生界奥陶系马家沟组马五段碳酸盐岩、上古生界二叠系石盒子组盒八段、山西组和太原组碎屑岩这两类沉积岩储层,总体表现为储层物性差、厚度薄、非均质性强、储量丰度低的特点。

1.1 储层物性差,为典型的低渗透—致密气藏

根据鄂尔多斯盆地内1230块密闭取心岩样的覆压分析试验,孔隙度主要介于5%~12%,渗透率介于0.03~0.60mD,属于低渗透—致密储层的范畴。如表1所示,GF1、GF2气田为致密气藏,储层物性差,覆压渗透率小于0.1mD,含气饱和度较低,平均约58%;GF3、GF4、GF5气田为低渗透气藏,孔隙类型以原生孔隙为主,储层物性相对较好,渗透率相对较高,覆压渗透率介于0.1~1.0mD,含气饱和度介于70%~80%。除GF1气田西部和东北部外,其他气田主体不产水,地层水以束缚水和层间滞留水为主。

1.2 储量主要分布在三套层系,致密气占主体

鄂尔多斯盆地具有广覆式生烃、连续性成藏的特点[10-11],含气面积大,超过$7\times10^4km^2$,然而由于储层物性差、有效储层厚度小,盆地内天然气藏的平均储量丰度约为$1\times10^8m^3/km^2$。纵向上,根据与烃源岩的关系,可划分出三套主力含气层系:二叠系下石盒子组盒八段—山西组山一段(源上组合);山西组山二段—太原组(源内组合);奥陶系马家沟组马五段(源下组合)。GF1气田主力产层为盒八段—山一段碎屑岩,GF2、GF3、GF4气田主力产层为山二段—太原组碎屑岩,GF5气田主力产层为马五段碳酸盐岩(表1)。按照储层物性划分,低渗透气藏储量占五个主力气田总储量的16%,致密气藏储量占84%。

表1 鄂尔多斯盆地主力气田储层参数表

名称	孔隙度(%)	覆压渗透率(mD)	含气饱和度(%)	有效厚度(m)	储量丰度($10^8m^3/km^2$)	有效储层分布层段	储层岩性	储层类型
GF1	7.6	0.088	58.5	10.1	1.06	盒八段、山一段	碎屑岩	致密
GF2	6.6	0.073	57.4	6.2	0.83	山二段、太原组	碎屑岩	致密
GF3	7.1	0.572	73.3	11.2	1.01	山二段	碎屑岩	低渗透
GF4	5.9	0.113	69.6	9.5	0.95	山二段	碎屑岩	低渗透
GF5	6.3	0.364	78.2	5.4	0.65	马家沟组马五段	碳酸盐岩	低渗透

1.3 多层协同供气和单层主力供气

鄂尔多斯盆地各气田储量丰度较接近,介于$0.6\times10^8\sim1.1\times10^8m^3/km^2$,但储层展布、供气模式差异较大,可以划分为多层协同供气和单层主力供气两种模式。致密气藏有效砂体呈多层叠置连片状分布,存在相对富集区;单个有效砂体的厚度介于2~5m,宽度介于300~500m,

长度介于500~700m,在空间上呈透镜状孤立分布,连续性差;由于垂向上发育多个有效砂体,钻遇3~5个有效砂体的单井居多;单层产气贡献率均低于30%,不存在明显主力层,致密气藏供气模式属于多层协同供气型。低渗透气藏虽然纵向上也发育多个小层,但主力层分布稳定,连续性好,气层连通范围可达2~3km;主力层单层产气贡献率在70%以上,供气能力较强,低渗透气藏供气模式属于单层主力供气型。

2 生产动态特征

低渗透—致密气藏单井泄气面积小且井间连通性差,分析气井开发指标是评价区块或气田开发效果的基础,其中关键指标包括单井日产气量及其递减率、动态储量、泄气面积、估算最终开采量(EUR)等。

2.1 单井日产气量低,且递减率高

低渗透—致密气藏只有经过储层压裂改造,气井才具有工业产能,且产气量普遍较低[12-13]。鄂尔多斯盆地投产气井超过$1.8×10^4$口,且以直井为主,依靠多井低产实现了低渗透—致密气藏的规模开发,其中87%的气井初期产气量介于$1×10^4$~$2×10^4 m^3/d$,13%的气井初期产气量介于$5×10^4$~$15×10^4 m^3/d$;另一方面,致密气井没有严格意义上的稳产期[14],气井投产之后即递减,生产曲线呈"L"形,气井初期产气量由有效砂体近井裂缝带提供,递减快,后期产气量由有效砂体远井端提供,此时产气量虽小但递减缓慢。致密气井的生产动态评价结果显示,直井的初期产量递减率平均为23.6%,前三年的产量递减率平均为22.0%,中后期气井的产量递减率逐渐降至13.5%。

2.2 单井泄气面积、动态储量及EUR小

由于低渗透—致密气藏储层渗流能力差,气井井控范围随生产的持续进行将逐渐扩大,若利用早期生产数据评价气井动态储量,其值通常偏小。因此,优选生产时间超过五年的老井来进行动态储量评价;同时,鉴于气井采用多层合采的方式进行开采,为了避免由于叠合有效厚度取值偏大而导致气井泄气面积计算值偏小的情况出现,尽量筛选发育1~2个气层的井以提高泄气面积评价结果的合理性。

动态储量计算方法包括物质平衡法、压降曲线法、产能不稳定法、生产曲线积分法等。在开发中后期,动态资料已较丰富,采用产能不稳定法和生产曲线积分法效果较好。因此,利用这两种方法,设定气井废弃条件为井口压力小于3MPa、日产气量小于$0.1×10^4 m^3$,计算单井动态储量。受储层地质条件、压裂改造效果、生产制度等因素的影响,单井EUR一般占气井动态储量的80%~90%。

计算结果表明,以低渗透储层为主的GF3、GF4、GF5气田井均泄气面积介于2~$5km^2$,动态储量介于$1.7×10^8$~$4.7×10^8 m^3$,井均EUR介于$1.4×10^8$~$4.0×10^8 m^3$。以致密储层为主的GF1、GF2气田井均泄气面积介于0.21~$0.23km^2$,动态储量介于$2400×10^4$~$2600×10^4 m^3$,井均EUR介于$2000×10^4$~$2100×10^4 m^3$。鄂尔多斯盆地单井泄气面积、动态储量及EUR整体偏低,低渗透气藏单井开发指标好于致密气藏。

3 储量分类评价及动用接替序列的建立

3.1 储量评价单元划分

鄂尔多斯盆地主力气田含气面积大,完钻井数多,合理划分储量评价单元是储量分类评价的基础。将地质与动态特征相近、管理模式一致、分布范围适中的区域划为同一个储量评价单元。储量评价单元范围不能太大,否则单元内部极强的非均质性将导致基于单井参数的统计值不具有代表性;储量评价单元也不能太小,以避免工作量徒增,同时也不利于形成规律性认识。

综合考虑开发管理区界限、所处的开发阶段、储层地质和动态特征,共划分出11个储量评价单元(图1)。GF5-1单元以下古生界马五段碳酸盐岩为开发对象,属于低渗透率储层,投入开发早,已进入稳产末期,开发管理上独立。GF3-1、GF3-2和GF4-1这三个单元主要以上古生界山二段碎屑岩为开发对象,也属于低渗透率储层,单井产气量较高,产气层集中在山二段下部,主力层产气贡献率超过70%。其中,GF3-1单元以丛式水平井开发为主,GF3-2以直井开发为主,GF4-1也以直井开发为主但单井产气量略低。这三个单元在管理上分属三个开发区块。GF1-1、GF1-2、GF1-3、GF1-4、GF2、GF4-2和GF5-2这七个单元以上古生界盒八段碎

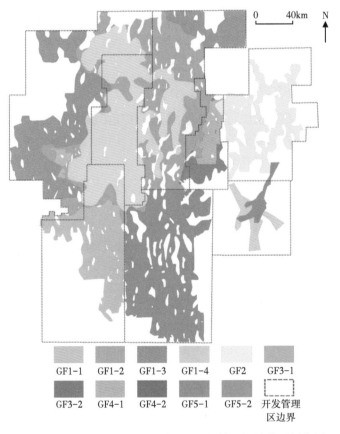

图1 鄂尔多斯盆地5个主力气田11个储量评价单元划分图

屑岩为开发对象,储层致密,具有多层协同供气的特点,单井产气量均较低,开发效益偏低。GF1-1、GF1-2、GF1-3 和 GF1-4 单元属于 GF1 气田,其中 GF1-3 单元受产水影响较大,另外三个单元的储层地质特征和气井产量差别较大。GF2、GF4-2 和 GF5-2 单元分属不同的开发管理区。由于单个储量评价单元内部储层的地质条件相近,气井生产动态特征相似,因此可以将每个储量评价单元视为相对均质体,通过求取评价参数的算术平均值来定量描述单元特点(表2),为储量分类评价奠定基础。

表2 鄂尔多斯盆地主力气田 11 个储量评价单元参数表

名称	埋深（m）	有效厚度（m）	孔隙度（%）	空气渗透率（mD）	含气饱和度（%）	储量丰度（$10^8 m^3/km^2$）	单井平均 EUR（$10^4 m^3$）
GF1-1	3300~3500	13.3	8.5	0.92	60	1.47	2352
GF1-2	3000~3300	11.2	6.6	0.58	55	1.10	1443
GF1-3	3400~3600	8.7	7.8	0.85	45	0.81	722
GF1-4	3500~3800	8.5	7.3	0.54	50	0.92	795
GF5-1	3100~3500	5.4	6.3	3.62	78	0.65	18413
GF5-2	2900~3300	9.0	6.8	0.62	56	1.06	1612
GF3-1	2700~3200	11.6	7.6	6.52	75	1.09	41735
GF3-2	2700~3200	10.5	6.5	4.85	71	0.93	25040
GF4-1	2500~2800	9.5	6.0	1.11	69	0.96	14505
GF4-2	2450~2750	10.5	6.7	0.58	60	0.75	1320
GF2	2000~2900	6.2	6.6	0.83	55	0.79	2012

3.2 储量分类综合评价

低渗透—致密气藏储层物性差,多数井需要经过储层改造才能获得工业气流,从而增加了开发成本,导致开发效益偏低,因此,储量能否有效动用,经济效益是一个关键的影响因素,同时也是低渗透—致密气藏储量分类评价的关键指标。由此,以内部收益率作为储量评价的核心参数,通过建立内部收益率与气井开发指标的关系,结合储层物性、含气饱和度等参数,综合构建储量分类评价体系。内部收益率(R)是国际上评价投资有效性的关键指标,是指资金流入现值总额与资金流出现值总额相等、净现值(NPV)为零时的折现率。由于气田开发前期投入大,收益相对支出小,因此现金流(V)为负,而随着生产的持续进行,V 逐渐为正。当式(1)中 NPV 为 0 时,根据历年 V 计算折现率,即为 R。由于低渗透—致密储层连通性差,在求取区块的内部收益率时常采用气井内部收益率的平均值。鄂尔多斯盆地主力气田的开发方案设计中设定单井生产年限介于 11~15 年。

$$NPV = \sum_{i=1}^{n} \frac{V_i}{(1+R)^i} \quad (1)$$

$$V_i = E_i - C_i \quad (2)$$

式中 NPV——净现值,万元;

V——气井年现金流,万元;
R——内部收益率;
i——年份;
n——气井总的生产年限;
E——气井年收益,万元;
C——气井年支出,万元。

如式(3)所示,由气井采气得到的收益(E)与商品率(a)、气价(P)和年产气量(Q)直接相关,而前二者在一定时期内是相对稳定的,因此,Q 是评价 E 的关键因素,而其与首年产气量及年递减率有关,如式(4)所示。将气井采气期内历年产气量累加即得到气井 EUR,如式(5)所示。需要指出的是,开发效益是有时间属性的,对应相同的气井 EUR,生产周期越长,E 越低。

$$E_i = aPQ_i \tag{3}$$

$$Q_i = Q_{i-1}(1 - D_i) \tag{4}$$

$$\text{EUR} = \sum_{i=1}^{n} Q_i \tag{5}$$

式中　a——商品率;
　　　P——气价,万元/$10^4 m^3$;
　　　Q——气井年产气量,$10^4 m^3$;
　　　D——年产气量递减率。

采气期内的支出主要包括气井综合成本(W)、生产经营成本(O)及销售税费及附加(F)、所得税(T)这四个部分,如式(6)所示。其中,W 是在气井投产之前的一次性投入,如式(7)所示,包括钻完井、储层压裂改造、地面配套等费用,与各区块储层埋深、岩石力学性质、地面交通条件、气藏开发管理模式等因素相关;O 包括操作费用、管理费用和销售费用,其中操作费用和管理费用与气井产气量线性相关,而销售费用与收益线性相关,如式(8)所示;F 包括城市建设维护费、资源税、教育附加税等,皆与 E 呈线性关系,相关税费与气井收益的相关系数为 0.0626 [(式(9)];T 为税前利润的15%,如式(10)所示。

$$C_i = W_i + Q_i + F_i + T_i \tag{6}$$

$$W_i = \begin{cases} W, i = 1 \\ 0, i = 2, 3, \cdots, n \end{cases} \tag{7}$$

$$Q_i = 0.1895 Q_i + 0.0138 E_i \tag{8}$$

$$F_i = 0.0626 E_i \tag{9}$$

$$T_i = 0.15 \left(E_i - \frac{W}{n} - F_i - Q_i \right) \tag{10}$$

式中　W——气井综合成本,万元;
　　　O——生产经营成本,万元;
　　　F——销售税金及附加,万元;

T——所得税,万元。

联立式(6)~式(10),得

$$C_i = W_i - 0.15\frac{W}{n} + 0.1611Q_i + 0.2195E_i \quad (11)$$

联立式(2)、式(3)、式(11),得

$$V_i = (0.7181P - 0.1611)Q_i - W_i + 0.15\frac{W}{n} \quad (12)$$

如式(12)所示,P 越高,Q 越高,W 越低,则 V 越高,对应的 R 越高。根据式(1)、式(5)、式(12),编制了不同 W(固定气价)、P(固定成本)下 R 与气井 EUR 的关系图版。在固定 P 的前提下(P 为 1.15 元/m³),若气井 EUR 为 2000×10⁴m³,W 为 800 万元时,R 可达 22%;而 W 为 1600 万元时,R 仅为 0(图2a)。在固定 W 的前提下(W 为 1000 万元),在现有技术水平条件下随着 P 升高或财税补贴,R 有较大程度提升(图2b);气井 EUR 为 1800×10⁴m³ 时,P 为 1.00 元/m³、1.15 元/m³、1.30 元/m³、1.50 元/m³、1.80 元/m³,R 分别为 3%、8%、14%、23% 及 40%。

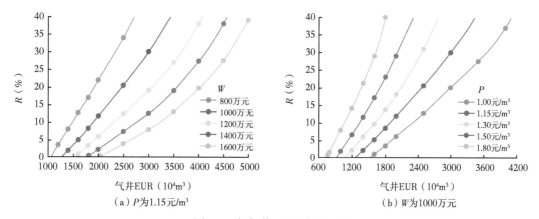

图 2　R 与气井 EUR 关系图版

鄂尔多斯盆地低渗透储层井均 EUR 在 1×10⁸m³ 以上,对应内部收益率普遍超过 30%。近年来,由于国家能源结构转型的推进,开发天然气资源的优惠政策不断落实,致密气开发的 R 下限由之前的 12% 降至 8%,未来还有望进一步降到 5%~6%。由此,综合考虑储层物性、含气性及现有开发技术条件下可以获得的气井累计产气量,以内部收益率 30%、8%、5% 为界,将 11 个储量评价单元划分为高效、效益、低效及难动用四种储量类型(表3)。

GF3-1、GF3-2、GF4-1、GF5-1 单元的储量为高效储量,属于低渗透率气藏类型,储量规模占五大主力气田总储量的 16.0%。储层物性较好,储层厚度尽管不大,但是分布稳定、连续性好,主力产层明显。气井生产稳定,气井平均 EUR 大于 1×10⁸m³,R 大于 30%。

GF1-1、GF2 单元的储量为效益储量,储量规模占五大主力气田总储量的 32.9%。有效砂体呈透镜状,连续性差,纵向多层叠合连片发育。气井平均 EUR 介于 0.2×10⁸~1.0×10⁸m³,R 介于 8%~30%。

GF1-2、GF5-2 单元的储量为低效储量,储量规模占五大主力气田总储量的 33.4%。储层相对致密，物性较差，储量丰度较低，气井平均 EUR 介于 $0.1×10^8$~$0.2×10^8 m^3$，R 介于 5%~8%。

难动用储量主要分布在 GF1-3、GF1-4 单元，其次在 GF3-2、GF4-2 单元的局部地区，储量规模占五大主力气田总储量的 17.7%。区内储层致密或含水，以低产气井和产水井为主，气井平均 EUR 低于 $800×10^4 m^3$，R 低于 5%，目前尚未实现有效开发。

表3 鄂尔多斯盆地主力气田储量分类表

储量分类	R (%)	单井 EUR ($10^8 m^3$)	空气渗透率 (mD)	含气饱和度 (%)	单井泄气面积 (km^2)	典型区块所在位置	储量占比 (%)
高效储量	>30	>1.0	>1.0	>65	1.0~5.0	GF3-1、GF3-2、GF5-1、GF4-1	16.0
效益储量	8~30	0.2~1.0	0.6~1.0	58~65	0.2~1.0	GF1-1、GF2	32.9
低效储量	5~8	0.1~0.2	<0.6	50~58	<0.2	GF1-2、GF5-2	33.4
难动用储量	<5	<0.1	<0.6	<50	<0.2	GF1-3、GF1-4、GF4-2	17.7

3.3 储量动用接替序列的建立

对于低渗透—致密气藏而言，单井 EUR 取决于可动用气层的厚度和连通范围，是气藏自身地质条件决定的，同时也受压裂改造工艺技术的影响。不同储量单元的储层条件、开发方式、递减规律和气井综合成本等不同，达到一定的 R 所对应的井均 EUR 下限差异较大，可以对比某储量单元的实际井均 EUR 和满足 R 为 8% 对应的井均 EUR，来评价储量的可动用性，在此基础上，建立储量动用接替序列。如图3所示，不同色块代表不同的储量评价单元，图中各单元块中间的蓝色线对应单元内的井均 EUR（蓝色数字），各单元块上边线对应单元内井的最

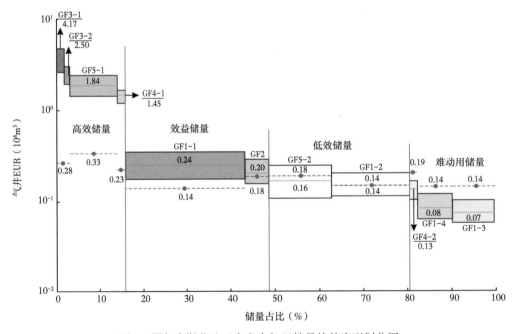

图3 鄂尔多斯盆地五个主力气田储量接替序列划分图

大 EUR，下边线对应单元内井的最小 EUR，图 3 中红色虚线对应 R 为 8% 的井均 EUR（红色数字）；各储量单元的储量占比越大，色块越长。若某储量单元实际井均 EUR 大于 R 取 8% 对应的井均 EUR 时，则该储量单元在现有条件下可以有效动用，反之则不能有效动用。

以井均 EUR 为依据的储量动用接替序列直观反映了储量开发的有效性，对于鄂尔多斯盆地天然气储量的开发次序、开发潜力及长期开发战略的制定具有指导意义。同时，该序列还具有较强的拓展性，一方面在现有序列的基础上可以更新各储量单元的储量动用比例，以体现盆地内储量的动用情况；另一方面，盆地未来新增的探明储量，也可补充到这一框架下，不断完善。

4 开发技术对策

4.1 储量动用程度评价

低渗透气藏储层连续性相对较好，井网一次性部署，后期局部调整，储量动用程度评价方法与常规气藏相同。致密气藏井间连通性差，气井泄气面积小，后期加密潜力大，提出了以井控法为核心的储量动用程度评价方法，关键步骤是确定单井泄气面积，并以动、静态储量比反映储量动用程度。鄂尔多斯盆地致密气藏分布面积广，直井、水平井均有[15]，根据泄气面积计算结果确定井网对储量的控制程度，可以分为密井网和稀井网。其中，密井网一般井网密度大于 2 口/km²，井网对储量控制程度高，井区内储量可以视为全部有效动用；稀井网一般井网密度小于 1 口/km²，井网对储量控制程度较低，可用区内所有井的动态储量之和作为区块已动用的储量。

根据方案实施情况和目前井网的完善程度进行测算，盆地内主力气田在现有井网下储量动用程度为 32%；已动用储量主要为高效或效益储量，分布在 GF2、GF5-1、GF4-1 及 GF1-1 单元；大量未动用储量以低效或难动用储量为主，主要分布在 GF5-2、GF1-2、GF1-3 和 GF2 单元。

4.2 开发技术对策

结合储量动用程度和目前开发的主要技术手段，提出不同类型储量的开发技术对策。

4.2.1 高效储量

该类储量单元储层品质较好，井网对储量的控制程度较高，已进入开发中后期。结合井网完善程度，测算各评价单元储量动用程度介于 68%～84%，平均为 76%，未动用储量规模小，主要分布在储层条件差的外围边角地带和局部富水区，后期开发的主要对策是增压开采和局部井网调整。数值模拟预测结果显示增压开采可以提高采收率 10% 左右，井网完善程度高的区域最终采收率可以达到 70%。

4.2.2 效益储量

效益储量单元以致密气藏为主，含气面积大，由于井控范围小，井网完善程度低，井间发育未动用储量，储量动用程度约为 42%，井网加密是提高该类储量动用程度的核心[16]。通过储层结构解剖、单井泄气面积计算、密井网试验区开发效果分析和不同井网密度数值模拟预测等方法，认为该类储量可以采取 3～4 口/km² 的加密井网进行开发[17]，结合生产制度优化、老井侧钻等配套措施，预计可以将采收率提高到50%左右[18]。

4.2.3 低效储量

低效储量单元相比于效益储量单元，储层物性变差、含气饱和度降低、开发效果更差，目前储量动用程度为15%，需要优选甜点区，滚动开发，逐步动用，降低开发风险。根据试气资料分析和地质精细解剖，提出低效储量的甜点区优选标准：在地质条件方面，要求有效砂体相对集中，连续性较好，单层厚度大于5m或者合采层厚度大于8m，储量丰度大于$1×10^8m^3/km^2$；在开发动态方面，要求测试产气量大于$2×10^4m^3/d$，无阻流量大于$5×10^4m^3/d$，EUR大于$1300×10^4m^3$。

4.2.4 难动用储量

难动用储量单元主要受储层致密或含水的影响，单井产气量低或产水，由于目前缺少有效的开发技术手段，仅动用极少量的"甜点区"，储量动用程度不足5%，长远来看该类储量是鄂尔多斯盆地潜在的可开发资源，需加大排水采气、储层改造等技术的攻关，大幅提高单井产气量，实现储量的有效动用。

5 结论

（1）鄂尔多斯盆地以低渗透—致密气藏为主，多层系含气，储层物性差，单井动态储量小、产气量低，产气类型可以划分为多层协同供气和单层主力供气两种。

（2）依据储层地质条件和单井动态特征相近的原则，结合开发管理区块的分布情况，将鄂尔多斯盆地划分为11个储量评价单元，并通过取评价参数的算术平均值来定量描述各个储量评价单元。

（3）以内部收益率为8%对应的井均EUR值为参照，与各个储量评价单元实际的井均EUR值进行对比，将11个储量评价单元进行排序，建立了储量经济有效动用接替序列。

（4）以经济效益为导向，将内部收益率30%、8%和5%作为界限，把11个储量评价单元划分为高效、效益、低效和难动用四种储量类型。

（5）高效储量以增压开采和局部井网调整为主，效益储量通过井网加密进一步提高储量动用程度，低效储量优选富集区实现滚动开发，难动用储量则需要加强富水区识别、排水采气工艺和精细压裂改造等技术攻关，力争实现效益开发。

参 考 文 献

[1] 孙龙德，方朝亮，李峰，等．中国沉积盆地油气勘探开发实践与沉积学研究进展[J]．石油勘探与开发，2010，37(4)：385-396．

[2] 邹才能，杨智，何东博，等．常规—非常规天然气理论、技术及前景[J]．石油勘探与开发，2018，45(4)：575-587．

[3] 马新华，贾爱林，谭健，等．中国致密砂岩气开发工程技术与实践[J]．石油勘探与开发，2012，39(5)：572-579．

[4] 谭中国，卢涛，刘艳侠，等．苏里格气田"十三五"期间提高采收率技术思路[J]．天然气工业，2016，36(3)：30-40．

[5] 卢涛，刘艳侠，武力超，等．鄂尔多斯盆地苏里格气田致密砂岩气藏稳产难点与对策[J]．天然气工业，

2015, 35(6): 43-52.
[6] 中华人民共和国国家质量监督检验检疫总局, 中国国家标准化管理委员会. GB/T 30501—2014 致密砂岩气地质评价方法[S]. 北京: 中国标准出版社, 2014.
[7] 孙玉平, 陆家亮, 唐红君. 国内外储量评估差异及经验启示[C]//2014年全国天然气学术年会, 贵阳, 2014.
[8] 王永祥, 张君峰, 段晓文. 中国油气资源/储量分类与管理体系[J]. 石油学报, 2011, 32(4): 645-651.
[9] 李忠兴, 郝玉鸿. 对容积法计算气藏采收率和可采储量的修正[J]. 天然气工业, 2001, 21(2): 71-74.
[10] 邹才能, 朱如凯, 吴松涛, 等. 常规与非常规油气聚集类型、特征、机理及展望: 以中国致密油和致密气为例[J]. 石油学报, 2012, 33(2): 173-187.
[11] 杨华, 付金华, 刘新社, 等. 鄂尔多斯盆地上古生界致密气成藏条件与勘探开发[J]. 石油勘探与开发, 2012, 39(3): 295-303.
[12] 吴凡, 孙黎娟, 乔国安, 等. 气体渗流特征及启动压力规律的研究[J]. 天然气工业, 2001, 21(1): 82-84.
[13] 何明舫, 马旭, 张燕明, 等. 苏里格气田"工厂化"压裂作业方法[J]. 石油勘探与开发, 2014, 41(3): 349-353.
[14] 冉富强, 李雁, 陈显举, 等. 致密油气藏储层评价技术[J]. 中国石油和化工标准与质量, 2017, 37(18): 177-178.
[15] 何东博, 贾爱林, 冀光, 等. 苏里格大型致密砂岩气田开发井型井网技术[J]. 石油勘探与开发, 2013, 40(1): 79-89.
[16] 贾爱林, 王国亭, 孟德伟, 等. 大型低渗—致密气田井网加密提高采收率对策——以鄂尔多斯盆地苏里格气田为例[J]. 石油学报, 2018, 39(7): 802-813.
[17] 郭智, 贾爱林, 冀光, 等. 致密砂岩气田储量分类及井网加密调整方法——以苏里格气田为例[J]. 石油学报, 2017, 38(11): 1299-1309.
[18] 郭建林, 郭智, 崔永平, 等. 大型致密砂岩气田采收率计算方法[J]. 石油学报, 2018, 39(12): 1389-1396.

原文刊于《天然气工业》, 2020, 40(3): 65-73.

Modeling and analyzing gas supply characteristics and development mode in sweet spots of Sulige tight gas reservoir, Ordos Basin, China

Ji Guang

(Research Institute of Petroleum Exploration & Development, Petrochina, Beijing 100083, China)

Abstract: In order to study the characteristics of the gas supply and development mode in sweet spots of Sulige tight gas reservoir in Ordos Basin, China, a mathematical model was developed for the typical lenticular reservoirs in tight gas reservoirs, and its analytical solution was obtained. The ideal model was calculated by using the analytical solution. Analysis of the production data indicated a clear boundary between the high-permeability and low-permeability regions of the lenticular reservoir, and the boundary will supply gas to the low-permeability region. The reliability of this finding was validated by real production data. The development mode of the lenticular reservoir was obtained, that is, the high-permeability area was first used during the initial production; when the pressure wave reached the boundary in the high-permeability region, the production showed a pseudo-steady state; further increase of the production pressure exceeding the threshold of the surrounding low-permeability region triggered the utilization of the low-porosity and low-permeability regions. The established model can provide useful guidance for the development of similar tight gas reservoirs.

Keywords: Ordos Basin; Tight gas reservoir; Lenticular reservoir; Gas supply characteristics; Sweet spots

1 Introduction

With the rapid development of industrial production, oil and natural gas resource demand has been increasing over the years. As conventional oil/natural gas resources have been gradually depleted, tight oil/ natural gas reservoirs have recently become significant sources and have become a hotspot for exploration and development. China and North America are setting off a boom in nonconventional oil/natural gas and other resources[16,24,35]. The non-conventional oil and natural gas resources that have been proven and put into production include shale gas, coal bed methane, tight oil and natural gas, etc[2-3,18,25,26].

Sulige gas field in Ordos Basin, China, is a typical compact terrigenous sedimentary sandstone tight gas reservoir. Since 2005, as a typical large unconventional gas field, it has been put into large-scale development and has gradually become the largest tight gas field in China[22,27,28]. This region contains a huge amount of resources. Development in a scientific and effective way will lay the foundation for the follow-up development for China's oil and gas industry[16,29].

Considerable work was conducted by experts during the early exploration stage. Lan et al.[13] studied the relationship between the sedimentary facies control, the diagenetic transformation, and the favorable paleogeography distribution and natural gas accumulation in the expulsion phase of source rocks, and discussed the geological factors controlling the distribution of the "sweet spot" regions that have relatively high porosity. Yang et al.[31] found that there were many similarities between the reservoir at the He-8 section in the Sulige region and the Xiangxi group reservoirs in the middle of Sichuan, and found that the Paleozoic Permian He-8 and Shan-1 gas reservoirs in Ordos Basin have the characteristics of large depth of burial, long burial time and high degree of diagenesis[15]. Those reservoirs are typical tight gas, high quality, developed reservoirs, and most of the natural gas is stored in these "sweet spot" layers[34].

Along with the continuous development of tight gas reservoirs in Ordos Basin, many experts and scholars have conducted in-depth studies on the accumulation mechanism of tight gas, the classification characteristics of oil and gas reservoirs and the distribution characteristics of high-efficient oil and gas reservoir sweet spots in this area. They have produced numerous key results and have successfully completed the early exploration and development tasks[18-19,30]. Ordos Basin is composed of lenticular reservoirs, and the effective sand bodies are small in scale[7,9]. The control reserves of wells are low, and the reservoir structure is the typical "sand-sand" binary structure[21]. The middle part of the layer, the lenticular reservoirs with higher porosity and permeability, is a coarse sand based, high-quality layer, while the surrounding and edge regions are a fine sand based, tight layer. The seepage characteristics of these layers are special [17] and are not applicable to any previous seepage mechanism. However, few experts have investigated the seepage mechanism of the reservoir layer that has binary composite structure. In order to fill this gap, it is necessary to conduct a thorough study on the seepage mechanism of the lenticular binary composite structure reservoir[5-6].

Study of reservoir fluid seepage provides the basis of reservoir development. Based on the sweet spots in Sulige tight gas reservoir, Ordos Basin, China, in order to study the seepage of lenticular reservoirs in a comprehensive and quantitative way, a mathematical model of the lenticular reservoir needs to be developed [1,12,20,32], and the analytical solution can then be solved. The analytical solution is used to calculate the ideal model and derive the seepage mechanism by analyzing the calculation result. The gas supply characteristics and development mode of sweet spots in this region can eventually be obtained.

2 Basic geological characteristics of the tight gas reservoir in Ordos Basin

The tight gas reservoirs in Ordos Basin are pore-type reservoirs in which the sand bodies are distributed in the lenticular type. The reservoir structure is the typical "sand-sand" binary structure (Fig. 1). The abundance of reserves is low, and the daily output of a single well and the accumulated yield of a single well are also low. However, the distribution range of the gas reservoirs is vast. There is no obvious gas boundary, and the overall reserve scale is large. It is the major component of the gas reservoir in tight sandstone in China[10,16,21,26].

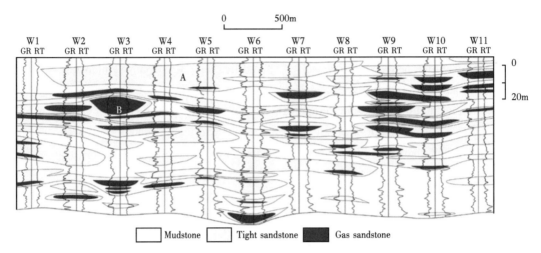

Fig. 1　Characteristic section of binary composite structure of the tight gas reservoir in Ordos Basin
(A)Channel,filled with middle-fine grained sand,matrix reservoir; (B)Discontinuous distributed channel bar,
grit rock,main gas-bearing sand bodies,discontinuous distribution

The lenticular reservoirs with relatively higher permeability widely distributed in tight sandstone is the most essential features of this kind of gas reservoirs. However, not all sandstones is effective reservoirs. The permeability of the lenticular reservoirs is more than 0.5mD, while the permeability of tight sandbody is less than 0.05mD. The hydrodynamic of the sediment in this type of river varies significantly, forming different sediments of coarse and fine sandstones. From the special cores analysis, the coarse sandstones formed a high-quality, gas-bearing sand body that has a porosity of 5% or more and a permeability of 0.5mD or more, which is the main producing layer of the gas reservoir. The fine sandstones, on the other hand, formed a tight layer with a porosity of 5.0% or less and a permeability of 0.1mD or less, which has limited contribution to production and is currently classified as a non-effective layer.

These types of sedimentation and diagenetic characteristics result in the small size of the effective sand bodies, scattered distribution, generally in the range of several tens to several hundreds of meters, and poor horizontal continuity and connectivity (Fig. 2).

However, these sand bodies are present in large quantities, have multi-layer structures and are distributed across a vast range. If all effective sand bodies are projected onto one layer, the gas-bearing area can reach more than 90%. The highly-scattered distribution of the lenticular reservoir determines the production characteristics of the gas reservoirs: due to the small scale of the effective sand body, the single well production is low, the stable production capacity is poor, and the single well control reserves and the final accumulated yield are low. Generally, the dynamic reservoir controlled by a single well is less than 30 million cubic meters, and the gas production rate can keep 10 thousand cubic meters for 3 years, then the production rate will keep decline for about 7 to 10 years until the reservoir pressure is too low to produce[4,11,33].

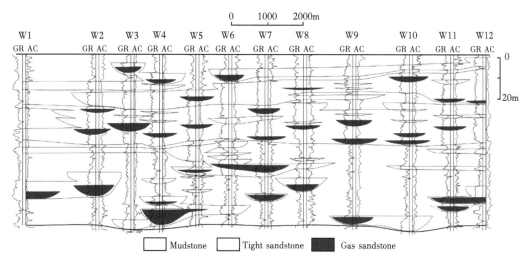

Fig. 2 Section of typical gas reservoirs in Sulige region, Ordos Basin

3 Mathematical model and its solution

3.1 Description of the physical model

As shown in Fig. 3, the so-called "sweet spot" refers to an effective reservoir that has relatively high quality, which is the main producing layer of the gas reservoir. After the wellbore opens, when the pressure wave propagates to the sweet spot boundary, it will continue to spread to the non-main producing layer, demonstrating the gas supply characteristics of the reservoir in the tight area.

In order to study the gas supply characteristics and development mode, a seepage model was established. Fig. 4 displays the schematic diagram of the seepage model for heterogeneous gas reser-

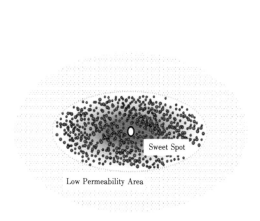

Fig. 3 Schematic diagram of reservoirs in heterogeneous gas reservoirs

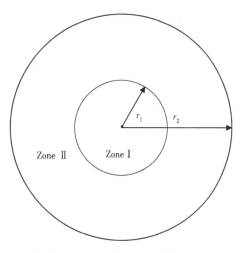

Fig. 4 Schematic diagram of the seepage model for heterogeneous gas reservoirs

voirs, which are divided into two zones due to different permeabilities. Zone I represents the "sweet spot" area that has relatively high permeability, and zone II represents the thicker, non-main producing layer. Set $r_2/r_1 = C$

3.2　Establishment of the mathematical model

According to the basic methods and steps of establishing the mathematical model of seepage flow, the unstable mathematical model of heterogeneous tight reservoir was established using single-phase compressible gas seepage as an example. The basic assumptions are as follows. The gas reservoir contains a single-phase homogeneous gas, which obeys the low-speed non-Darcy law; the gas compression and the stress sensitivity of reservoir are considered; the seepage process is isothermal.

The basic components of the seepage mathematical model are the continuity equation (mass conservation equation), the equation of motion (momentum conservation equation) and the state equation. In actual production, radial flow is generally presented in the vicinity of the bottom of each well. Thus, the mathematical model can be written in the coordinate system[14,23].

The continuity equation is given as

$$-\frac{\partial(\rho_g \vec{v})}{\partial r} = \frac{\partial(\rho_g \phi)}{\partial t} \qquad (1)$$

The equation of motion is given as

$$\vec{v} = \frac{K}{\mu}\left(\frac{dp}{dr} - G\right) \qquad (2)$$

The threshold pressure gradient is as follows,

$$G = 0.0874 K^{-0.459} \qquad (3)$$

The stress sensitivity of the gas reservoir is

$$K = K_0 e^{-\alpha(p_e - p)} \qquad (4)$$

The state equation is

$$\begin{cases} \rho_g = \dfrac{T_{sc} Z_{sc} \rho_{gsc}}{\rho_{sc}} \times \dfrac{p}{TZ} \\ C_p = \dfrac{-\dfrac{dV}{V}}{dp} = -\dfrac{1}{V} \cdot \dfrac{dV}{dp} = \dfrac{1}{p} - \dfrac{1}{z} \cdot \dfrac{dZ}{dp} \end{cases} \qquad (5)$$

The pseudo-pressure function is defined as

$$m^* = 2\int_{p_a}^{p} \frac{p e^{-\alpha(p_e - p)}}{\mu(p) Z(p)} dp \qquad (6)$$

The initial conditions are

$$\begin{cases} r = r \\ t = 0 \\ m^* = m_e^*(p = p_e) \end{cases} \quad (7)$$

Considering the boundary conditions in steady production:

$$\begin{cases} r = r_w, m^* = m_w^*(p = p_w) \\ r = r_c, m^* = m_c^*(p = p_c) \\ r = r_e, m^* = m_e^*(p = p_e) \\ r \dfrac{\partial m^*}{\partial r} = \dfrac{q_m T p_{sc}}{\pi K h Z_{sc} T_{sc} \rho_{gsc}} \end{cases} \quad (8)$$

The pseudo-threshold pressure gradient is defined as:

$$\dot{G} = e^{-\alpha(p_e - p)} G \quad (9)$$

Substituting Eqs. (2)-(5) into Eq. (1), Eq. (1) can be simplified as

$$\frac{\partial^2 m^*}{\partial r^2} + \frac{1}{r}\frac{\partial m^*}{\partial r} - C_\rho \dot{G} \frac{\partial m^*}{\partial r} = \frac{1}{\eta}\frac{\partial m^*}{\partial t} \quad (10)$$

in which

$$\eta = \frac{K}{\phi \mu(p) C_\rho} \quad (11)$$

Dimensionless parameter β is set as a disturbance factor,

$$\beta = 1 - C_\rho \dot{G} r \quad (12)$$

The governing equation Eq. (10) can be derived as:

$$\frac{\partial^2 m^*}{\partial r^2} + \frac{\beta}{r}\frac{\partial m^*}{\partial r} = \frac{1}{\eta}\frac{\partial m^*}{\partial t} \quad (13)$$

where, the pressure gradient at pseudo-threshold $\dot{G}$ is considered in the governing equation; interior boundary conditions can be derived using Darcy's law. Given the Boltzmann transformation,

$$u = \frac{r^2}{4\eta t} \quad (14)$$

Eq. (13) becomes

$$u \frac{d^2 m^*}{du^2} + (\beta + u)\frac{dm^*}{du} = 0 \quad (15)$$

3.3 Solution of the mathematical model

Considering the steady production of the inner boundary, Zone I that surrounds the well satisfies

Eq. (15); the permeability of zone I: $k_0 = k_{01}$.

Solving Eq. (15) gives:

$$m^* = C_2 + C_1 Ei\left(-u\left(1 - \frac{(1-\beta)}{2u} \cdot r\right)\right) = C_2 - C_1 \int_{u\left(1-\frac{(1-\beta)}{2u}\cdot r\right)}^{\infty} e^{-t}/t \, dt \qquad (16)$$

Considering boundary condition Eq. (7) gives

$$C_1 = \frac{q_m T p_{sc}}{\pi k h Z_{sc} T_{sc}} \cdot \frac{e^{\frac{1}{2}(-r_w + 2u_w + r\beta_w)}(r_w - 2u_w - r_w \beta_w)}{\left(1 - \beta_w - 2\frac{\partial u}{\partial r}\Big|_{r=r_w} - r_w \frac{\partial \beta}{\partial r}\Big|_{r=r_w}\right) r_w} \qquad (17)$$

Substituting Eq. (17) into Eq. (16), C_1 can be solved based on external boundary conditions:

$$C_2 = m_c^* - C_1 Ei\left(-u_c\left(1 - \frac{(1-\beta_c)}{2u_c} \cdot r_c\right)\right) \qquad (18)$$

where, $u_w = \frac{r_w^2}{4\eta_1 t}$, $u_c = \frac{r_c^2}{4\eta_1 t}$

Similarly, zone II surrounding the well satisfies Eq. (15). The permeability of zone II satisfies $K_0 = K_{02}$.

The pressure at $r = r_c$ is the pressure of the exterior boundary of zone I. The pseudo-pressure functions at this position between zone II and zone I can be expressed as

$$m_c^* = C_4 + C_3 Ei\left(-u_c\left(1 - \frac{(1-\beta_c)}{2u_c} \cdot r_c\right)\right) \qquad (19)$$

where

$$C_3 = \frac{(m_e^* - m_c^*)}{Ei\left(-u_e\left(1 - \frac{(1-\beta_e)}{2u_e} \cdot r_e\right)\right) - Ei\left(-u_c\left(1 - \frac{(1-\beta_c)}{2u_c} \cdot r_c\right)\right)} \qquad (20)$$

$$C_4 = m_e^* - C_3 Ei\left(-u_e\left(1 - \frac{(1-\beta_e)}{2u_e} \cdot r_e\right)\right) \qquad (21)$$

where

$$u_c = \frac{r_c^2}{4\eta_2 t}, u_e = \frac{r_e^2}{4\eta_2 t},$$

4　Results and discussion

4.1　Characteristics of the gas supply

The parameters used in the analysis are listed as follows: formation pressure is 25.0MPa; porosity of sweet spot zone is 10% while the porosity of low permeability is 3%; zone I is the high-permeability area (500m radius, 0.5mD permeability, 0.030MPa^{-1} stress sensitivity coefficient of reservoir); zone II is the low-permeability area (0.03mD permeability, 0.015MPa^{-1} stress sensitivity coefficient of reservoir); gas reservoir thickness is 10m; pressure relief radius is 1000m; wellbore radius is 0.1m; threshold pressure gradient is 0.003MPa/m; standard state pressure is 0.1MPa; gas compression factor is 0.89; gas isothermal compression coefficient is 0.03MPa^{-1}; standard state temperature is 273K; formation temperature is 396K; gas viscosity is 0.027mPa·s; gas production is $5.0 \times 10^4 m^3/d$.

Fig. 5 shows the variation of the heterogeneous formation pressure with distance at different production times. It can be seen that with increasing production time, the formation pressure decreases, and the pressure gradient of the formation changed obviously in the boundaries of zone I and II.

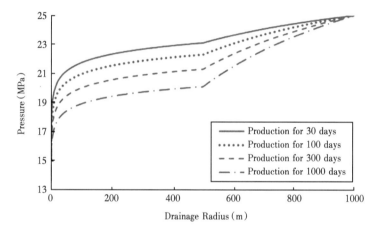

Fig. 5 Pressure at different production times

Fig. 6 shows the variation of the pressure gradient under heterogeneous formation with distance under different production times. With the increase in production time, the pressure gradient of the

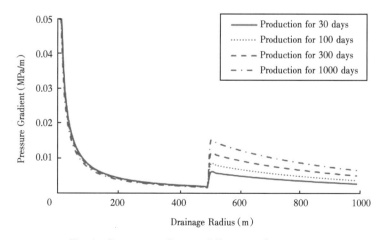

Fig. 6 Pressure gradient at different production times

formation in zone I did not change significantly in the high-permeability region but suddenly increased at the boundary point. The pressure gradient in zone II increased. It can be seen that with increasing time, the low permeability region supplies gas to the high-permeability region.

Fig. 7 shows the variation of formation pressure versus pressure drop radius under different conditions of gas production. It can be seen that under the steady production of the interior boundary, the formation pressure is gradually reduced with the increase in gas production, and the influence of heterogeneity on the pressure distribution increases.

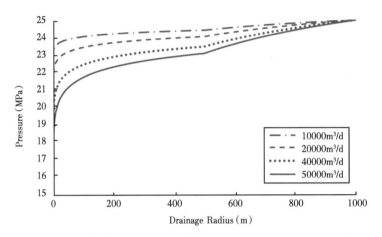

Fig. 7　Pressure at different production times

Fig. 8 shows the variation of the heterogeneous formation pressure gradient versus pressure drop radius under different production conditions. It can be seen that the pressure gradient dramatically increased at the interface between the high-permeability and low-permeability regions. The higher the daily gas production, the larger the formation pressure gradient.

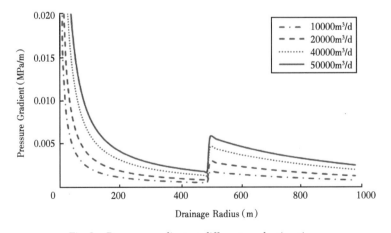

Fig. 8　Pressure gradient at different production times

4.2 Development mode of the sweet spot

The high-permeability region is first utilized in gas production. When the pressure reaches the boundary of the highly permeable region, the production shows a sharply decreasing. With the increase in pressure difference beyond the threshold pressure of the exterior low permeability region, the reservoir in the low permeable region begins to be utilized. The gas supply from the exterior low-permeability region is also observed in real production.

Fig. 9 displays the production curve of Su-XX well. Based on the gas well production, Su-XX well cumulatively produced $44.85 \times 10^6 \mathrm{m}^3$ of natural gas by January 2010. According to the dynamic reserves of $50.60 \times 10^6 \mathrm{m}^3$, the recovery factor of dynamic reserves reached 88.6%, approaching the condition of abandonment. However, the average daily production of this well in August 2015 was still above $1 \times 10^6 \mathrm{m}^3$ in reality, and the cumulative gas production was $59.06 \times 10^6 \mathrm{m}^3$ which is more than the reserves of sweet spot area.

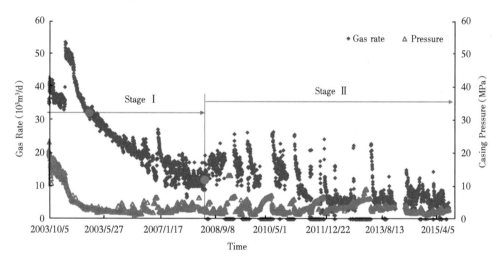

Fig. 9 Production curve of Su-XX well

(In stage I, only the sweet spot area supply the gas; In stage II, low permeably begin to supply gas, but the production is discontinuous because of low reservoir pressure)

Fig. 10 displays the fit of the dynamic control reserves of Su-XX well in the period of sweet spot supply. As shown in Fig. 10, gas wells were in continuous stable production. According to this relatively stable production data, the exponential decreasing method was used to predict the dynamic control reserves of Su-XX well to be $54.6 \times 10^6 \mathrm{m}^3$.

Fig. 11 displays the fit of the dynamic control reserves of Su-XX well in the late period of intermittent production because of low reservoir pressure. As shown in Fig. 11, intermittent production stage was used, which resulted in a large change in production since July, 2006. The trend of the production data in this period is different from that in the sweet spot supply period. According to the decreasing trend in the late stage of intermittent production, the dynamic control reserves of Su-XX well are predicted to be $64.5 \times 10^6 \sim 72.0 \times 10^6 \mathrm{m}^3$.

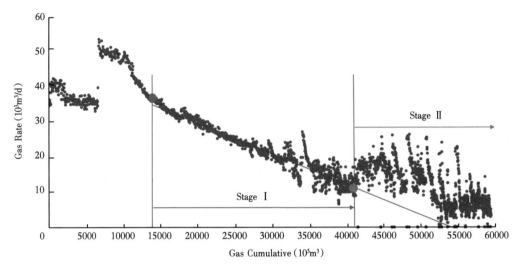

Fig. 10 Fit of dynamic control reserves of Su-XX well in the period of steady production
(In stage Ⅰ, only the sweet spot area supply the gas; In stage Ⅱ, low permeably begin to supply gas)

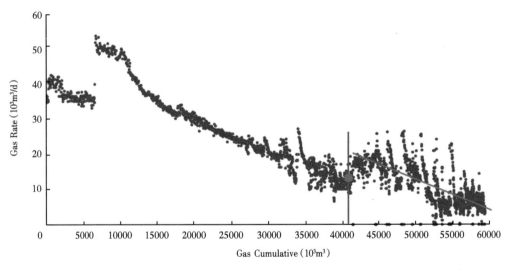

Fig. 11 Fit of dynamic control reserves of Su-XX well in the late period of intermittent production
(After the point, low permeably begin to supply gas)

From the comparison of the dynamic control reserves predicted by the low permeability area production data after the sweet spot supply gas state, the peripheral low-permeability area was supplying gas to the gas well. The comprehensive analysis shows that there was gas supply to the gas well in the low-permeability reservoirs of the Su-XX wells, and the gas supply was about $9.9 \times 10^6 \sim 17.4 \times 10^6 m^3$.

5 Conclusions

(1) Based on the mass conservation equation, the low-speed non-Darcy motion equation for gas and the real gas equation of state, and considering the reservoir stress sensitivity, a mathematical model of the low-speed non-Darcy flow in the sweet spot zone was established. The derived analytical solution of radial flow under boundary conditions provides a theoretical basis for studying the development characteristics of these typical tight gas reservoirs.

(2) With the increase in production time, the formation pressure gradient did not change significantly in the high-permeability region but suddenly increased at the boundary region, and gradually increased in the low-permeability region. With the increase in gas production, the formation pressure was gradually reduced, and the impact of heterogeneity on the pressure distribution increased.

(3) The pressure gradient of the heterogeneous formation under different gas production conditions showed a clear trend with the change in the pressure drop radius. The pressure gradient of the formation increased abruptly at the boundary of the high- and low-permeability regions. The higher the daily gas production, the greater the gradient of the formation pressure.

(4) The binary structural characteristics of the reservoirs in Ordos basin determined its production characteristics. By analyzing the production curve and dynamic control reserves, it can be seen that the gas production first occurred in the high-permeability region. When pressure reached the boundary of the high-permeability region, the production characteristics of the pseudo-steady state appeared. With the increase in the production pressure difference that exceeded the threshold pressure, the low-permeability region began to be used. Based on the real production trend, the phenomenon that the exterior low-permeability region supplied gas to the high-permeability region was also observed.

Declaration of conflicting interests

The author(s) declared no potential conflicts of interest with respect to the research, authorship, and/or publication of this article.

References

[1] Baziar S, Tadayoni M, Nabi-Bidhendi M, et al. Prediction of permeability in a tight gas reservoir by using three soft computing approaches: A comparative study[J]. Journal of Natural Gas Science and Engineering, 2014, 21: 718-724.

[2] Chew K J. The future of oil: unconventional fossil fuels. Philosophical transactions. Series A, Mathematical, Physical, and Engineering Sciences, 2014, 372 (2006): 1-32.

[3] Ding X, Yang P, Han M, et al. Characteristics of gas accumulation in a less efficient tight-gas reservoir, He 8 interval, Sulige gas field, Ordos Basin, China[J]. Russian Geology and Geophysics, 2016, 57(7): 1064-1077.

[4] Fan J, Qu X, Wang C, et al. Natural fracture distribution and a new method predicting effective fractures in tight oil reservoirs in Ordos Basin, NW China[J]. Petroleum Exploration and Development, 2016, 43(5): 806-814.

[5] Freeman C M, Moridis G, Ilk D, et al. A numerical study of performance for tight gas and shale gas reservoir systems[J]. Journal of Petroleum Science and Engineering, 2013, 108: 22-39.

[6] Ghanizadeh A, Clarkson C R, Aquino S, et al. Petrophysical and geomechanical characteristics of Canadian tight oil and liquid-rich gas reservoirs: I. Pore network and permeability characterization[J]. Fuel, 2015, 153: 664-681.

[7] Guo Z, Jia A, Bo Y. Effective sand body distribution and its main controlling factors in tight sandstone gas reservoirs[J]. Petroleum Geology and Experiment, 2014, 36(06): 684-691.

[8] Guo Y, Pang X, Li Z, et al. The critical buoyancy threshold for tight sandstone gas entrapment: Physical simulation, interpretation, and implications to the Upper Paleozoic Ordos Basin[J]. Journal of Petroleum Science and Engineering, 2017, 149: 88-97.

[9] He D, Jia A, Ji G, et al. Well type and pattern optimization technology for large scale tight sand gas, Sulige gas field[J]. Petroleum Exploration and Development, 2013, 40(1): 79-89.

[10] He D, Wang L, Ji G, et al. Well spacing optimization for Sulige tight sand gas field[J]. Petroleum Exploration and Development, 2012, 39(4): 458-464.

[11] Jang H, Lee W, Kim J, et al. Novel apparatus to measure the low-permeability and porosity in tight gas reservoir [J]. Journal of Petroleum Science and Engineering, 2016, 142: 1-12.

[12] Jiang R, Han G, Wand Y. Research on special seepage mechanism and prediction method for stable productivity of low permeability gas pools[J]. Oil Drill Production Technology, 2015, 37(04): 67-71.

[13] Lan C, He S, Zhang J. Discussion on the factors of controlling the distribution of the reservoir "sweet spots" of Sulige Gasfield[J]. Journal of Xi'an Shiyou University (Natural Science Edition), 2007, 22 (01): 45-48.

[14] Liu P, Li W, Xia J, et al. Derivation and application of mathematical model for well test analysis with variable skin factor in hydrocarbon reservoirs[J]. AIP Advances, 2016, 6 (6): 1-11.

[15] Liu P, Zhang X. Enhanced oil recovery by CO_2-CH_4 flooding in low permeability and rhythmic hydrocarbon reservoir[J]. International Journal of Hydrogen Energy, 2015, 40 (37): 12849-12853.

[16] Lu T, Liu Y, Wu L, et al. Challenges to and countermeasures for the production stabilization of tight sandstone gas reservoirs of the Sulige Gasfield, Ordos Basin[J]. Natural Gas Industry B, 2015, 2(4): 323-333.

[17] Luo S, Peng Y, Wei X. Characteristics and classification of gas-water relative permeability curves of tight sandstone reservoirs in Sulige Gas Field[J]. Journal of Xi'an Shiyou University (Natural Science Edition), 2015, 30(06): 55-61.

[18] Pan S, Horsfield B, Zou C, et al. Statistical analysis as a tool for assisting geochemical interpretation of the Upper Triassic Yanchang Formation, Ordos Basin, Central China[M]. International Journal of Coal Geology, 2017, 173: 51-64.

[19] Ran X, Li A. Development theory of Sulige gas field[M]. Beijing: Petroleum Industry Press, 2008.

[20] Ren J, Zhang L, Ezekiel J, et al. Reservoir characteristics and productivity analysis of tight sand gas in Upper Paleozoic Ordos Basin China[J]. Journal of Natural Gas Science and Engineering, 2014, 19: 244-250.

[21] Santagiuliana R, Fabris M, Schrefler B A. Subsidence above depleted gas fields[J]. Engineering Computations, 2015, 32 (3): 863-884.

[22] Shad S, Holmgrün C, Calogirou A. Near wellbore thermal effects in a tight gas reservoir: Impact of different reservoir and fluid parameters[J]. Journal of Unconventional Oil and Gas Resources, 2016, 16: 1-13.

[23] Stehfest H. Numerical inversion of laplace transform[J]. Communications of the ACM, 1970, 13 (1): 47-49.

[24] Tan Z, Lu T, Liu Y, et al. Technical ideas of recovery enhancement in the Sulige gasfield during the 13th five-year plan[J]. Natural Gas Industry B, 2016, 3(3): 234-244.

[25] Wang G, Chang X, Yin W, et al. Impact of diagenesis on reservoir quality and heterogeneity of the upper Triassic Chang 8 tight oil sandstones in the Zhenjing area, Ordos Basin, China. Marine and Petroleum Geology, 2017,

83: 84-96.

[26] Wang T, Dong S, Wu S, et al. Numerical simulation of hydrocarbon migration in tight reservoir based on Artificial Immune Ant Colony Algorithm: A case of the Chang 81 reservoir of the Triassic Yanchang Formation in the Huaqing area, Ordos Basin, China[J]. Marine and Petroleum Geology, 2016, 78: 17-29.

[27] Wu L, Zhu Y, Liu Y, et al. Development techniques of multi-layer tight gas reservoirs in mining rights overlapping blocks: A case study of the Shenmu gas field, Ordos basin, NW China[J]. Petroleum Exploration and Development, 2015, 42(6): 904-912.

[28] Yang H, Liu X S, Meng P L. New development in natural gas exploration of the Sulige gas fields[J]. Natural Gas Industry, 2011, 31(02): 1-8.

[29] Yang H, Xi S, Wei X. Analysis on gas exploration potential in sulige area of Thrordos basin[J]. Natural Gas Industry 2006, 26(12): 45-48.

[30] Yang H, Liang X, Niu X, et al. Geological conditions for continental tight oil formation and the main controlling factors for the enrichment: A case of Chang 7 Member, Triassic Yanchang formation, Ordos basin, NW China [J]. Petroleum Exploration and Development, 2017, 44(1): 11-19.

[31] Yang X, Zhao W, Zou C. Comparison of formation condition of 'sweet point' reservoir in Sulige gas field and Xiangxi group gas field in the central Sichuan basin[J]. Natural Gas Industry, 2007, 27(01): 1-7.

[32] Zhang Y, Zeng J, Qiao J, et al. Experimental study on natural gas migration and accumulation mechanism in sweet spots of tight sandstones[J]. Journal of Natural Gas Science and Engineering, 2016, 36: 669-678.

[33] Zhao J, Pu X, Li Y, et al. A semi-analytical mathematical model for predicting well performance of a multistage hydraulically fractured horizontal well in naturally fractured tight sandstone gas reservoir[J]. Journal of Natural Gas Science and Engineering, 2016, 32: 273-291.

[34] Zhou Y, Ji Y, Xu L, et al. Controls on reservoir heterogeneity of tight sand oil reservoirs in Upper Triassic Yanchang Formation in Longdong Area, southwest Ordos Basin, China: Implications for reservoir quality prediction and oil accumulation[J]. Marine and Petroleum Geology, 2016, 78: 110-135.

[35] Zou C N, Zhai G M, Zhang G Y, et al. Formation, distribution, potential and prediction of global conventional and unconventional hydrocarbon resources[J]. Petroleum Exploration and Development, 2015, 42(01): 13-25.

原文刊于《Enery Exploration & Exploitation》,2018,36(4):895-909.

致密砂岩气藏储渗单元研究方法与应用
——以鄂尔多斯盆地二叠系下石盒子组为例

郭建林　贾成业　闫海军　季丽丹　李易隆　袁贺

(中国石油勘探开发研究院，北京 100083)

摘要：精细表征储层特征和储层结构是致密砂岩气藏开发中后期的主要技术需求。基于不同类型砂体的相似孔隙度、渗透率特征将辫状河沉积体系中河道充填和心滩砂体聚类为储渗单元，提出了储渗单元研究概念和研究思路，开展了辫状河沉积体系储渗单元发育模式研究。通过露头观测和测井相标志，识别出辫状河沉积体系中储渗单元发育心滩叠置型、河道充填叠置型、心滩和河道充填叠置型、心滩或河道充填孤立型四种储渗单元叠置模式；基于储渗单元发育模式，提出了河流相致密砂岩气藏开发井型的适应性，指出辫状河沉积体系中河道叠置带是叠置型储渗单元发育的有利部位，是水平井开发的有利目标，辫状河沉积体系中的过渡带和洼地主要发育孤立型储渗单元，适合直井或丛式井组开发。鄂尔多斯盆地二叠系下石盒子组野外露头研究和苏里格气田加密井区井间干扰试验表明，辫状河沉积体系中储渗单元发育规模为顺古水流方向长 600 m 和垂直水流方向宽 400 m 左右；表明证实苏里格气田具备进一步加密到 400 m×600 m 的条件，预计可提高采收率 15%~20%。

关键词：致密砂岩气藏；辫状河沉积体系；储渗单元；叠置模式；开发井型；采收率

致密砂岩气藏是指地层条件下覆压渗透率小于 0.1 mD(不包含裂缝渗透率)的砂岩储层，一般情况下没有自然产能或自然产能低于工业标准，需要采用增产措施或特殊工艺井才能获得商业气流[1]。致密砂岩气藏是中国主要的天然气藏类型之一，在天然气产业地质储量和年产量中占相当大的比重。截至 2016 年 12 月，致密砂岩气藏探明储量占全国天然气总探明储量的 35%，致密砂岩气藏年产量占国内天然气总年产量的比例为 22%。近十年来，随着苏里格、大牛地、榆林、子洲、神木、米脂等一批致密砂岩大气田投入开发，致密砂岩气探明储量和年产量实现了快速双增长[2-3]。但上产生产高峰期之后，主力致密砂岩气田将相继进入开发中后期，开发调整、稳产挖潜和提高采收率是该期致密砂岩气藏开发中面临的主要技术问题。与国外海相—海陆过渡相致密砂岩储层不同，河流相砂体是中国致密砂岩气藏的主要储层类型，进一步精细表征河流相沉积体系特征，开展河流相沉积体系中沉积微相和微相组合精细描述是致密砂岩气藏开发中后期的主要研究方向。

Martin(1993)、Collinson(1996)和 Miall(1996)对河流相沉积体系的研究表明[4-6]，辫状河砂体具有较好的渗透率、孔隙度和较高的净毛比，是品质较好的油气藏储集体。同时，由于辫状河体系内部渗透性差异形成的低渗单元是阻碍流体流动、制约波及系数的主要因素，是储层表征和油气藏开发中面临的主要技术挑战[7-12]。Hearn(1984)提出了流动单元的概念[13]，流动单元研究以一致的岩石学和水动力学特征为基础，将具有不同特征的沉积微相划分为不同级别的流动单元,预测剩余油分布规律。Miall(1996)依据河流沉积层序中的不同级次界面和结构单元建立了储层内部建筑结构[6],将不同岩性界面划分为五级界面,为表征河流相储层非均质性和河流沉积学研究提供了有益的研究思路。在前人对河流相沉积体系储层构型和流动单元划分的基础上,结合多年的研究与实践,笔者提出了储渗单元研究思路,以河流相沉积边界和储层非均质性差异为标志,针对河流相沉积体系中高渗透率、低渗透率储层单元开展识别和分析,建立不同渗透性特征的储集体空间分布模式,指导河流相致密砂岩气藏开发实践。储渗单元是指受岩性或物性边界约束的、具有相似储集性能和渗流特征的沉积亚(微)相或亚(微)相组合。由于天然气的流动性远高于原油,通常气藏开发中压降波及范围内的天然气可采储量均可实现商业开发,因此储集体内部储集和渗流特征评价是气藏开发评价的研究重点。与流动单元不同,储渗单元研究以阻流边界(通常为岩性或物性边界)识别为基础,将阻流边界控制范围以内、分布连续、具有相似物性特征的沉积微相和微相组合划分为不同品质的储渗单元。从本质上,流动单元研究是对沉积微相按流动特征的分级分类,而储渗单元研究是将不同类型的沉积微相按渗透性聚类,通过不同沉积微相的叠置关系建立储渗单元内部结构模式。本文通过鄂尔多斯盆地二叠系下石盒子组辫状河沉积体系露头和实钻储层特征分析,识别储渗单元和建立发育模式,以期为致密砂岩气藏开发中后期加密部署和水平井开发提供技术思路。

1 储渗单元特征与发育模式

1.1 储渗单元边界类型

Lynds 和 Hajek(2006)对美国内布拉斯加州奈厄布拉勒河和北卢普河现代河流沉积的研究表明[10],受水深的控制,同期河道沉积的顶界为泛滥平原或溢岸沉积;随着河道改道作用的影响,河道沿侧向移动,河道对泛滥平原或溢岸沉积的泥岩产生切割,在下一期河道底部形成于泥岩衬里。由于泛滥平原或溢岸沉积是河道滞留沉积的横向而不是纵向伴生亚相,所以其与河道沉积属同一期河流沉积的侧向沉积物。因此,不同期次河道沉积界限为泛滥平原或溢岸泥岩相,即辫状河储层建筑结构中的四级构型界面[6],多期河道侧向往复改道,形成纵向和横向上多期河道砂体、泛滥平原或溢岸泥岩相互叠置形成复合河道带,鄂尔多斯盆地二叠系下石盒子组气藏即为多期辫状河河道叠置而成的大型复合河道带。

储渗单元研究的首要任务是识别储渗单元内部和外部边界。储渗单元识别的基础是不同储渗单元与内、外部边界的岩性和物性差异。河流相致密砂岩气藏有效储层成因与岩石组构、成岩作用密切相关。以鄂尔多斯盆地二叠系下石盒子组气藏为例,辫状河沉积体系中心滩和河道充填底部粗砂岩分选差、大粒径矿物颗粒形成岩石骨架结构,石英类刚性矿物含量高、抗压实能力强,有利于原生孔隙的保存和孔隙流体的流动,溶蚀作用相对发育,粒径大于0.5mm的粗砂岩孔隙度普遍大于5%,渗透率与孔隙度呈正相关关系,整体上粗砂岩平均孔隙度为9.1%、平均覆压渗透率为0.098mD[14-17],渗流条件好(图1),是有效储层发育的有利岩相;河道充填中—上部的中细砂岩中火山岩屑等塑性颗粒含量高、分选好,呈致密压实相,不利于孔隙流体的流动和溶蚀作用的发生,孔隙度和渗透率均较低,有效储层不发育。同时,废弃河道、泛滥平原、心滩内部不同期次的单元坝间泥岩夹层和落淤层等泥岩相渗透率极低。储渗单元研究中将心滩、河道底部充填等渗透率高、物性条件好的沉积微相或微相组合归为储渗单元，而心滩和河道底部充填受岩性和物性边界的隔挡，形成阻流边界。溢岸、心滩侧向加积的坝内粉砂质泥岩夹层和落淤层是储渗单元研究中的内部边界，废弃河道、泛滥平原是储渗单元研究中的外部边界。

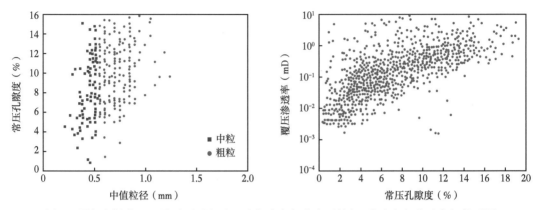

图 1 鄂尔多斯盆地二叠系下石盒子组致密砂岩气藏岩石粒径、孔隙度与渗透率间关系图

泛滥平原和溢岸泥岩（图2），即同期河道复合砂体的顶界面，是储渗单元的标志性顶底界面，也是储渗单元在纵向上的主要识别标志。泛滥平原和溢岸泥岩厚度一般为数十厘米到数米不等，延伸范围较广。废弃河道泥岩位于储渗单元顶面附近，同属于储渗单元的外部边界，是河道水流侧向运动或河流水动力减小形成，一般河流上游部位废弃河道泥岩规模发育较小，河流下游由于水动力减弱废弃河道泥岩发育规模增大，但整体上废弃河道泥岩位于同期河道充填沉积范围以内。河道充填沉积顶部由于水流的沉积分异作用，沉积物分选好，粒径相对较小，通常在细粉砂级别，受压实作用影响大渗透性较差。由于河道顶部充填物性较差，形成储渗单元的外部物性边界。河道充填沉积顶部与底部间的岩性界面是河流相储层建筑结构的三级界面[6-7,18]。

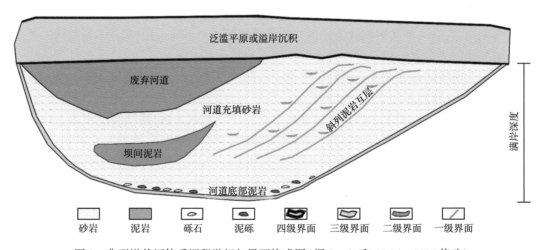

图 2 典型辫状河体系沉积微相与界面构成图（据 Lynds 和 Hajek，2006 修改）

单元坝是构成心滩的基本单元,河道水流受多个心滩单元坝阻隔发育坝间次级水道,或称串沟,次级水道水动力相对较弱,粉砂质泥岩、泥岩在该区域易沉降,形成坝间泥岩。坝间泥岩为储渗单元内部边界,通常规模相对较小、物性差,一般厚度小于0.5m,是河流相储层建筑结构的二级界面。

坝内泥岩是心滩侧向加积作用产生的粉砂质、泥质夹层,通常厚度较薄,纹层或夹层级,分选好;落淤层泥岩则形成于季节性洪水期的泥质沉积物,一般侧向延伸宽度有限,发育规模较小。坝内泥岩、落淤层或统称为斜列泥岩互层(图2),同属储渗单元的内部边界,是层组或单个层系界面,属建筑结构中的一级界面。

1.2 储渗单元发育模式

野外露头观测和致密砂岩气藏开发中密井网区精细解剖可识别不同类型的储渗单元,从而建立储渗单元叠置模式。笔者采用储渗单元研究思路,对鄂尔多斯盆地南缘柳林地区二叠系下石盒子组露头剖面观测,结合气井钻遇砂体分析,将储渗单元划分为:心滩与河道底部充填叠置型、河道充填叠置型、心滩叠置型和心滩或河道底部充填孤立型四种发育模式(图3)。

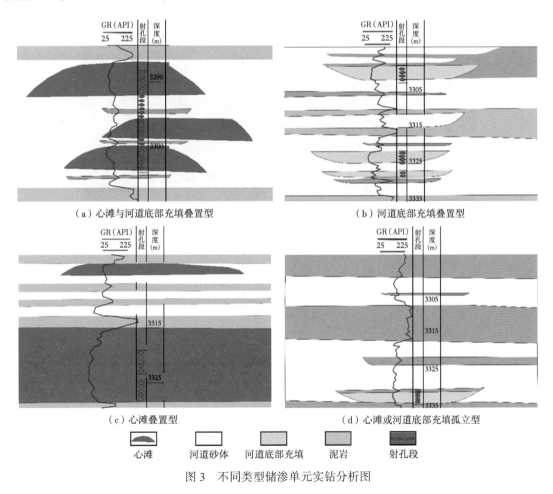

图3 不同类型储渗单元实钻分析图

1.2.1 心滩与河道底部充填叠置型

心滩整体上一般为正粒序,底部为砾岩或粗砂岩相,向上过渡为粗、中砂岩相。由于心滩边部水动力作用较弱,通常为斜层状的泥岩或粉砂质泥岩(落淤层),同时侧向加积作用剧烈,形成心滩复合砂体内部夹杂泥岩或粉砂质泥岩夹层,即储渗单元的内部阻流边界。该阻流边界通常规模较小,呈纹层状,从露头剖面和水平井实钻轨迹中可识别出该类型储渗单元内部阻流边界。心滩与河道底部充填型储渗单元内部落淤层阻流边界纵向发育规模较小,一般小于0.5m(图4),水平井钻井过程中通常沿心滩侧向钻进过程中钻遇薄层状泥质夹层即为落淤层边界。心滩与河道底部充填叠置型储渗单元底部边界为上期河道消亡时沉积的泛滥平原或废弃河道泥岩、粉砂质泥岩,属岩性边界;其上部边界为河道顶部充填沉积形成的中、细砂岩相,分选较好,物性较差,形成物性边界。该类型储渗单元通常为厚层状,纵向上厚度较大,一般在10~15m之间。

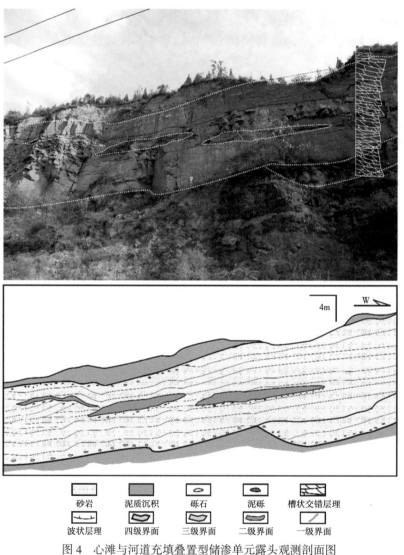

图4 心滩与河道充填叠置型储渗单元露头观测剖面图

1.2.2 河道底部充填叠置型

两期或多期河道底部充填呈垂向叠置,河道带砂体底部发育明显的冲刷界面,一般呈不规则下凹状,底部以含砾粗砂岩、粗砂岩为主,单期河道砂体内部呈正粒序旋回。受河道迁移、改道作用的影响,在不同期次河道充填的顶部形成废弃河道或泛滥平原泥岩、粉砂质泥岩的互层,是该类型储渗单元的内部岩性边界(图5)。由于河道底部充填的冲刷作用,泥岩或粉砂质泥岩互层较薄,通常在0.5~1m之间。单个河道带砂体厚度4~7m,河道底部充填叠置型储渗单元通常由3~5个河道砂体垂向叠置而成。因此,该类型储渗单元厚度较大(6~10m)。

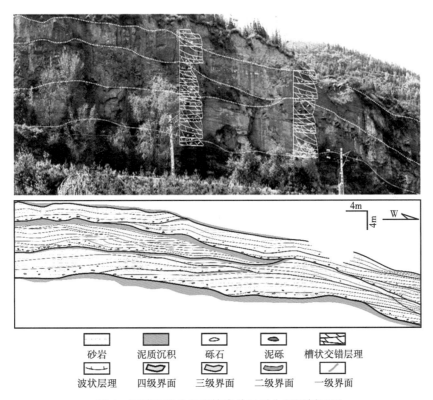

图5 河道充填叠置型储渗单元露头观测剖面图

1.2.3 心滩叠置型

辫状河发育带中两期或多期心滩垂向叠置形成规模较大的储渗单元,该类型储渗单元厚10~20m(图6)。不同期次心滩砂体间由于辫状河道的改道作用频繁,通常夹薄层状泛滥平原或废弃河道泥岩,一般厚度在3~5m之间,形成储渗单元内部岩性边界,整体上该类型储渗单元发育频率较低。心滩砂体一般呈块状,具有纵向上粒度逐渐变细的特征,但心滩砂体内部由于侧向加积作用,内部常见倾斜状泥岩夹层(落淤层)。

1.2.4 心滩或河道充填孤立型

与上述叠置型储渗单元不同,该类型储渗单元由于改道作用的影响,不同期次辫状河道砂体侧向变化距离较大,在辫状河体系过渡带或体系间洼地,心滩或河道充填砂体沉积频率较低,从而纵向上心滩或河道底部充填砂体呈孤立状(图7、图8)。心滩砂体顶部偶见侧向加积

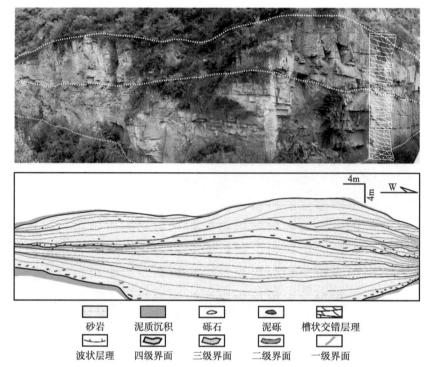

图 6　心滩叠置型储渗单元露头观测剖面图

图 7　心滩孤立型储渗单元露头观测剖面图

形成的倾斜状泥岩,呈薄层状,通常在 0~0.5m 之间。河道充填孤立型储渗单元由于河道体系过渡带或河道体系间水动力较弱,储渗单元顶部通常为河道顶部充填砂体,粒度较细,呈中、细砂岩,渗透性较差;同时,随着该期河道的改道或消亡,河道充填顶部向上为废弃河道或泛滥平原沉积泥岩或粉砂质泥岩。单个心滩厚度一般为 5~8m,河道底部充填砂体厚度一般为 3~5m,因此该类型储渗单元与叠置型相比,具有发育规模较小、侧向上连续性和连通性差的特点。

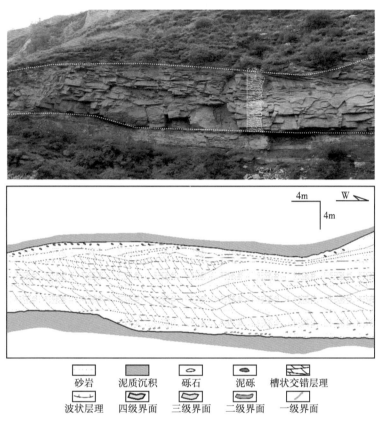

图 8 河道充填孤立型储渗单元露头观测剖面图

1.3 储渗单元发育规模

1.3.1 储渗单元井间识别

通过密井网井间干扰试验可进一步精细识别和表征井间储渗单元形态和边界类型。以密井网开发试验区早期投产井苏 38-16-5 和加密井苏 6-j21 为例,两口井均射开二叠系下石盒子组盒八段砂体,纵向上射孔层段对应一致,因此相应层位连续性和连通性可对比性强。苏 38-16-5 井射孔层段测井曲线形态底部呈齿化箱形、上部呈平滑箱形,储渗单元类型为河道与心滩叠置型;苏 6-j21 射孔层段测井曲线形态底部为平滑箱形与齿化箱形垂向分布,储渗单元类型为心滩与河道充填叠置型,上部为平滑箱形零星分布,储渗单元类型为心滩孤立型。苏 38-16-5 井投产于 2003 年 10 月,经过 6 年的生产,到 2009 年 9 月地层压力已降至 6MPa,而苏 6-j21 加密井在 2009 年 9 月投产时,地层压力仍高达 30.45MPa,仍维持原始地层压力。结合两口井射孔层段的一致性,可判断出苏 38-16-5 井泄流范围小于两口井间距即对应层段储渗单元间存在外部边界,根据对应的部位分别为心滩和河道充填砂体,两口井储渗单元边界类型为坝间泥岩相形成的岩性边界(图9)。

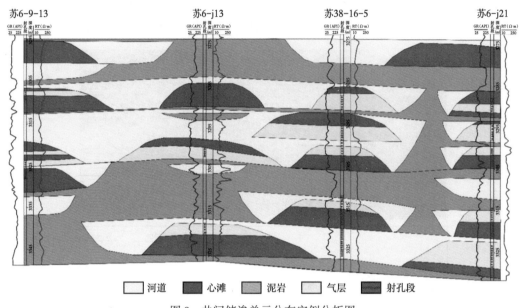

图 9　井间储渗单元分布实例分析图

1.3.2　储渗单元发育规模

同一期河道充填的满岸深度与储渗单元发育规模呈正相关。沉积序列完整的心滩微相代表了河流的满岸深度,山西柳林地区二叠系下石盒子组露头剖面观测表明,单个心滩厚度为 3~6.5m。据此,苏里格地区辫状河体系的河流满岸深度为 3~6.5m。而河流相沉积的宽厚比通常在 40~70 之间[19-20],因此河流相沉积体系中储渗单元宽度范围在 200~400m 之间。不同学者对现代河流沉积砂体的研究表明:高坡降辫状河的心滩坝微相不发育,主要微相单元是河道亚相;低坡降辫状河的心滩微相发育明显,顺物源方向心滩砂体长宽比一般为 2~5,长宽比最大可达到 10:1[21-25]。据此,苏里格气田下石盒子组心滩砂体长度分布范围为 400~1500m,由于顺河道水流方向心滩砂体形态通常呈弯曲状,沿南北向心滩砂体长度为 400~900m[26-28]。

密井网开发试验是井间砂体解剖的最直接方式,同时配合干扰试井分析是判识井间渗透性和砂体连通性的最有效途径。通过大量加密井井间地层压力测试,对比不同批次加密井原始地层压力,若投产前地层压力接近原始地层压力,则表明储渗单元不连通或连通性较差,同时,通过储渗单元测井相分析识别出储渗单元类型,从而建立储渗单元发育规模知识库。

通过对投产时间长的气井井底压力和新投产井原始地层压力的监测,可判断出井间储渗单元发育规模。以苏里格气田苏 6 加密井区为例,该井区开发层位均为二叠系下石盒子组致密砂岩储层,如图 10 所示,作为苏里格气田最早的开发试验区该井区在初期一次骨架井网 1200m×800m 的基础上,通过不断加密部署,逐渐形成井距 366~800m、排距约 600m 的加密井网。骨架井网原始地层压力约 30MPa,随着投产时间的不断增加,投产井井底压力不断降低,最早投产的苏6井、苏38-16、苏38-16-2、苏38-16-3、苏38-16-4和苏38-16-5井井底压力已分别下降至4.99MPa、6.02MPa、9.35MPa、13.78MPa、6.03MPa及9.67MPa。同一井排上加密井井距分布366~466m不等,投产前原始地层压力除苏6-j5井外,均呈现较大幅度的压降;而另外两排加密井井距分布范围为417~820m,新投产井原始地层压力23.60~30.72MPa。表明当井距低于400m时,井间已出现明显压力干扰;当井距超过400m,井间压力干扰的概率大幅度降低,储渗单元井间东西向规模小于400m。同时,对相邻两排井投产前原始地层压力监测表明,上下两排气井与首批投产井排间排距约600m,投产前地层压力均未发现压力降低,未受到早期投产气井生产的影响。因此,储渗单元南北向发育规模范围小于600m。

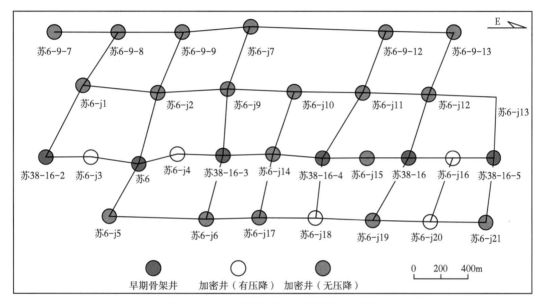

图10 苏里格气田苏6加密井区井位图

2 储渗单元研究与开发部署

2.1 储渗单元与气田加密部署

井网加密是大规模致密砂岩气藏开发中后期提高采收率的主要技术手段之一。位于美国得克萨斯州西南部的奥卓拉气田(Ozona)是致密砂岩气藏开发的典型气田。该气田于1960年代投入开发,开发初期一次部署井网面积为1.29km²;1995年以后,通过加密部署和井间干扰试验,采用地质统计分析,开展气井泄流面积和井间剩余储量评价,以井间压力干扰为加密部署极限的判识标准。随着地质认识逐渐加深,砂体规模小于井网密度、井间砂体连续性有限,进一步对储渗单元泄流能力评价,气井泄流面积0.16~0.32km²,平均0.24km²;将单井控制面积逐步加密到0.65km²和0.32km²,主力开发区单井控制面积0.16km²(占总井数52%)、井网密度平均6口/km²,最大井网密度可达12口/km²[29]。

苏里格气田自2003年投入开发,规模开发初期开展"甜点"区预测、富集区优选和地质建模研究,基于辫状河体系主河道带复合砂体长1000~1800m、中值1200m,宽600~1000m,在气田中部富集区采用800m×1200m骨架井网,气井平均可采储量2000×10⁴m³,实现了气田有效开发;到2009年,通过井位优化和加密部署在原有一次井网基础上,通过主河道带内单砂体精细刻画和一次井网下剩余储量分布研究,局部加密到600m×800m,在保障气井可采储量不下降的前提下,进一步扩大气田产能;2010年以后,气田实现整体开发,通过滚动评价、扩边生产,开发区域逐步扩展到东区、中区和西区,直井开发区块内全面实现600m×800m井网,气田开发规模进一步上升,且较一次井网下采收率提高幅度为10%,成为中国储量和产能规模最大的天然气田[2-3,26,27]。通过储渗单元研究,鄂尔多斯盆地辫状河沉积体系中心滩和河道充填砂体形成的储渗单元呈现不同叠置模式,不同类型储渗单元南北向(顺古水流方向)和东西向(垂直古水流方向)发育规模分别为600m和400m以内,表明苏里格气田具备进一步加密的条件,达到400m×600m井网,预计提高气田采出程度15%~20%。

2.2 储渗单元与水平井部署

致密砂岩气藏储渗单元研究以不同类型砂体间叠置关系研究为主体,将具有相似渗流特征的砂体聚类为储渗单元,结合复合砂体的形态和辫状河体系发育特征,辫状河体系主河道叠置带内心滩和河道充填砂体富集,心滩叠置型、心滩与河道叠置型,以及河道充填叠置型储渗单元发育。整体上辫状河体系砂体发育规模较小,不适合大面积采用水平井开发;但辫状河主河道叠置带河道继承性发育,较强的水动力条件有利于孔渗条件好的粗砂岩形成和富集,砂地比普遍大于70%,纵向上多期叠置型储渗单元富集,叠置型储渗单元总厚度通常超过10~15m,空间分布稳定,侧向上储渗单元连续性好(图11)。因此,叠置型储渗单元发育带即辫状河体系的主河道叠置带是水平井部署的有利目标区,可优选井位、部署水平井。

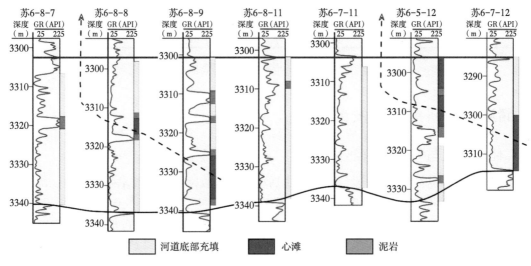

图11 辫状河沉积体系河道叠置带储渗单元发育特征

对于辫状河体系过渡带和体系间洼地,辫状河体系过渡带虽然发育因洪水期改道短期内形成的河道充填,沉积速率快、粒度粗,形成单层厚度较大的粗砂岩,但整体上砂地比较低,一般在30%~70%(图12);辫状河体系间洼地,砂地比更低一般小于30%,以泥岩、粉砂质泥岩为主,夹薄层状中细砂岩(图13)。辫状河体系过渡带和体系间洼地储渗单元类型主要为零星分布的心滩或河道充填孤立型,叠置型储渗单元零星分布,不利于水平井整体部署,宜采用直井或直井丛式井组开发。

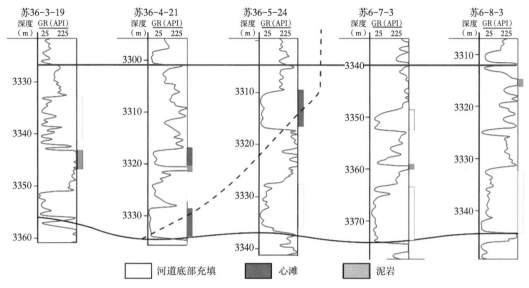

图 12 辫状河沉积体系过渡带储渗单元发育特征

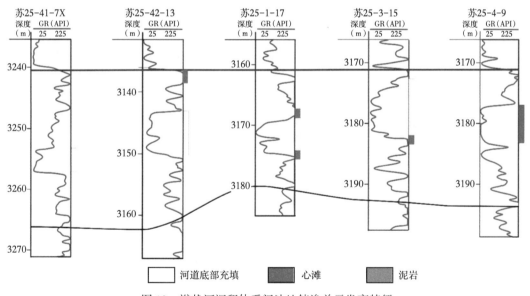

图 13 辫状河沉积体系间洼地储渗单元发育特征

3 结论

(1)将具有相似储渗特征的地质体或沉积微相聚类为储渗单元,通过储渗单元发育特征、叠置模式、露头识别研究,建立储渗单元空间分布和发育规模等地质认识,明确不同类型储渗单元分布规律,可有效指导天然气藏开发中井网和井型、井距优选、开发中后期加密部署等关键开发技术政策的制定,为天然气藏高效开发和提高采收率提供技术支撑。

(2)采用储渗单元研究技术思路和方法,根据不同类型砂体的储渗特征,将孔隙度和渗透率高的粗砂岩相砂体聚类为储渗单元,心滩和河道充填砂体是主要的储渗单元类型;溢岸、心滩侧向加积的坝内粉砂质泥岩夹层和落淤层是储渗单元研究中的内部边界,废弃河道、泛滥平原形成储渗单元研究中的外部边界。

(3)河流相致密砂岩气藏储渗单元发育四种叠置模式:心滩叠置型、河道充填叠置型、心滩或河道充填叠置型,以及心滩或河道充填孤立型。叠置型储渗单元主要发育在辫状河沉积体系的河道叠置带内,是致密砂岩气藏水平井开发的有利目标;孤立型储渗单元主要发育在辫状河沉积体系过渡带和体系间洼地,适宜采用直井或直井丛式井组开发。通过露头观测和加密井区井间干扰试验,对储渗单元发育规模定量开展表征,储渗单元空间发育规模为顺古水流方向600m和垂直古水流方向400m以内;苏里格气田在当前600m×800m井网的基础上,仍具备井网加密的条件,预计进一步加密到400m×600m井网可提高采收率15%~20%。

参 考 文 献

[1] 国家能源局. 中华人民共和国石油和天然气行业标准(SY/T 6832—2011):致密砂岩气地质评价方法. 北京:石油工业出版社,2011.

[2] 戴金星,倪云燕,吴小奇. 中国致密砂岩气及在勘探开发上的重要意义[J]. 石油勘探与开发,2012,39(3):257-264.

[3] 马新华,贾爱林,谭健,等. 中国致密砂岩气开发工程技术与实践[J]. 石油勘探与开发,2012,39(5):572-579.

[4] Martin J H. Areview of braided fluvial hydrocarbon reservoirs:The petroleum engineer's perspective[G]//Best J L and Bristow,C,S,Braided rivers,Geological Society (London) Special Publication,1993,75,333-368.

[5] Collinson,J D. Alluvial Sediments[G]// Reading H G. Sedimentary Environments:Processes,facies and stratigraphy,Oxford,United Kingdom,Blackwell Science,1996:37-82.

[6] Miall A D. The Geology of Fluvial Deposits[M]. Springer Verlag,New York,1996:75-178.

[7] Bridge,J S,Mackey S D. A theoretical study of fluvial sandstone body dimensions[G]// Flint S S and bryant I D et al. Geological modeling of hydrocarbon reservoirs. International Association of Sedimentologists,Special Publication 15,Utrecht,Netherlands,1993:213-236.

[8] Leclair S F,Bridge J S,Wang F,et al. Preservation of crossstrata due to migration of subaqueous dunes over aggrading and non-aggrading beds:comparison of experimental data with theory[J]. Geoscience Canada,1997,24(1):55-66.

[9] Lane S N. Approaching the system-scale understanding of braided river behavior[G]// Smith G S,Best J,Bristow C et al. Braided Rivers:Process,deposits,ecology and management,blackwell publishing,London,2006:107-135.

[10] Lynds R,Hajek E. Conceptual model for predicting mudstone,dimensions in sandy braided-river reservoirs[J]. AAPG Bulletin,2006,90(8):1273-1288.

[11] Labourdette R. Stratigraphy and static connectivity of braided fluvial deposits of the Lower Escanilla Formation,South Central Pyrenees,Spain[J]. AAPG Bulletin,2011,95(4):585-617.

[12] Lunt I A,Smith G H,Best J L,et al. Deposits of the sandy braided South Saskatchewan River Implications for the use of modern analogs in reconstructing channel dimensions in reservoir characterization[J]. AAPG Bulle-

tin,2013,97(4):553-576.
- [13] Hearn C L, Ebanks W J, Tye R S, et al. Geological factors influencing reservoir performance of the Hartzog Draw Field, Wyoming[J]. Journal of Petroleum Technology, 1984, 36(8): 1335-1344.
- [14] 何东博,贾爱林,田昌炳,等. 苏里格气田储集层成岩作用及有效储集层成因[J]. 石油勘探与开发, 2004, 31(3): 69-71.
- [15] 李易隆,贾爱林,何东博,等. 致密砂岩有效储层形成的控制因素[J]. 石油学报, 2013, 34(1): 71-82.
- [16] 王国亭,何东博,王少飞,等. 苏里格致密砂岩气田储层岩石孔隙结构及储集性能特征[J]. 石油学报, 2013, 34(4): 660-666.
- [17] Jia C Y, Jia A L, Zhao X, et al. Architecture and quantitative assessment of channeled clastic deposits, Shihezi sandstone (Low Permian), Ordos Basin, China[J]. Journal of Natural Gas Geoscience, 2017, 2(1): 11-20.
- [18] Kelly S. Scaling and hierarchy in braided rivers and their deposits: examples and implications for reservoir modeling[G]// Smith G S, Best J, Bristow C et al. Braided rivers: Process, deposits, ecology and management, blackwell Publishing, London:2006:75-106.
- [19] 廖保方,张为民,李列,等. 辫状河现代沉积研究与相模式——中国永定河剖析[J]. 沉积学报, 1998, 16(1): 34-39.
- [20] 吴胜和. 储层表征与建模[M]. 北京:石油工业出版社,2010.
- [21] 钟建华,马在平. 黄河三角洲胜利I号心滩的研究[J]. 沉积学报,1998,16(2):38-42.
- [22] 于兴河,马兴祥,穆龙新,等. 辫状河储层地质模式及层次界面分析[M]. 北京:石油工业出版社,2004.
- [23] 何顺利,兰朝利,门成全,等. 苏里格气田储层的新型辫状河沉积模式[J]. 石油学报, 2005, 26(6): 25-29.
- [24] 金振奎,杨有星,尚建林,等. 辫状河砂体构型及定量参数研究——以阜康、柳林和延安地区辫状河露头为例[J]. 天然气地球科学, 2014, 25(3): 311-317.
- [25] 雷卞军,李跃刚,李浮萍,等. 鄂尔多斯盆地苏里格中部水平井开发区盒8段沉积微相及砂体展布[J]. 古地理学报, 2015, 17(1): 91-105.
- [26] 何东博,王丽娟,冀光,等. 苏里格致密砂岩气田开发井距优化[J]. 石油勘探与开发, 2012, 39(4): 458-464.
- [27] 何东博,贾爱林,冀光,等. 苏里格大型致密砂岩气田开发井型井网技术[J]. 石油勘探与开发, 2013, 40(1): 79-89.
- [28] 卢涛,刘艳侠,武力超,等. 鄂尔多斯盆地苏里格气田致密砂岩气藏稳产难点与对策[J]. 天然气工业, 2015, 35(6): 43-52.
- [29] Cipolla C L, Mayerhofer M. Infill drilling and reserve growth determination in lenticular tight gas sands[C]// SPE Annual Technical Conference & Exhibition, Denver, Colorado: Society of Petroleum Engineers, 1996: 533-554.

原文刊于《高校地质学报》,2018,24(3):412-424.

致密砂岩气田储量分类及井网加密调整方法
——以苏里格气田为例

郭 智 贾爱林 冀 光 宵 波 王国亭 孟德伟

(中国石油勘探开发研究院,北京 100083)

摘要:苏里格气田是中国致密砂岩气田的典型代表,储层物性差,有效砂体规模小,分布频率低,非均质性强,区块之间差异明显。依靠 600m×800m 的主体开发井网难以实现储量的整体有效动用,采收率仅为 30%左右,需要开展储量分类评价,针对各类储量区分别实施井网加密调整。优选气田中部苏 14 区块为研究区,通过密井网区精细解剖、干扰试井分析明确了储层的发育频率及规模;以沉积相带为约束,结合储量丰度值、储层叠置样式、差气层影响和生产动态特征,将气田储量分成五种类型。从 I 类—V 类,储层厚度减小,连续性变差,储量品位降低,单井产量变低。依据密井网实际生产数据与数值模拟结果,针对各储量类型,研究了井网密度、干扰程度、采收率的关系,论证了合理井网密度下的单井开发指标。在现有的经济及技术条件下,各类储量区合适井网密度为 2~4 口/km²,气田最终采收率约为 50%。通过系统研究确定了致密砂岩气田复杂地质条件下的储量构成,为开发中后期加密调整方案的编制提供了地质依据。

关键词:致密砂岩气;苏里格气田;储量分类评价;储层叠置样式;井网加密调整

苏里格气田是国内发现的一个特大型低渗透致密砂岩气田,气田地质条件表现为低孔隙度、低渗透率、低丰度[1],储层连续性及连通性差[2],非均质性强,开发难度大。气田经过十几年的科研及生产攻关,研发形成了一系列低成本开发的特色技术[3],无论从储量还是产量规模来看,都已成为国内最大的气田[4]。气田地质储量 $4.77×10^{12} m^3$,于 2014 年提前达到 $249×10^8 m^3$ 的规划年产能(约为全国天然气年产量的 1/5),进入稳产阶段。苏里格这一超大气田的稳产,对于"陕京线"向京津冀地区长期平稳供气具有战略意义,在油价持续低迷、天然气业务地位不断上升的背景下,也是中国石油上游产业链保持较高盈利水平的重要举措。

然而由于地质条件和开发技术约束,气田稳产及提高采收率面临着严峻的挑战:随着开发的深入,优质储量区不断减少,储层品质逐步降低,开发对象日益复杂;单井泄气面积小,产量低,生产压差下降快,单井及区块递减率高;气田现有的 600m×800m 主体开发井网对储量控制不足,采收率仅在 30%左右;气藏多层段含气,水平井开发虽然能提高采气速度,但从长远来看,不可避免地造成部分层段储量漏失,影响最终采收率[5]。国内外开发实践表明,采用直井井网加密是致密砂岩气藏提高储量动用程度和气田采收率的最有效手段[6]。苏里格气田年产能综合年递减率 20%~22%,若稳产每年需弥补递减 $50×10^8$~$60×10^8 m^3$,新钻直井 1000~1500 口。井网加密一是要解决部署区域(优质储量筛选)的问题,二是要解决加密方式(合理井网密度)的问题。长久以来,长庆油田公司与中国石油勘探开发研究院就气田合理井网密度存在着争议。长庆油田公司着眼于保证I类+II类井比例和单井开发效益,认为井网密度应控制在 3 口/km² 以内;中国石油勘探开发研究院从气田整体开发有效、提高采收率入手,认为可以接受一定程度的井间干扰,可整体加密至 4 口/km²。由于储层的强非均质性,各区块之间甚至同一区块内部差异明显,井网加密密度还不能一概而论,需要在落实储量规模的基础上,明确不同类型储量构成、分布及动用程度,分类形成各储量区加密调整对策,论证合理动用顺序。

目前主流的储量分类方法主要依靠储量丰度这一个参数,虽然能表现各区带储量规模及平面分布情况,在勘探及开发评价早期阶段也发挥了一定的作用,但无法反映出储层叠置样式、气体流动性等复杂地质信息,导致地质静态与生产动态关联度低,满足不了开发中后期的需要,例如较高比例的生产井表现出低丰度高产、高丰度低产的特征。为此,本研究充分结合钻井、测井和生产动态资料,深刻剖析动、静态的内在关系,探索影响开发效果的关键地质因素,构建储量分类的多参数划分标准,开展适合于开发中后期的储量分类综合评价研究。

1 气田概况

1.1 砂体与有效砂体呈"砂包砂"二元结构

苏里格气田位于鄂尔多斯盆地伊陕斜坡的西北侧(图1),主要产层为二叠系下石盒子组盒八段和山西组山一段。主体沉积环境为陆相辫状河,在宽缓的构造背景下,河道多期改道、

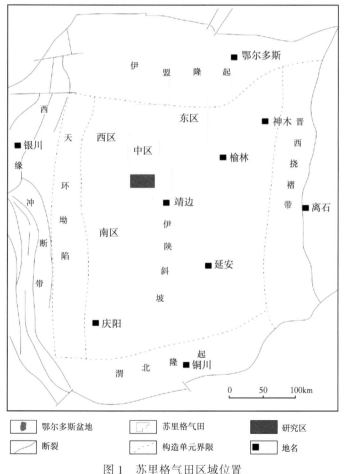

图 1 苏里格气田区域位置

叠置,形成几千至数万平方千米的大规模砂岩区[7],呈片状连续分布。储层沉积后遭受强烈的压实和胶结等成岩作用变得致密[8],原生孔隙所剩无几,孔隙类型以次生孔隙为主。在普遍低渗透致密砂岩背景下,孔隙度、渗透率值高(孔隙度大于5%,渗透率大于0.1mD),含气性好(含气饱和度大于45%)的砂体被称为"有效砂体",是探明储量计算的主体对象和产能的主要贡献者。不同于砂体的大规模连续分布,有效砂体发育规模小、频率低,在空间上呈甜点状分散分布[9],与连片的致密砂体呈"砂包砂"二元结构,有效累计厚度仅占砂体厚度的1/3~1/4,空间预测难度大,需要开展储层精细解剖,明确有效砂体分布特征。

1.2 储层非均质性强,各区块差异明显

受物源、沉积水动力、成岩作用改造和后期构造运动等多因素影响[10],储层非均质性强[11]。气田勘探面积约为 $5\times10^4 km^2$,主要分为中区、西区、东区和南区等(图1)。中区储层质量相对较好,投产时间长,开发效果最好;西区储层物性与中区接近,但大面积含水[12],造成超过 $7000\times10^8 m^3$ 的储量难于动用;东区岩屑含量高,储层较致密,北部部分层段见水;南区距离沉积物源远,岩石粒度细,储层厚度薄[13],仅有部分区块建产,基本还处在开发评价阶段。各区块之间特征差异明显,同属于区域的多个开发区块之间甚至同一区块内部,地质特点及开发效果也不尽相同。需要在抽提储层共性的基础上对储量进行分类评价[14]。

1.3 单井产量低,递减快,管理难度大

气田平均储量丰度低($1.2\times10^8 m^3/km^2$),地层压力低(压力系数0.8~0.9),直井泄气面积小,单井产量低(平均日产气量 $1\times10^4 m^3$),经济有效开发难度大。低渗透致密气藏气井没有严格意义上的稳产期,投产之后即递减,前三年平均递减率22.7%。直井按照所处的地质条件和开发效果可分为Ⅰ类、Ⅱ类和Ⅲ类,三年期平均日产量分别为 $1.95\times10^4 m^3$、$0.97\times10^4 m^3$ 和 $0.51\times10^4 m^3$,井数所占比例分别为20%、40%和40%。目前产量低于 $0.5\times10^4 m^3/d$ 的低产井5000余口,占气田总井数的一半以上,并呈增加趋势。大部分Ⅲ类井本身就是低产井,Ⅰ类井和Ⅱ类井投产5—6年后,随着产量递减亦成为低产井。大量低产、低效井的存在,加大了气田效益开发的难度。需要进一步筛选优质储量,提升Ⅰ类+Ⅱ类井的比例。

1.4 储量规模落实,优质储量比例小,动用程度低

气田地质储量为 $4.77\times10^{12} m^3$,受矿权、保护区等因素影响的地质储量为 $0.85\times10^{12} m^3$,受含水区影响的地质储量为 $0.72\times10^{12} m^3$,开发区内地质储量为 $2.99\times10^{12} m^3$,其中探明储量 $1.23\times10^{12} m^3$。虽然储量基数大,但优质储量比例小,开发区探明储量仅占气田地质储量的26%。气田整体开发井网由早期的600m×1200m调整为600m×800m,储量动用程度由20%提升至30%左右。但现有井网依然对有效砂体控制不足,导致最终采收率偏低,与处在开发中后期阶段的国内外其他气田相比,仍然有较大的提升空间。需要系统研究合理井网密度[15],论证不同类型储量动用顺序。

2 储量分布特征及动用程度评价

从苏里格气田众多的开发区块中优选苏14区块作为研究区(图1)。这是因为:(1)研究区位于气田中区,面积850km²,储层条件较好;(2)区块于2006年投产,开发时间长,是苏里格

气田最早投产的几个区块之一,动态资料可靠;(3)区内共有井数 646 口(直井 570 口,水平井 76 口),加密区六个,动、静态资料完备,适合开展综合研究。

2.1 有效砂体分布规律

有效砂体刻画是储量计算、分类评价的基础。气田有效砂体连续性差,空间分布零星。目前主体开发井网井距大于有效砂体的规模尺度,使得有效砂体精细描述难度大。2008—2015 年,在苏 14、苏 6 区块和苏 36-11 区块等打了多排加密井并进行了 42 个井组的井间干扰实验,井网密度为 2.5~5.0 口/km²,是有效砂体解剖的宝贵资料。气田井距方向为东西向,排距方向为近南北向,与砂体展布方向基本一致(图 2)。干扰试验表明:井距在 300~800m,排距在 500~900m,随着井、排距增加,干扰概率逐渐降低,反映有效砂体的连通概率降低。在井距方向大于 600m 条件下试验了三组,未见干扰;在排距方向大于 800m 条件下试验了五组,一组受到干扰,干扰概率为 20%(图 3),可以判断有效砂体主体规模小于 600m ×800m。

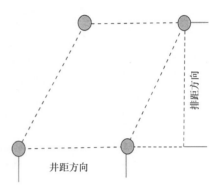

图 2 储层展布方向与井、排距方向

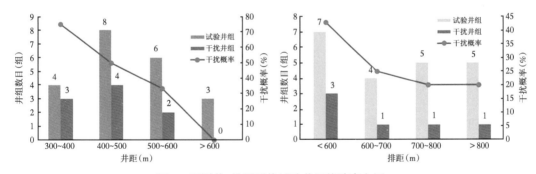

图 3 不同井、排距干扰试验井组统计直方图

以干扰试井分析为依据,结合野外露头观测、沉积物理模拟获得的长宽比(1.5~4.0)、宽厚比(50~120)数据[16],通过密井网区精细地质解剖,明确了研究区有效砂体的规模、叠置样式及分布频率。有效单砂体厚度为 1~5m,宽度为 100~500m,长度为 200~600m,1km² 地层内平均发育有效砂体 20~30 个。气田 80%的有效砂体呈孤立分布,规模小,平均小于 400m×600m;气田的 20%有效砂体通过垂向叠置、侧向搭接,规模较大,储量占总储量的比例达到 45%。

2.2 开发储量评价

储量核算垂向上刻画到单砂体,平面上区分砂层组,具体方法为:(1)以测井解释为依据,针对研究区所有直井以单砂体为单元计算储量丰度,提高了储量计算的精度和准确性;(2)考虑有效砂体在平面的连续性,以砂层组为单位,圈定各储量丰度区间的含气面积,计算砂层组的地质储量;(3)将各砂层组储量累加,得到区块地质储量。

储量评价结果为:研究区地质储量为 $1288.6×10^8m^3$;多层叠合后,研究区含气区块平均储量丰度为 $1.52×10^8m^3/km^2$。在各层段中,盒八段上亚段、山一段含气面积大,储量规模大,集中了区块70%以上的储量(表1)。受主河道控制,研究区东部储量相对富集。

表1 苏14区块各层段储量参数表

层段	含气面积 (km^2)	地质储量 (10^8m^3)	储量占比 (%)	含气区储量丰度 ($10^8m^3/km^2$)
下石盒子组八段上亚段	261.6	108.7	8.4	0.42
下石盒子组八段下亚段	825.2	632.3	49.1	0.77
山西组一段	737.9	282.3	21.9	0.38
山西组二段	266.4	95.6	7.4	0.36
其他	515.0	169.7	13.2	0.33

2.3 储量动用程度

结合产量不稳定法(Blasingame法、流动物质平衡)和递减曲线分析法评价每口单井的动态储量和泄气范围。区块直井为570口,平均动储量为 $3005×10^4m^3$,平均泄气范围为 $0.20km^2$;水平井为76口,平均动用储量为 $7283×10^4m^3$,平均泄气范围为 $0.62km^2$。累加得到区块已动用地质储量为 $226.7×10^8m^3$,已动用面积 $161.9km^2$,分别占区块储量的18%和面积的19%,整体动用程度较低,通过井网加密提高储量动用程度和采收率的潜力较大,但针对何种储量区采用何种井网密度进行加密还需要进一步的论证。

3 储量分类综合评价

3.1 生产动态主控地质因素

3.1.1 储量丰度

较高的储量丰度是气井高产与区块效益开发的物质基础,井的产量与储量丰度有一定的正相关性。随着储量丰度的增加,高产量井比例逐渐增高。但在较高的储量丰度条件下,仍然对应一定比例的低产井。在储量丰度大于 $2.5×10^8m^3/km^2$ 区域范围内,依然有13%的井EUR(预测最终累计产量)小于 $1300×10^4m^3$。储量丰度是影响气井产能的重要因素之一,并连同储量规模、储量构成(纯气层、差气层比例)、储层垂向叠置样式、平面连续性等多参数共同影响了气井的最终开发效果。

3.1.2 单层厚度

当有效厚度、储量丰度接近时,储层分布样式不同,各井累计产气量差异较大。有效储层为地下三维地质体,具有一定长宽比和宽厚比[17]。单层厚度越大,则优质储层在平面的延伸规模越大,储层的连续性和连通性越好。若厚砂体厚度为薄砂体的两倍(图4),则其体积($8\pi ab$)可达到薄砂体体积(πab)的8倍。假设有两口井,A井钻遇了一个厚层,B井钻遇了两个薄层(图4),两井的累计有效厚度均为 $2h$,A井所控制储层的延伸面积为B井的4倍。

约80%的区块有效单砂体小于4m,将大于4m的单砂体定为厚砂体。厚砂体平均厚度为5.58m,薄砂体平均厚度为2.46m。前者平均厚度为后者的2.27倍,在同样累计厚度下,延伸面积约为薄砂体的5.2倍。

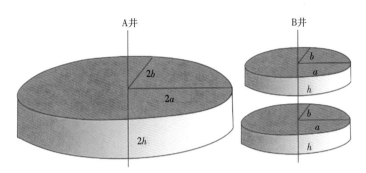

图4 储层分布对储量影响模式图

3.1.3 差气层比例

气田有效砂体根据物性及含气性差异可分为差气层及纯气层两类。差气层孔隙度大于5%,渗透率大于0.1mD,含气饱和度大于45%;纯气层一般孔隙度大于7%,渗透率大于0.3mD,含气饱和度大于55%。气田表现为多层含气的特点,单井平均钻遇4~5个有效砂体,差气层及纯气层各占约50%。差气层与纯气层相比,物性差,含气饱和度小,含水饱和度高,储层内气体相对渗透率小,流动性差(图5)。

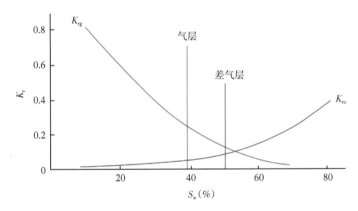

图5 气水两相渗流模式图

为了对比差气层与纯气层的产能差异[18],在研究区570口直井中挑选出仅钻遇差气层井23口,仅发育气层井15口。在储量丰度接近及有效厚度接近的条件下,仅发育差气层的气井平均EUR为$1873×10^4 m^3$,仅发育纯气层的气井平均EUR为$2492×10^4 m^3$,差气层产能仅为纯气层的75%。

3.2 储量分类评价

除了储量丰度,厚层发育程度、气层发育比例也是控制气井产能的重要地质参数。提出了厚层系数、气层系数等概念,对原有的储量丰度进行了修正。厚层系数 F_h 是指单井钻遇的若干个有效砂体中,折算厚层占累计有效厚度的比例,反映了储层的叠置程度及平面的连通性。由于在同样累计厚度下,厚储层延伸面积平均为薄储层的 5.2 倍,因此薄层折算厚度为薄层有效厚度除以 5.2。气层系数 F_g 是指单井累计有效厚度中折算纯气层的比例,与物性及气相的流动性密切相关。由于同样储量丰度及有效厚度条件下,差气层的产能仅为纯气层 75%,因此差气层的折算厚度为差气层累计有效厚度乘以 75%。由于一口井钻遇的若干有效砂体中,全都为厚层或全都为纯气层的概率微乎其微,故大多数情况下 F_h、F_g 皆小于 1,极个别情况等于 1。

$$F_h = \frac{(h_h + h_b/5.2)}{h_s} \tag{1}$$

式中　F_h——厚层折算系数;
　　　h_h——厚层累计厚度;
　　　h_b——薄层累计厚度;
　　　h_s——有效砂体累计厚度。

$$F_g = \frac{(h_g + h_{pg} \times 0.75)}{h_s} \tag{2}$$

式中　F_g——气层折算系数;
　　　h_g——纯气层累计厚度;
　　　h_{pg}——差气层累计厚度。

$$I = (a \times F_h + b \times F_g) \times A_s \tag{3}$$

式中　I——修正储量丰度;
　　　A_s——原始储量丰度;
　　　a、b——相关系数,之和为 1。

通过试算,当 $a=0.8$,$b=0.2$ 时,单井累计产量与修正储量丰度相关性最高。

分别统计各井的累计有效厚度及累计纯气层、差气层、厚层、薄层有效厚度等几个参数,通过式(1)—式(3),得到修正储量丰度。拟合修正储量丰度与单井预测最终累计产量的关系可以看出,修正后的储量丰度与生产动态相关性显著提升,R 由原始储量丰度的 0.5 提升至修正后的 0.8(图 6)。反映出修正后的储量丰度能够表现地质与生产动态的内在关联,可以较准

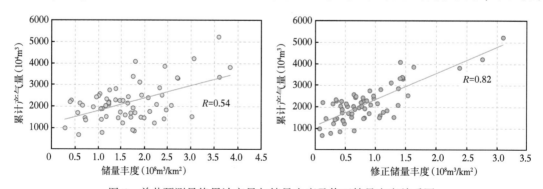

图 6　单井预测最终累计产量与储量丰度及修正储量丰度关系图

确地区分出不同类型的储量,从而可以作为储量分级的主要依据。

综合考虑储量规模、储层分布特征、差气层影响,以修正储量丰度为主要依据,结合动、静态多个参数,建立了储量综合分类标准(表2)。根据储量综合分类标准,将研究区储量分为五种类型。从Ⅰ类—Ⅴ类储量区,储量品质变差,储量趋于分散,有效厚度逐渐变薄,气层比例逐渐减小,单井累计产量逐渐变低。Ⅰ类、Ⅱ类及部分Ⅲ类储量对应优质储量,丰度大于 $1.5×10^8 m^3/km^2$。Ⅲ类及Ⅳ类储量的丰度有一定程度的重叠,其差别主要体现在储层叠置样式及差气层比例,进一步验证了储量丰度不是储量分类的唯一参数。各类储量区分布面积比例分别为17%、24%、30%、15%和14%,储量规模分别为 $328.9×10^8 m^3$、$362.6×10^8 m^3$、$363.2×10^8 m^3$、$170.1×10^8 m^3$ 和 $63.8×10^8 m^3$。由于综合了多因素,各类储量分布与储量丰度值在大趋势上有一定相关性,在局部细节又有诸多差异(图7)。

表2 储量综合分类标准

储层类型	修正储量丰度 ($10^8 m^3/km^2$)	原始储量丰度 ($10^8 m^3/km^2$)	累计有效厚度 (m)	单层有效厚度 (m)	气层比例 (%)	单井预测累计产量 ($10^4 m^3$)	投产井类型
Ⅰ类	≥1.4	≥2.0	≥18	≥5	≥70	≥3500	Ⅰ类井为主
Ⅱ类	0.8~1.4	1.5~2.0	14~18	≥4	50~70	2500~3500	Ⅰ类+Ⅱ类井
Ⅲ类	0.5~0.8	1.1~1.5	10~14	≥3	40~50	1900~2500	Ⅱ类井为主
Ⅳ类	0.2~0.5	1.1~1.4	6~12	<3	30~40	1350~1900	Ⅱ类+Ⅲ类井
Ⅴ类	<0.2	<1.1	<6	<3	<30	<1350	Ⅲ类井

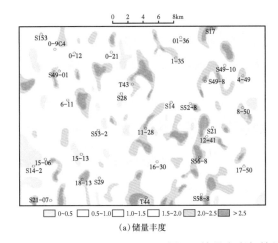
(a)储量丰度

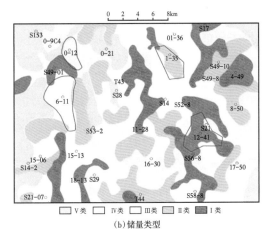
(b)储量类型

图7 储量丰度与储量类型分布平面对比图

Ⅰ类储量区位于高能主砂带主体,砂地比高,储层连续性强,有效砂体规模大,块状厚层型、多层叠置型比例高,有效砂体分布集中,纯气层发育数目占有效砂体中的比例大于70%(图8),是研究区开发潜力最好的一类储量。区内平均有效厚度19.7m,平均储量丰度 $2.3×10^8 m^3/km^2$,投产井以Ⅰ类井为主,井均预测最终累计产气量 $5117×10^4 m^3$,井均泄气面积 $0.29km^2$。

Ⅱ类储量区位于高能主砂带翼部及次高能主砂带主体,砂地比较高,储层连续性较强,有效砂体规模较大,区内平均有效厚度17.1m,平均储量丰度$1.76×10^8m^3/km^2$,井均预测最终累计产气量$3045×10^4m^3$,泄气面积$0.23km^2$。与Ⅰ类储量相比,块状厚层比例低,多期叠置比例高,纯气层比例减少,为30%~50%。

Ⅲ类储量区主要分布在低能砂带,是分布最广泛的一类储量区,分布面积$254km^2$。有效砂体较为孤立,局部为多层叠置型,平均有效厚度13.8m,平均储量丰度$1.43×10^8m^3/km^2$,纯气层比例40%~50%,投产井以Ⅱ类井为主,井均预测最终累计产气量$2228×10^4m^3$,井均泄气面积$0.21km^2$。与Ⅰ类、Ⅱ类储量区相比,储层连续性差,气层比例减小。

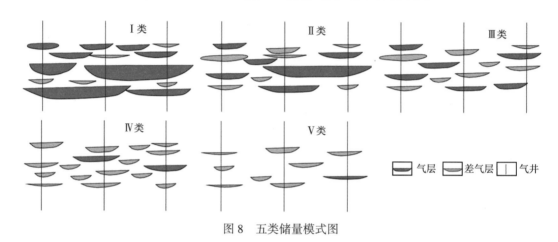

图8 五类储量模式图

Ⅳ类储量区主要分布在砂带边部,砂地比低,有效砂体基本为孤立型,厚度薄,储层物性差,净毛比低,纯气层比例进一步减小至30%~40%。区内平均有效厚度13.1m,平均储量丰度$1.35×10^8m^3/km^2$,井均预测最终累计产气量$1880×10^4m^3$,仅有边界效益,平均泄气面积$0.16km^2$。

Ⅴ类储量区主要分布在区块的边部及砂带间,有效砂体厚度薄,规模小,在空间零星分布。单层有效单厚度一般小于3m,累计有效厚度小于6m,井均发育有效砂体1~2层,平均储量丰度$0.53×10^8m^3/km^2$,单井平均预测最终累计产气量$880×10^4m^3$,泄气面积$0.12km^2$。在现有的经济及技术条件下,单井累计产量达不到钻完井及储层改造成本所需的下限累计产量($1277×10^4m^3$),没有加密潜力,建议暂不开发。故之后的加密调整研究主要针对Ⅰ类—Ⅳ类储量区。

不同类型储量区表现出不同的生产动态特征,这正是分类进行井网加密而不采用均一井网加密的原因。需要指出的是,本文的储量分类方法虽然能反映出差别较大的不同储量类型储层地质特点和生产动态特征,但由于地质条件的复杂性及储层的强非均质性,在同一类型的储量区内,储层也不是均匀分布的。

4 井网加密调整对策

一般随着井网密度增加,井间干扰变得严重,单井平均产量降低,井组增加产量的速度变慢,采收率增加幅度变小。井网过稀,储量得不到有效动用,采出程度低;井网太密,受地质条件和产能干扰,单井累计产量降低,影响开发效益(图9)。确定合理井网密度要兼顾技术和经

济条件,应遵循以下原则:

(1)同时保证较高的单井产量规模和区块采收率,允许一定程度的干扰,又要避免干扰严重,即 a(产生干扰井网密度)<适宜井网密度<b(严重干扰时井网密度);

(2)骨架井及加密井整体达到12%内部收益率(最终累计产量≥$1780×10^4m^3$),即适宜井网密度≤c(经济极限井网);

(3)每口加密井能够不亏本(最终累计产量≥$1277×10^4m^3$),即适宜井网密度≤d(最大收益井网密度)。

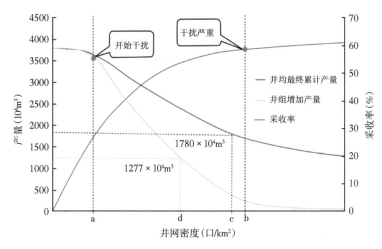

图9 产量、采收率与井网密度关系图

4.1 各类储量区加密调整研究

为了研究储层连通性,评价单井开发指标,气田在苏14、苏6、苏36-11等区块开展了不同储量类型、变井网密度下的八个密井网区的加密试验,井网密度2.1~5.0口/km²,涵盖Ⅰ类—Ⅳ类储量区(表3)。以这八个密井网区的实际生产数据(单井累计产量、区块采收率)为依据,针对Ⅰ类—Ⅳ类储量区,分别建立地质模型、进行数值模拟,兼顾经济和技术因素,开展加密调整研究,论证各储量类型的适宜井网密度及开发指标。

表3 密井网区开发参数表

密井网区	井数(口)	井网(m×m)	井网密度(口/km²)	储量丰度($10^8m^3/km^2$)	修正储量丰度($10^8m^3/km^2$)	储量类型	预测最终累计产量(10^4m^3)	井组采收率(%)
苏36-11试验区	13	400×500	5.0	2.17	1.76	Ⅰ	2524	58.2
苏14三维区E	8	600×700	2.4	2.21	1.18	Ⅱ	2732	29.7
苏14加密区	18	500×700	2.9	1.74	0.79	Ⅲ	2336	39.8
苏6试验区	13	400×600	3.9	1.67	0.75	Ⅲ	1906	44.5
苏14三维区A	11	600×600	2.8	1.30	0.46	Ⅳ	1499	32.3
苏14三维区B	8	500×700	2.9	1.28	0.50	Ⅳ	1602	36.3
苏14三维区C	7	650×750	2.1	1.35	0.35	Ⅳ	1749	27.2
苏14三维区D	7	500×800	2.5	1.32	0.49	Ⅳ	1594	30.2

首先针对各类储量优选建模区(建模区位置见图 7b),建立地质模型。要确保建模精度,须使每类建模区拥有足够多的井数(>10 口)。由于各类储量区在平面上变化较快,很难保证每一类建模区只对应唯一类型的储量,但应要求尽量以相应的储量区为主。建模中通过基于目标的模拟方法建立砂体模型[19],在相控下建立有效砂体模型[20],建模结果与实际井的物性、储层规模、储量规模统计结果较一致,说明模型可靠,能较准确地表现出储层"砂包砂"二元结构。

在分别建立各类储量区地质模型的基础上,利用数值模拟方法模拟井网密度由 1 口井/km² 到 8 口井/km² 的生产过程,预测气井开发指标和生产期末最终采收率。变更井网密度时,先将老井抽离,重新布新井,打井的位置也就发生了变化。由于模拟是基于实际地质模型,储层非均质性极强,井网密度改变前后即使没有井间干扰,单井的最终累计产量也可能发生变化,这是与概念均质模型不同的地方。

为了区别产量的减少是由储层变差引起的还是井间干扰造成的,这里引入"新老井产量比"的概念,是指井网密度每增加 1 口/km²,井组增加的产量(等效于新井的最终累计产气量)与老井平均最终累计产气量之比。分析认为,新老井产量比的下降幅度较缓且大于 50% 时,主要是储层变差引起的产量降低;当新老井产量比呈"断崖式"迅速下降且小于 30% 时,意味着干扰严重,井网再加密,对采收率提升有限。

以 I 类储量区为例开展分析。前文已述,根据生产动态数据计算得到 I 类储量区的气井平均泄气面积为 0.29km²,推算井网密度为 3~4 口/km² 时产生井间干扰。根据数值模拟结果(表4),从 2 口/km² 增加到 3 口/km² 时,井组增加产量为 2936×10⁴m³(等效新钻井最终累计产量),2 口老井平均最终累计产量为 4466×10⁴m³,则新钻井与老井平均产量比 2936/4466 = 65.7%(>50%),说明井网密度为 3 口/km² 时,干扰不严重。井网密度从 3 口/km² 增加到 4 口/km² 时,新老井产量比为 28.6%(<30%),干扰严重,故 I 类储量区井密度应小于 4 口/km²。

表 4 I 类储量加密指标参数

井网密度 (口/km²)	采出程度 (%)	平均单井最终累计产量 (10⁴m³)	井组增加产量 (10⁴m³)	新老井产量比 (%)
1	23.6	5296	5296	—
2	38.6	4466	3636	68.7
3	51.7	3956	2936	65.7
4	57.0	3250	1132	28.6
5	59.8	2753	765	23.5
6	62.6	2392	587	21.3
7	64.6	2130	558	23.3
8	66.9	1923	474	22.3

I 类储量区储层质量相对较好,从区块整体效益来看,达到 8 口井/km² 时,区块仍具经济效益,井均最终累计产量为 1923×10⁴m³(大于 1780×10⁴m³)。从新钻井自保的角度,当井网密度达到 3 口井/km²,新钻井最终累计产量为 2936×10⁴m³(大于 1277×10⁴m³),进一步加密到 4 口/km²,新钻井最终累计产量为 1132×10⁴m³(小于 1277×10⁴m³),故井网密度应不大于 4 口/km²。

综合考虑避免严重干扰、所有井整体有效和新钻井能自保等三条加密原则(图10a)，Ⅰ类储量区适宜井网密度为3口/km²。

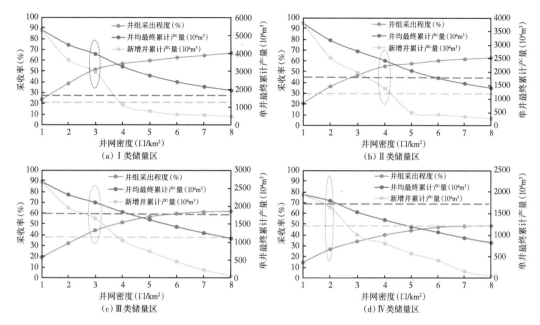

图10 各类储量区产量、采收率与井网密度关系

表5 各类储量区加密井指标

储量类型	未动用面积 (km²)	建模区平均储量丰度 (10^8m³/km²)	严重干扰井网密度 (口/km²)	整体有效井网密度 (口/km²)	新井自保井网密度 (口/km²)	适宜井网密度 (口/km²)	加密后单井平均最终累计产量 (10^4m³)	井组采出程度 (%)
Ⅰ	101.5	2.30	4	≤8	≤3	3	3956	51.7
Ⅱ	147.0	1.76	5	≤5	≤4	4	2407	55.1
Ⅲ	195.7	1.43	6	≤4	≤3	3	2086	43.7
Ⅳ	87.6	1.35	7	≤2	≤2	2	1782	26.4

Ⅱ类—Ⅳ类储量区井均控制范围分别为0.23km²、0.21km²、0.16km²，结合"新老井产量比"分析，认识到Ⅱ类—Ⅳ类储量区严重干扰时对应的井网密度分别为5口/km²、6口/km²和7口/km²。再根据所有井整体有效、加密井自保原则，研究了Ⅱ类—Ⅳ类储量区适宜井网密度(图10)。从Ⅰ类—Ⅳ类储量，随着储层品质依次降低：储量规模不断减小，平均储量丰度从2.30×10^8m³/km²降到1.35×10^8m³/km²；储层连续性逐渐下降，严重干扰井网密度从4口/km²上升到7口/km²；经济效益逐步减少，整体有效井网密度从8口/km²降到2口/km²；新井可自保的井网密度受干扰程度和储量质量双重影响，处在2~4口/km²范围之间(表5)。储量品质高，较少的井就可以有效控制储量，再打加密井，效益受损；储量品质低，经济效益差，多打加密井，开发风险大。这就产生了一种引人注意的现象：区块Ⅱ类储量区新井自保对应的最大井网密度能到达到4口/km²，而储量品质更好的Ⅰ类储量区、品质稍差的Ⅲ类储量区仅

为 3 口/km²。

明确了Ⅰ类—Ⅳ类储量区适宜井网密度,分别为 3 口/km²、4 口/km²、3 口/km² 和 2 口/km²。Ⅰ类—Ⅳ类储量区加密后井均预测最终累计产量分别为 $3956×10^4m^3$、$2407×10^4m^3$、$2086×10^4m^3$ 和 $1782×10^4m^3$,相比于各类储量区 1 口/km² 井网条件下最终累计产量分别减少了 23%、12%、3% 和 2%。加密后优质储量区平均井网密度为 3.7 口/km²,证实了中国石油研究总院"气藏可整体加密至 4 口/km²"的判断是基本可靠的,也通过一系列分析基本消除了研究单位与油田现场对于合理井网密度的历史争议。

4.2 稳产及动用顺序论证

研究区按照井网对储量的控制程度可将井网分成三类:井网完善区、井网基本控制区和井网未控制区。井网完善区主要指水平井网及密度大于 3 口/km² 的直井井网,对储量控制比较完善,下一步不再加密调整;井网基本控制区井密度 1~2 口/km²,储量类型多样,包涵 600m×800m,600m×1200m 多种井网,整体加密困难,适合甜点式加密;井网未控制区井密度小于 1 口/km²,面积和储量均占研究区总面积和总储量的 70% 以上,是加密调整的重点目标,建议根据各类储量适宜井网密度整体布井一次成型。

区块于 2014 年进入稳产阶段,规划年产能 $18×10^8m^3$,根据历年投产井的递减率按井数加权得到区块综合递减率为 22%,确定今后每年需弥补递减 $3.96×10^8m^3$。考虑管道、集气站、天然气处理厂等地面建设完善程度[21],秉承"有质量、有效益、可持续"开发原则,按照"开发一批、储备一批、攻关一批"的思路,建议针对井网基本控制区、井网未控制区Ⅰ类、Ⅱ类、Ⅲ类、Ⅳ类储量依次加密。综合各类储量区剩余储量、加密井密度、开发井指标预测及递减率分析,认为区块还能稳产 22 年,井网基本控制区、井网未控制区Ⅰ类—Ⅳ类储量区分别能够支撑区块稳产 2 年、6 年、7 年、6 年和 1 年,整个稳产阶段还需打新井 1862 口,区块最终采气量 $635.92×10^8m^3$,采收率 49.3%。

5 结论

(1)致密砂岩气田储层结构复杂,非均质性强。根据储量规模、储层叠置样式、差气层影响、生产动态特征,将气田储量分为五类,避免了仅依据储量丰度这一单因素分类所造成的误差。兼顾技术和经济因素,明确了各类储量区适宜井网密度,Ⅰ类—Ⅳ类储量分别为 3 口/km²、4 口/km²、3 口/km² 和 2 口/km²,Ⅴ类储量区单井产量低,达不到经济下限,暂不开发。

(2)储量分类及井网加密调整方法推广到气田的其他区块时,储量分类标准不变,只是其他区块五类储量的分布比例与研究区有所不同,各类储量区加密指标与研究区略有差异。考虑矿权、保护区等地面影响因素,其他区块的稳产年限会有所下降。通过井网加密,气田各区块可平均再稳产 15~20 年,采收率由 30% 提高至约 50%。

(3)气田现有的密井网试验区存在一些问题,导致开发指标具有一定的不确定性:(1)八个密井网试验区全部位于气田中区,储层品质好于气田的平均水平,缺乏代表性;(2)井网密度偏小(仅 36-11 试验区达到 5 口/km²,其余皆小于 4 口/km²),井距、排距一般大于 400m,对于识别长度和宽度在 400m 以下的有效砂体难度大。建议在气田的西区、东区加强密井网开发区先导实验,进一步论证单井的开发指标、明确各类储量区适宜井网密度及合理动用顺序。

参 考 文 献

[1] 何光怀, 李进步, 王继平, 等. 苏里格气田开发技术新进展及展望[J]. 天然气工业, 2011, 31(2): 12-16.
[2] 张文才, 顾岱鸿, 赵颖, 等. 苏里格气田二叠系相对低密度砂岩特征及成因[J]. 石油勘探与开发, 2004, 31(1): 57-59.
[3] 卢涛, 刘艳侠, 武力超, 等. 鄂尔多斯盆地苏里格气田致密砂岩气藏稳产难点与对策[J]. 天然气工业, 2015, 35(6): 43-52.
[4] 马新华, 贾爱林, 谭健, 等. 中国致密砂岩气开发工程技术与实践[J]. 石油勘探与开发, 2012, 39(5): 572-579.
[5] 李建奇, 杨志伦, 陈启文, 等. 苏里格气田水平井开发技术[J]. 天然气工业, 2011, 31(8): 60-64.
[6] 何东博, 贾爱林, 冀光, 等. 苏里格大型致密砂岩气田开发井型井网技术[J]. 石油勘探与开发, 2013, 40(1): 79-89.
[7] 杨华, 付金华, 刘新社, 等. 鄂尔多斯盆地上古生界致密气成藏条件与勘探开发[J]. 石油勘探与开发, 2012, 39(3): 295-303.
[8] 赵文智, 汪泽成, 朱怡翔, 等. 鄂尔多斯盆地苏里格气田低效气藏的形成机理[J]. 石油学报, 2005, 26(5): 5-9.
[9] 闵琪, 付金华, 席胜利, 等. 鄂尔多斯盆地上古生界天然气运移聚集特征[J]. 石油勘探与开发, 2000, 27(4): 26-29.
[10] 何东博, 贾爱林, 田昌炳, 等. 苏里格气田储集层成岩作用及有效储集层成因[J]. 石油勘探与开发, 2004, 31(3): 69-71.
[11] 付金华, 魏新善, 任军峰. 伊陕斜坡上古生界大面积岩性气藏分布与成因[J]. 石油勘探与开发, 2008, 35(6): 664-667.
[12] 孟德伟, 贾爱林, 冀光, 等. 大型致密砂岩气田气水分布规律及控制因素——以鄂尔多斯盆地苏里格气田西区为例[J]. 石油勘探与开发, 2016, 43(4): 607-614.
[13] 郭智, 贾爱林, 薄亚杰, 等. 致密砂岩气藏有效砂体分布及主控因素——以苏里格气田南区为例[J]. 石油实验地质, 2014, 36(6): 684-691.
[14] 王永祥, 张君峰, 段晓文. 中国油气资源/储量分类与管理体系[J]. 石油学报, 2011, 32(4): 645-651.
[15] 严谨, 史云清, 郑荣臣, 等. 致密砂岩气藏井网加密潜力快速评价方法[J]. 石油与天然气地质, 2016, 37(2): 125-128.
[16] 贾爱林, 何东博, 何文祥, 等. 应用露头知识库进行油田井间储层预测[J]. 石油学报, 2003, 21(6): 51-53.
[17] 贾爱林, 程立华. 数字化精细油藏描述程序方法[J]. 石油勘探与开发, 2010, 37(6): 623-627.
[18] 计秉玉, 王春艳, 李莉, 等. 低渗透储层井网与压裂整体设计中的产量计算[J]. 石油学报, 2009, 30(4): 578-582.
[19] 贾爱林. 中国储层地质模型20年[J]. 石油学报, 2011, 32(1): 181-188.
[20] 郭智, 孙龙德, 贾爱林, 等. 辫状河相致密砂岩气藏三维地质建模[J]. 石油勘探与开发, 2015, 42(1): 76-83.
[21] 武力超, 朱玉双, 刘艳侠, 等. 矿权叠置区内多层系致密气藏开发技术探讨——以鄂尔多斯盆地神木气田为例[J]. 石油勘探与开发, 2015, 42(6): 826-832.

原文刊于《石油学报》, 2017, 38(11): 1299-1309.

致密砂岩气藏多段压裂水平井优化部署

位云生　贾爱林　郭　智　孟德伟　王国亭

(中国石油勘探开发研究院,北京 100083)

摘要:国内致密砂岩气藏普遍存在含气砂体分布零散、储集体内非均质性强的特征,含气砂体准确预测的难度较大。对于水平井开发方式而言,两口相向水平井靶点 B 之间留有较大间距,将造成储量平面控制和动用程度降低。本文从国外致密气开发实践调研入手,基于苏里格致密砂岩气藏有效砂体空间展布、规模尺度及开发动态特征分析,结合数值模拟方法,论证水平井在不同部署方式下的开发效果,提出水平井的优化部署方案。研究表明:相向两口水平井靶体 B 点接近重合、压裂段等间距部署,可以大幅度提高水平井对储量的控制和动用程度,同时有效提高气田整体开发经济效益。研究成果在致密气实际开发中具有可操作性和推广应用前景。

关键词:致密砂岩气藏;多段压裂水平井;井距设计;数值模拟;采出程度;经济效益

致密砂岩气藏在地层条件下平均渗透率小于 0.1mD(不包括裂缝渗透率),气井没有自然产能或自然产能低于工业标准[1],分段压裂水平井是目前国内外最常用的增产增效及提高采收率的技术手段[2-5]。鉴于国内致密气藏普遍存在含气砂体分布零散、储集体内非均质性强的特点,水平井设计和部署时,需首先开展有效储层或有效砂体预测,进而在有效储层预测基础上钻遇水平井并选择性实施压裂改造,针对非有效砂体段一般不实施压裂改造,受有效砂体分布预测可靠性的限制,对于相向的两口相邻水平井而言,靶体 B 点之间留有较大的间距必将造成储量平面控制和动用程度的降低,这也是国内致密砂岩气藏水平井开发方式下采收率较低的主要原因之一。本文通过借鉴国外致密气开发实践经验,分别从苏里格致密砂岩气藏有效砂体空间分布及规模尺度、生产动态特征(静、动态两个层面)分析水平井井距存在的合理性,同时利用数值模拟手段预测两种水平井部署方式对提高采收率及经济效益两个方面的效果,最终提出致密砂岩气藏水平井井距和压裂间距部署新思路,为提高致密气藏水平井对储量的控制和动用程度及降低开发评价成本提供技术支撑。

1 致密砂岩气藏有效储层分布特征

国内致密砂岩气藏普遍具有含气面积大,主力含气砂体分布零散,储集体非均质性强的特征。苏里格致密气田属陆相辫状河或辫状河三角洲沉积体系,砂体大面积连片分布,数万平方千米范围内整体含气,但有利相带心滩微相为非连续相,导致气藏内部存在很强的储层非均质性,平面上有效砂体呈透镜状零散分布,表现为"孤立甜点"特征,甜点规模一般小于 2km^2;垂向上有效储层呈多层叠置或分散分布,盒八上、盒八下、山一段、山二段均有发育,单层厚度1~5m,总厚度 10m 左右。储集体具有"砂包砂"二元结构特征,心滩相沉积的主力含气砂体孤立分布于连续发育的辫状河河道充填沉积的基质储层中[6-7]（图1），总体上，天然气富集甜点规模小、厚度薄、高度分散、横向连通性差、预测难度大。针对这一特点，国内普遍采用较为稳妥的水平井部署方法，即首先利用直井作为骨架井预测有效砂体展布，进而以骨架井作为出发井和目标井，在井间预测的有效砂体展布位置部署水平井，根据钻遇有效砂体气测显示，进行非均匀压裂设计，即只对钻遇气层实施压裂，而对含气层、差气层和泥岩干层不压裂。

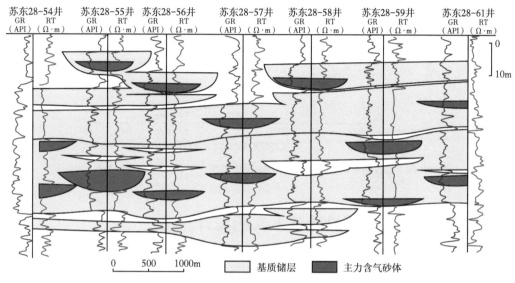

图1 苏里格气田东部试验区储层"二元"结构分布剖面

2 国外致密砂岩气藏水平井开发设计思路

致密砂岩气藏水平井开发方式下，储层流体渗流机理复杂[8-9]，难以建立适用的产能评价模型。同时，致密气藏普遍整体含气、主力含气砂体预测难度大，因此，国外水平井开发普遍采用均匀分段压裂措施进行增产[4-5,10,11]，且两口水平井的末端靶点 B 尽可能靠近或重合（图2），进而通过单压裂段产能（图3）和压裂段数来评价气井及整个气藏的产能[12]，即气藏

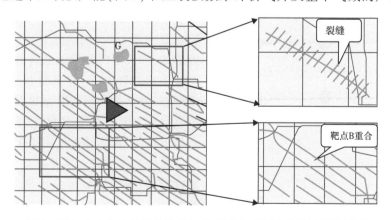

图2 西加拿大盆地某区块致密气藏水平井井距及分段压裂部署图

的可采储量(EUR)=单压裂段最终可采储量×总压裂段数。该评价方式避免了以气井为评价单元时所导致的不同压裂段数气井产能差异大的问题。从气藏整体压力分布和储量动用程度的角度考虑,人工裂缝在水平段和气藏中均匀分布最为合理。

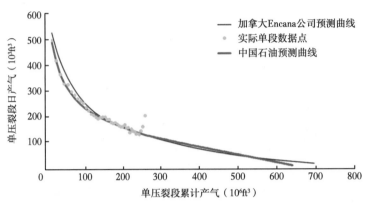

图3 西加拿大盆地致密气产能评价的单压裂段典型生产曲线

3 目前国内水平井井距设计与本文的设计对比

目前国内致密气藏水平井井排距设计思路仍沿用直井的部署思路[13-17],如苏里格致密砂岩气藏南北向水平井水平段长1000m,东西向排距600m,两口相向水平井的靶点B间距与直井排距一致,为800m(图4)。这对于苏里格气田致密气藏有效砂体规模小,横向连通性差,发育频率低,空间上以孤立分布为主的特征,势必会造成储量的遗漏。通过实钻井钻遇有效砂体解剖,苏里格气田有效单砂体宽度范围100~500m,长度范围300~700m(图5)。从最小井距400m的苏6区块加密井排精细解剖来看,在可识别的有效储层个体范围内,600m井距仅可动用其中的61%(图6),可见,在两口相向水平井靶点B间距800m的情况下,将有一定数量的有效砂体,即天然气储量无法有效动用。同时,针对水平井部署方式,鉴于钻井和压裂施工条件限制,后期剩余储量挖潜也将存在很大困难,无论通过水平井还是直井进行井网加密,均会面临较大挑战,最终导致气田整体的采收率降低。

鉴于致密气藏水平井产量主要靠多段裂缝改造贡献,本文研究提出在目前经济技术和装备条件下,A点的靶前距尚不可避免外,但两口相向水平井的靶点B间距应尽量缩短或直至到

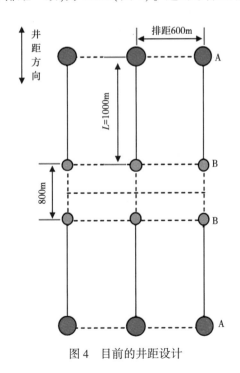

图4 目前的井距设计

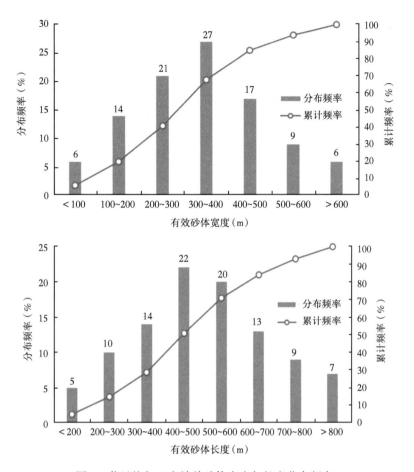

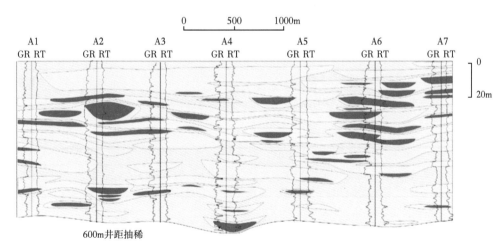

图 5 苏里格气田有效单砂体宽度与长度分布频率

图 6 苏里格气田苏 6 区块加密井排有效砂体动用分析

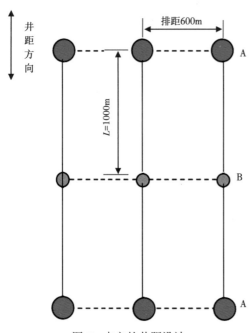

图7 本文的井距设计

0,即B点重合(图7),进而通过设置合理的裂缝间距有效动用原间距下的天然气储量。理由有以下两点:(1)靠近B点的末端裂缝在B点方向上压降最大波及范围为一个压裂间距,国内外研究成果表明:致密气藏水平井压裂间距,范围75~200m,苏里格气田气井动态分析显示,气井泄流范围主要分布在0.15~0.25km²,即等效圆形的平均泄流半径在250m左右。(2)由于水平井筒和压裂工具限制,靠近B点的末端垂直裂缝不可能与B点重合,一般与B点仍有几十米至上百米的距离。因此,即使两相向水平井的靶点B重合,靠近B点的两条末端裂缝之间仍有一定距离,这个距离与裂缝间距相等时即为最优。

4 实例分析与对比

以苏里格致密气田为例,定量分析两种井距设计产生的差异。苏里格致密气田有效储层具有明显的"二元"结构,以苏里格某区块实际的井组地质模型为基础开展水平井开发数值模拟分析,模型面积2.64km²,在模型中部署两口相向钻进的水平井,水平段长度均为1000m,均匀压裂六段,压裂间距为160m,靶前距为400m。以两种井距设计思路进行部署(表1):一种是两口水平井靶点B间距为800m;另一种是B点重合。结合苏里格地区水平井的实际开发情况,设定气井配产5×10⁴m³/d,稳产期三年,经数值模拟预测:靶点B间距800m的条件下,平均单井累计产量为7963×10⁴m³;B点重合条件下,平均单井累计产量7744×10⁴m³,单井累计产量降低约2.8%(图8、图9)。

表1 地质模型及数值模拟参数表

网格尺寸 (m)	地质模型参数			水平井部署方式	数模预测 EUR ($10^4 m^3$)
	平均孔隙度 (%)	平均渗透率 (mD)	平均含水饱和度 (%)		
I:50 J:50 K:1	含气砂体:11.7 基质:6.5	含气砂体:0.05 基质:0.001	含气砂体:31.5 基质:46.8	靶点B间距800m	7963
				B点重合	7744

苏里格致密气田水平井部署区平均地质储量丰度约为1.3×10⁸m³/km²,靶点B间距800m部署方式单井控制面积1.08km²,水平井开发层段采出程度约为56.7%;B点重合部署方式单井控制面积0.84km²,水平井开发层段采出程度约为70.9%,较靶点B间距800m部署方式提高了14.2%,从提高气田采收率的角度看,效果十分明显。从中国致密气藏巨大的储量规模

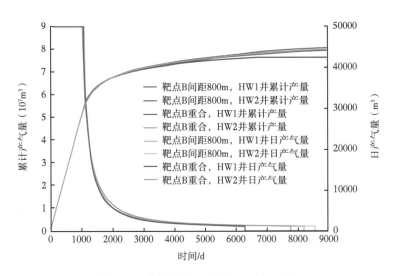

图 8 两种井距设计思路气井产量对比

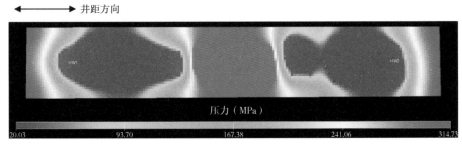

（a）靶点B间距800m数值模拟开采末期压力等值图

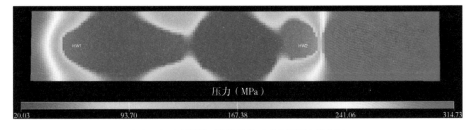

（b）B点重合设计下数值模拟开采末期压力等值图

图 9 两种井距设计下生产期末压力分布图

（中国石油四次资评全国致密气可采资源 $10.92×10^{12}m^3$）来考虑，该井距设计方式的实施将为气田长期稳产及未来持续上产提供有力的技术支持。从单井效益角度看，本文井距设计下的经济效益稍差，但从整体效益角度考虑，苏里格致密气田含气面积约 $35000km^2$，去除富水区和当前经济条件尚无法经济开发的低丰度区，按10%面积部署水平井，单井投资2400万元、气价1000元/10^3m^3，采用静态法计算经济效益，结果表明：本文井距设计思路的整体效益有所上升（表2），且随着气价的提高，经济效益将更加明显。

表 2　不同井距设计下的经济效益对比（静态法）

不同井距	靶点 B 间距 800m	B 点重合
可布井数（口）	3241	4167
投入（亿元）	778	1000
累计产气（$10^8 m^3$）	2581	3227
收入（亿元）	2581	3227
毛利润（亿元）	1803	2227
投入产出比	0.308	0.302

5　结论

基于致密气藏普遍整体含气、主力含气砂体预测难度大的特点，从提高采收率的角度出发，借鉴国外致密砂岩气藏水平井设计经验，创新性提出国内致密砂岩气藏水平井井距设计的新思路，即在裂缝间距合理的前提下，两口相向水平井的靶点 B 间距缩短至零，即 B 点重合。

以中国最为典型的致密气田——苏里格气田为例，综合考虑有效储层静态规模尺度和动态泄气范围，同时借助数值模拟手段，分析对比两种水平井井距部署方式在提高气田采收率和整体开发效益两个角度的优劣，本文所提出的 B 点重合井距设计思路既可以避免天然气储量遗漏后期挖潜难的问题，有效提高气田最终采收率，又可提高气田开发整体经济效益，同时对同类气藏规模水平井部署支撑气田稳产及上产具有指导意义。

参 考 文 献

[1] Stephen A Holditch. Tight gas sands[A]. SPE 103356, 2006.

[2] Giger F M. Low-permeability reservoirs development using horizontal wells[C]. SPE 6406, 1985.

[3] Joshi S D. Horizontal well technology[M]. Tulsa：Penn Well Publishing Company, 1991.

[4] Bruce R Meyer, Lucas W, Bazan R, et al. Optimization of multiple transverse hydraulic fractures in horizontal wellbores[C]. SPE 131732, 2010.

[5] Rasheed O Bello, Robert A Wattenbarger. Multi-stage hydraulically fractured shale gas rate transient analysis [C]. SPE 126754, 2010.

[6] 贾爱林，唐俊伟，何东博，等. 苏里格气田强非均质致密砂岩储层的地质建模[J]. 中国石油勘探，2007,（1）：12-16.

[7] 何东博，贾爱林，冀光，等. 苏里格大型致密砂岩气田开发井型井网技术[J]. 石油勘探与开发，2013, 40(1)：79-89.

[8] 李军诗，侯建锋，胡永乐，等. 压裂水平井不稳定渗流分析[J]. 石油勘探与开发，2008, 35(1)：92-96.

[9] 李树松，段永刚，陈伟. 中深致密气藏压裂水平井渗流特征[J]. 石油钻探技术，2006, 34(5)：65-69.

[10] 王瑞和，张玉哲，步玉环，等. 射孔水平井产能分段数值计算[J]. 石油勘探与开发，2006, 33(5)：630-633.

[11] 吴晓东，隋先富，安永生，等. 压裂水平井电模拟实验研究[J]. 石油学报，2009, 30(5)：740-748.

[12] 位云生,贾爱林,何东博,等. 致密气藏分段压裂水平井产能评价新思路[J]. 钻采工艺,2012,35(1):32-34.
[13] 何东博,王丽娟,冀光,等. 苏里格致密砂岩气田开发井距井网优化[J]. 石油勘探与开发,2012,39(4):458-464.
[14] 刘月田. 各向异性油藏水平井开发井网设计方法[J]. 石油勘探与开发,2008,35(5):619-624.
[15] 凌宗发,王丽娟,胡永乐,等. 水平井注采井网合理井距及注入量优化[J]. 石油勘探与开发,2008,35(1):85-91.
[16] 王振彪. 水平井地质优化设计[J]. 石油勘探与开发,2002,29(6):78-80.
[17] 刘月田. 水平井整体井网渗流解析解[J]. 石油勘探与开发,2001,28(3):57-66.

原文刊于《天然气地球科学》,2019,30(6):919-924.

Technical strategies for effective development and gas recovery enhancement of a large tight gas field: A case study of Sulige gas field, Ordos Basin, NW China

Ji Guang Jia Ailin Meng Dewei Guo Zhi
Wang Guoting Cheng Lihua Zhao Xin

(Research Institute of Petroleum Exploration & Development,
PetroChina, Beijing 100083, China)

Abstract: Based on the analysis of influencing factors of tight gas recovery and reservoir geological characteristics, the types of remaining tight gas reserves in the Sulige gas field are summarized from the perspective of residual gas genesis to estimate residual gas reserves of different types and provide corresponding technical strategies for enhancing gas recovery. The residual gas reserves in the Sulige gas field can be divided into four types: well pattern uncontrollable, horizontal well missing, imperfect perforation, blocking zone in composite sandbodies. Among them, the uncontrolled remaining gas of well pattern and blocking zone in composite sandbodies are the main body for tapping potential and improving recovery factor, and well pattern infilling adjustment is the main means. Taking into account reservoir geological characteristics, production dynamic response and economic benefit requirements, four methods for infilling vertical well pattern, i.e., quantitative geological model method, dynamic controlled range of gas well method, production interference method and economic and technical index evaluation method, as well as a design method of combined vertical well pattern with horizontal well pattern are established. Under certain economic and technological conditions, the reasonable well pattern density of enrichment zone of gas field is proved to be 4 wells per square kilometers, which can increase the recovery rate of the gas field from 32% to about 50%. Meanwhile, five matching techniques for enhancing gas recovery aimed at interlayer undeveloped residual gas have been formed, including tapping potential of old wells, technological technology optimizing of new wells, rational production system optimizing, drainage and gas producing, and reducing waste production, which could increase the recovery rate for 5% based on well pattern infilling. The research results provide effective support for the long-term stable production of $230 \times 10^8 m^3/a$ of the Sulige gas field and production growth in the Changqing gas area.

Keywords: Ordos Basin; Sulige gas field; tight gas; remaining reserves; well pattern infilling; enhancing gas recovery; matching technologies

Introduction

The tight sandstone gas field in the Sulige area of Ordos Basin is typical tight gas in China. The reservoirs are highly heterogeneous, with poor physical properties, high irreducible water saturation, and high gas percolation resistance; and wells in the reservoir often see fast decline in energy, small effective sweeping range, and low recovery of reserves. The Sulige gas field was discovered in 1996 and put into development in 2005. It is now the largest natural gas field in China with the largest reserves and production. In the gas enrichment area, the production capacity of $250×10^8 m^3$ was constructed at the end of 2014, and the annual output reached $230×10^8 m^3$. The enrichment area generally refers to the high-quality reserve area with abundance of more than $1.5×10^8 m^3/km^2$, relatively high reserve concentration, and final cumulative well production of more than $2 000×10^4 m^3$. Since 2015, the Sulige gas field has kept a stable production, and will keep so for more than 20 years according to the development plan. Different from conventional gas reservoirs, tight gas wells have hardly any stable production period, and to keep long-term stable production, new wells need to be drilled to make up for the decline. There are two ways to maintain stable production. One is to build new production blocks for successive production, and the other is to increase the producing reserve and gas recovery in the developed enrichment area through infilling the well patterns. From the current development situation of the Sulige gas field, most of the unproduced blocks are either lower in reserve abundance or higher in water saturation, and the wells there have generally lower production than infilling wells in the enrichment area. Therefore, it is more economically effective to increase the recovery in the enrichment area, and keep the unproduced blocks as resource reserve for long-term stable production.

Development practices both in China and abroad show that infilling well pattern is one of the effective measures to improve gas recovery. Regarding the analysis of rational well pattern density, in the early stage, the well pattern density was designed as no more than 3 wells/km^2[1] for the following reasons: focusing on the matching of well pattern and reservoir distribution, and keeping each effective sand body controlled by only one well so as to avoid interference as much as possible; ensuring the rate of class I+II wells and single well development efficiency; the designed well pattern was the optimal technological well pattern. Presently, for the maximum efficiency of gas field development and increasing gas recovery, a certain extent of inter-well interference is accepted when profitable, the well pattern of the best economic efficiency is designed, and four methods to estimate the rational well pattern density, i.e. quantitative geological modelling, dynamic drainage area, production interference and economic and technical index evaluation have been proposed. With progressive demonstration, it is concluded that in the enrichment area the 600m×800m framework well pattern can be infilled to 4 wells/km^2, which will enhance the recovery rate from 32% to about 50%. In addition, other matching measures such as vertical well sidetracking, repeated fracturing, and water drainage gas recovery can also tap the remaining reserves of old wells to some extent, and increase gas recovery by about 5%. Centering on the production demand of improving gas recovery in tight gas enrichment area, this paper presents in details the research results on factors affecting recovery rate, description of remaining reserves, optimization of well patterns and matching measures for improving gas recovery.

1 Basic geological characteristics of tight gas reservoirs

The Sulige gas field is located on the northwest side of the Yishan slope in the Ordos Basin.

The main pay zones are the Permian He 8 and the Shan 1 members. They are continental facies braided river sediments. With the board and gentle tectonic setting, the river channel experienced multiple diversion and stack, forming large-scale sandstone areas of several thousand to over ten thousand square kilometers distributed continuously in patches. After intense compaction and cementation, etc., the sandstone layers become tight reservoirs, with largely secondary pores. Under the general low permeability-tight sandstone background, the sand bodies with higher porosity, permeability and gas content are "effective sand bodies", which are the main targets of proven reserves calculation and contributors to production. Different from the continuous distributed large-scale sand bodies, the effective sand bodies are small in scale, and appear as multiple lenticular beds. The huge continuous sandstone patches and the small effective ones form a binary structure of "sand-in-sand".

The tight gas reservoirs have complex pore throat structure, poor physical properties, low gas filling degree and relatively high water saturation. The average water saturation of Sulige gas field is about 40%, and the formation water occurs in free water, retained water and irreducible water. Except the west and the northern part of the east area of Sulige, the free water is generally low in proportion. The formation water in the main blocks of the gas field is retained water and irreducible water, coexisting with the natural gas in the form of gas-water layer or a gas-bearing water layer, which is different from the "upper gas-lower water" gas-water separation mode of the conventional gas reservoir. In production, the production of water in tight gas wells is quite common. Due to rapid decline of formation energy in the near-well zone, the gas well production is low, with poor liquid carrying capacity, and liquid is very liable to accumulate in the well bottom without taking water drainage measure, blocking gas production.

2 Factors affecting tight gas recovery

From the matching relationship of effective sand body scale and well pattern macroscopically and influence of pore structure and fluid percolation characteristics on gas production microscopically, the factors affecting tight gas recovery can be attributed to the following three. (1) Reservoir heterogeneity: If the reservoir is heterogeneous, the gas-bearing sand bodies could be poor in continuity and connectivity, and thus the well pattern is likely not dense enough to control all the reserves. (2) Reservoir permeability: The reservoir is tight, with poor pore connectivity and weak percolation capacity. (3) Gas-water two-phase flow: Gas and water two-phase flow in the reservoir would cause high percolation resistance to gas, and thus low production and weak liquid carrying capacity of gas well, and the wellbore fluid accumulation causes the gas well abandonment pressure to rise.

2.1 Reservoir heterogeneity

The sedimentary environment of the Sulige tight gas reservoirs is mainly continental facies river sedimentary system featuring great variation of hydrodynamic conditions, so the single-stage river channels are small in scale[2] and complex in stack pattern, and the effective sand bodies are mostly in the gritstone at the bottom of the river channel and in the mid-bottom of the central bar, forming a binary structure of "sand-in-sand" with the matrix sand body. About 80% of the effective sand bodies in the gas field are single-stage and isolated, with big difference in scale[3]. The effective sand bodies are 1~10m thick, mostly 1.5~5.0m, and 50~1000m wide and long. The fitting calculation of gas reservoir engineering shows that the drainage area controlled by single well is 0.15~0.30 km^2, obviously smaller than the 0.48km^2 of the single well in the 600m×800m well pattern in present Sulige gas field, clearly, the development well pattern isn't dense enough to control all the reserves.

The frequent migration of braided river channels in multiple stages resulted in gas-bearing sand bodies of small scale in multiple vertical sections[4], which cross laterally and stack with each other to form large complex effective sand bodies. These sand bodies are generally heterogeneous inside[5], in which the tight and fine or argillaceous interlayers deposited at weak hydrodynamic conditions become "blocking zones", blocking the fluid percolation channel, and making it difficult to produce the reserves in the complex effective sand body fully and the gas recovery reduce. According to the outcrop anatomy of the modern braided river-Yongding River, on the downstream sedimentary section, multiple siltation intercalations are formed in the upper part of the central bar and at the intersection of the braided channel and the central bar[6]. Therefore, it is difficult to fully control the gas-bearing sand bodies with single well high-output pattern for the tight gas reservoir[7-10], and the recovery rate is often low.

2.2 Reservoir permeability

The tight gas reservoirs are low in porosity and permeability, so the pressure transmission is much weaker than that in the conventional gas reservoirs. The fluid-rock adsorption make the reservoirs have a starting pressure, and when the production pressure difference is small and the flow rate is low, the gas cannot overcome the starting pressure gradient and flow to the wellbore, resulting in almost no natural production capacity after well is completed. To obtain industrial gas flow the reservoirs must be fractured, and the fracture network connects the near-well reservoir to improve reservoir permeability, increase reservoir producing degree, single well production, and realize the efficient development. However, as the scale of reservoir fracturing is limited, the recovery rate of tight gas reservoirs is still lower than that of conventional gas reservoirs.

2.3 Gas-water two-phase flow

Due to rock wettability and capillary pressure[11-12], the water occupies the small pore throat and pore wall in priority in the rock; when the gas flows in water-bearing pores, it first flows into the large pores, and gradually drives the water in the small throats or makes the water film on the pore wall thinner with the increase of the flow pressure difference. The water saturation in the core varies with the flow of the gas. When the flow rate is low, the gas flow increases non-linearly with the increase of the pressure difference. The gas front jumps forward and is easily blocked by water. Therefore, when the gas flows in the water-bearing pores, a certain starting pressure (critical flow pressure) is also required; the higher the water saturation, the bigger the starting pressure will be needed.

The relationship between the gas flow pressure difference and the flow rate of the core sample is shown in Fig. 1. The starting pressure of the rock samples with different initial water saturations can be obtained by data fitting. The rock samples with the initial water saturation of 66.34%, 52.69% and 39.96% have a starting pressure of 0.08640, 0.00973, and 0.00239 MPa, respectively. The ratio of the starting pressure to the core length (4.5m) is the starting pressure gradient of the core, which is 0.01920, 0.00210, 0.00053 MPa/m respectively for these samples. As the water saturation decreases, both the starting pressure and starting pressure gradient decrease (Table 1). For reservoirs with active formation water or high water saturation, the gas flow percolation resistance increases due to the production of water. When the production reduces to a certain level at which it fails to carry the liquid out of the well, the liquid begins to build up in the wellbore, which makes the production reduce further, and the well production enters into a vicious circle. In such circumstance, water drainage has to be carried out to prevent flooding of gas wells. Liquid loading will increase the gas well abandonment pressure and decrease the recovery rate. At the same time, taking water drainage measures will increase the development cost and reduce economic efficiency.

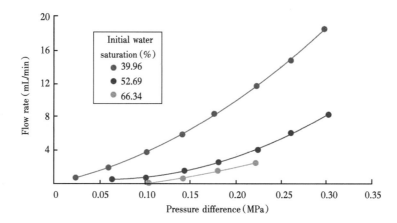

Fig. 1 Flow rate-pressure difference curves at different water saturations.

Table 1 Starting pressure gradient under different water saturations.

Water saturation (%)	Average permeability (10^{-3} μm^2)	Inverse permeability (10^{-3} μm^2)	Starting pressure (MPa)	Starting pressure gradient (MPa/m)
66.34	0.3152	3.173	0.08640	0.01920
52.69	0.4560	2.193	0.00973	0.00210
39.96	0.6240	1.603	0.00239	0.00053

3 Evaluation of remaining reserves in developed areas of tight gas fields

Due to the special structure of tight sandstone gas reservoirs, it is difficult to fully control the gas-bearing sand bodies of different scales in order to maximize the production of vertical wells, resulting in incomplete production of reserves. The development practice proves that the main well pattern of 600m×800m in present Sulige gas field is only able to control the main gas-bearing sand bodies, and difficult to control the smaller-scale gas-bearing sand bodies, which leads to remaining reserves between the wells and between layers. By using geology, geophysics, gas reservoir engineering and other methods, the remaining reserves in blocks, wells and layers have been estimated carefully. Combined with gas production technology, the remaining reserves in the developed areas can be classified into four types: that uncontrolled by well pattern, that in blocked zone of complex sand body, that missed due to incomplete perforation and horizontal wells (Fig. 2).

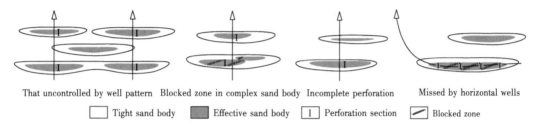

Fig. 2　Different types of remaining reserves.

3.1　Reserves uncontrolled by well pattern

The effective sand bodies in the tight gas reservoirs of Sulige gas field are small in scale, poor in horizontal connectivity, low in development frequency, and mostly isolated. In the early stage of gas field development, a 600m×1200m well pattern was set up, and the well pattern density is bigger than that of the conventional gas reservoir of 1~3 km. With the deepening of gas field development, the main well pattern was adjusted from 600m×1200m to 600m×800m after 2010, and the well density was adjusted from 1.4~2 wells/km^2, as a result, the producing reserve was greatly enhanced. But still the gas-bearing sand bodis were not fully controlled. Calculated with the final cumulative output of $2400×10^4 m^3$ for a single well and the reserve abundance of $1.5×10^8 m^3/km^2$, the current recovery rate is only 32%. The uncontrolled isolated reserves account for 50%~60% of the total remaining reserves, and is the main part for remaining reserves.

3.2　Reserves in blocked zone in complex sand body

The horizontal well trajectory geological section shows that the complex sand body is not connected inside, with several "blocked zones" extending in perpendicular to the water flow direction, with a width of 10~30m and an interval of 50~150m. The gas test data indicates that there exist flow boundaries in the sand bodies in a vertical well, confirming that the "blocked zone" affects the percolation capacity of the complex sand bodies and the reserves producing degree of vertical wells, and thus a certain amount of gas would be left. The remaining reserves in the blocked zones of complex sand bodies account for 25% of the total remaining reserves. Multi-stage fracturing of horizontal wells is able to overcome the influence of the blocked zones.

3.3 Incomplete perforation

Effective sand bodies can be classified into two groups according to differences in physical properties and gas content: poor gas layer and pure gas layer. In comparison, the poor gas layer has thinner effective sand body of $1\sim3m$, poorer physical properties with porosity of $5\%\sim7\%$, and burden pressure permeability of $0.01\times10^{-3}\sim0.10\times10^{-3}\mu m^2$; lower gas saturation of $45\%\sim55\%$; higher water saturation of more than 45%, lower gas relative permeability and poor fluidity. In the early stage of development, limited by the vertical well fracturing technology, residual gas was formed due to imcomplete perforation or fracturing of poor gas layers. Based on the drilling and logging data of 1200 wells in the Sulige gas field, the quantity, thickness, porosity, gas saturation, etc., of the effective sand bodies drilled in a single well were counted to screen out the reservoirs with incomplete perforation. Together with the geological peremeters such as width-to-thickness ratio and length-to-width ratio, the reserves in the reservoirs of incomplete perforation and their ratios in the whole remaining reserves were estimated. The results show that on average the perforation incomplete remaining reserves account for 14% of the well-controlled reserves, and these reserves are mainly in a few development wells and evaluation wells put into operation in the early stage. From 2008, with the advancement in the separate-layer fracturing technology, such remaining reserves basically no longer occur any more. Therefore, this type of remaining reserves can be taken as potential target in certain wells, but are not significant to the the overall recovery improvement of gas reservoirs.

3.4 Reserves missed by horizontal wells

In the Sulige gas field, there are multiple tight gas reservoirs, and the reserves in major reservoirs He 8 and Shan 1 account for about 80% of the geological reserves. By increasing the contact area between horizontal well and the reservoir, and using multi-stage fracturing to break the blocked zone, the producing degree of the reserves in the main sections can be enhanced. However, the geological feature of multiple gas-bearing layers makes it inevitable for the horizontal well to miss some reserves vertically. According to the statistics of drilled effective reservoirs in more than 1300 horizontal wells, the horizontal wells can control $60\%\sim70\%$ of the geological reserves, and leave off $30\%\sim40\%$ of the reserves. Estimated at the average single well control area of $1\ km^2$, the total remaining reserves missed by the horizontal wells is $600\times10^8\sim800\times10^8 m^3$, and the average abundance of the remaining reserves is $0.5\times10^8 m^3/km^2$. In this case, the economic efficiency of the measures to tap remaining reserves can be guaranteed.

4 Main technologies to improve the recovery of tight gas reservoirs

Subtracting the accumulated producing reserves of the completed well from the geological reserves of different blocks in the Sulige gas field, the general remaining reserve is obtained. According to the anatomical analysis on 39 gas reservoir sections through connected wells, the various types of remaining reserves were calculated, and then the ratio of various remaining reserves to the remaining general reserves were worked out. For remaining reverses in the isolated gas-bearing sand bodies not controlled by well pattern or blocked in the complex sand bodies, the essential reason of their being is that the well pattern is not able to control them, so the two are classified as inter-well unproduced reserves. For the reserves missed by incomplete perforation or horizontal wells, as detained in the longitudinal layers, they are classified as inter-layer unproduced reserves. The inter-well reverses accounted for about 82%, and the inter-layer for 18% (Table 2). Therefore, optimizing and infilling the well pattern to improve the producing degree of inter-well remaining reserves is the main measure to increase gas recovery. To be more specific, there are two measures: vertical well pattern infilling and combination of vertical and horizontal wells.

Table 2 Statistics on remaining reserves in the enrichment area of the central Sulige gas field.

Remaining reverse	Genetic type	Ratio
Inter-layer remaining reverse	Thin layer or gas-bearing layer left by incomplete perforation in vertical wells	9%
	Minor layers missed by horizontal wells	9%
Inter-well remaining reverse	Isolated gas-bearing sand bodies uncontrolled by well patterns	57%
	Gas blocked in complex sand bodies	25%

4.1 Vertical well pattern infilling to improve gas recovery

The vertical well pattern infilling is applicable to the area where mutiple gas-bearing layers scatter, and the core is to work out the economically effective well pattern density and optimize the well pattern geometry. Tight sandstone gas reservoirs are characterized by extensive hydrocarbon generation and continuous reservoir formation, with large gas-bearing area and poor physical properties. The reservoir structure exhibits strong heterogeneity microscopically. The author's team has long devoted to the research on tight gas stable production and recovery enhancement. In the early-stage, our coginitions mainly included: in the reservoir geological evaluation, not only the reserve scale should be studied, the distribution pattern of the reservoirs and the influence of gas content on the output also needed to be investigated[1]. For the indexes of infilling, multiple parameters should be optimized and integrated, scientific infilling principles and systematic evaluation indexes should be set up[13]. One of the characteristics of the early stage research was to infill the well pattern according to reserve types, i.e. different types of reserves corresponding to different well pattern densities.

In recent years, with the deepening of the gas field development, we have got better understanding on the gas field differing from the early stage: by making use of the geological feature of multiple isolated sand bodies of different stages stacked vertically, the strong heterogeneity in microstructure of reservoirs can be taken as the equivalent of homogeneity on macroscopic distribution of the reservoirs in the enrichment zone, which is testified by the feature of "every well produces gas but is hardly high in production". Based on this development concept, the distribution of the enrichment zone was firstly defined based on seismic data interpretation and sedimentary facies constraints, and the enrichment zone was taken as a relatively homogeneous whole, and developed by fractory-like large scale well deployment and one-off well pattern building. Doing so makes well position optimization unnecessay, drilling cost reduce, post stage management easier, and development efficiency improve. From 2008 to 2015, dense well pattern was tested in the enrichment areas of SU 6, SU 14 and SU 36-11 of the Sulige gas field, providing information for well pattern infilling analysis. Another advantage of reserch on the well infilling without grouping reserves is that the data analysis samples increase significantly (previously the He 8 well pattern areas were divided into five types of reserves), improving the reliability and accuracy of the research.

Based the previous researches, and after further sorting, summarizing and refining, 4 well pattern density evaluation methods have been come up, namely quantitative geological modelling, dynamic drainage ranging, output interference rating and economic-technic index evaluation. Of the 4 methods, the factors considered increase in turn, and the constraints also increase. Comprehensive research showed that in the enrichment area of Sulige large-scale tight sandstone gas field the well pattern density could be infilled to 4 wells/km^2. Compared with the previous research results, the progresses made this time include: (1) Combined with the analysis on output interference rate, 4 stages of recovery variation with the well pattern density have been identified, and the specific well pattern density of each stage has been determined; (2) the theoretical connotation of the final cumulative output of single well, gas production increase volume of infilled well and well pattern infilling, have been deenpened, and the original conceptual models have been gradually quantified, making them increase in application value in field; (3) following the development of tight gas, the economic evaluation has been strengthened, and suitable well pattern density under conditions of different gas prices has been discussed as the internal rate of return of tight gas was lowered from 12% to 8%.

4.1.1 Quantitative geological modelling

The core of the quantitative geological modelling is to define the scale and distribution frequency of the effective single sand bodies. According to the main data (thickness, width, length, etc.) of the effective single sand bodies, the class and reserve producing degree of sand bodies effectively controlled by the current well pattern were evaluated. Fine description of core is an important means to analyze the thickness of the effective single sand body. Based on the calibration of the rock-electricity relation and in combination with the logging data, the effective single sand body thickness of non-coring wells were estimated carefully. The results shows that the isolated effective single sand bodies in Sulige gas field are mostly 1.5~5.0m thick. The width and length of the effective single sand body can be estimated by dissecting the infilled well pattern, or by getting the width-to-thickness ratio and length-to-width ratio of the sediment in the corresponding sedimentary environment from outcrop observation and sedimentary physical simulation, and then with the thickness of the sand body, the length and width of the effective sand body can be calculated. The research results show that in the Permian He 8 and Shan 1 of Ordos Basin the width-to-thickness ratio of the river channel filling is 50~120, the length-to-width ratio is 1.2~4.0. In Sulige gas field, the isolated effective single sand bodies are largely 200~500m wide, accounting for 65%; and largely 300~700 m long, accounting for 69% (Fig. 3). The effective single sand bodies are mostly 0.08~0.32km^2, 0.24km^2 on average. Calculated with the average abundance of reserves and size of effective single sand bodies, there are 20~30 effective sand bodies developed in 1km² stratum. 80% of the effective sand bodies are isolated and small in scale, at less than 400m×600m on average; the rest 20% are vertically stacked and laterally overlapped, and larger in scale, holding 45% of the total reserves.

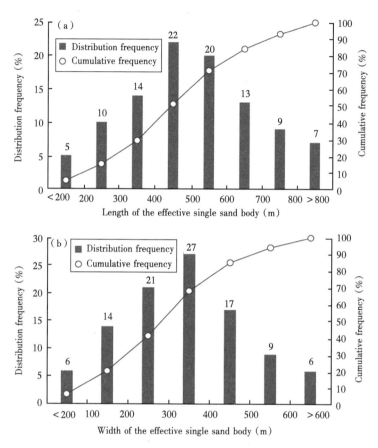

Fig. 3 The width and length distribution frequency of the effective single sand bodies in the Sulige gas field.

With the present 600m×800m main development well pattern, the development area covered by one well is 0.48km², which is twice the average size of effective single sand body. A large number of effective sand bodies are missed between the wells, which causes low producing degree of reserves. According to the quantitative geological model analysis, the well pattern density needs to reach 4 wells/km² (the reciprocal of 0.24km²).

4.1.2 Dynamic drainage ranging

The presence of starting pressure in tight gas at low flow rates would make the open flow capacity of the gas well lower, the abandonment pressure of the gas reservoir rise, and the recovery drop. The rational well space was estimated according to the core experiment analysis and the relationship between the starting pressure gradient and the formation permeability[14] before: when the original formation pressure was 40MPa and the formation permeability was 0.1×10⁻³μm², the maximum well space was 88.6m, i.e. the well drainage radius didn't exceed 44.3m. The successful development of tight gas is inseparable from the progress in the reservoir stimulation technologies. The tight gas reservoir is also called "artificial gas reservoir", of which the permeability includes matrix permeability and artificial permeability. The value and distribution of the artificial permeability are difficult to measure accurately, which results in a large difference between the rational well spacing obtained by the previous researchers and the actual situation.

The pressure and output data during development is reliable basis for drainage range analysis. In fact, during gas well production, the presence of starting pressure gradient causes changes in gas well pressure and production. In dynamic drainage ranging method, the gas wells with production time of more than 500 days and basically reaches the quasi-stable state, are chosen, with their pressure and output, and in comprehensive consideration of the parameters of artificial fractures and physical properties of the reservoir, important indexes such as the dynamic drainage ranging, producing reserve, final gas well cummulative output, etc. are fitted and determined, the distribution frequency of the dynamic drainage range is calculated and analyzed, thus the reserve producing degree of the present well pattern is evaluated. Through the fitting with Blasingame and the flowing material balance methods, it is found that the drainage area of the vertical wells in the gas field differ greatly, from the minimum of less than 0.1km^2 to the maximum of above 1.0km^2, mostly between 0.1 ~ 0.5km^2, and 0.27km^2 on average. The above data is consistent with the conclusion of geological analysis, and reflects the insufficiency of reserves control by the existing 600m × 800m framework well pattern. According to the analysis of the gas drainage range, the well pattern density needs to be 3.7 wells/km^2. In fact, the gas well drainage range is affected by multiple factors, including reservoir scale, reservoir stacking pattern, blocked zone location, and artificial fracture form. As the well pattern is generally an irregular polygon, i.e. when the well pattern density is less than 3.7 wells/km^2, the drainage areas of wells could overlap, so the wording "drainage radius" is not accurate.

4.1.3 Output interference rating method

Whether the well space is rational is evaluated mainly according to the interference test on site. In interference test, the well production regime is adjusted and excited by shut-in and opening, and the changes of pressure and output in the observation wells are tracked to find out whether there is inter-well interference in the tested well group. One gas well usually drills an average of 3~5 effective sand bodies. A new well drilled can encounter new reservoirs, but some effective sand bodies of larger scale and better connectivity might by controlled by two or more wells, increasing the interference possibility between wells. Meanwhile, most isolated reservoirs of smaller scales have no interference yet. In addition, in the Sulige gas field, separated-layer fracturing, commingled production and downhole throttling are adopted. However, it is difficult to carry out layer-by-layer output test in the interference test, and is not able to determine the inter-well connectivity layer by layer. Therefore, it is impossible to truly reveal how much the output is affected by the interference well ratio, so the application of this method in the Sulige gas field is obvious limited.

To solve this problem, the index of output interference rating is proposed to quantitatively characterize the influence of well pattern infilling on the average production of gas wells in a certain area of tight gas reservoir, and to reasonably evaluate the feasibility of well pattern infilling. The output interference rate is defined as the ratio of the difference of average single well cumulative output before and after the well pattern infilling in a certain area to the average single well cumulative output before infilling.

$$I_R = \frac{\Delta Q}{Q} \tag{1}$$

where I_R——output interference rate, %;
ΔQ——difference of average single well cumulative output before and after infilling, $10^4 m^3$;
Q——average single well cumulative output before infilling, $10^4 m^3$.

The interference tests of 42 well groups in the gas field show that when the well pattern density reaches $4/km^2$, about 60% of the gas wells have interference. The interference would be taken as very serious in traditional view, but the actual interference rate is only 20%~30%, which mean the interference is slight. By selecting typical blocks and combining geological modelling and numerical simulation, the relationships between reservoir abundance, well pattern density and output interference rate of Sulige tight gas reservoirs were studied (Fig. 4). The results show that the output interference rate increases with the increase of well pattern density. When the well pattern density is 2.5~4.5 wells/km^2, the output interference rate increases rapidly, indicating that the gas drainage range of most gas wells is 0.22~0.40km^2, and testifying the previous conclusion. When the well pattern density reaches 4.5 wells/km^2, the output interference rate increases slowly. Generally, the larger the average abundance of the block, the more the reservoirs and the larger the cumulative thickness of reservoirs, the stronger the continuity of sandbodies between wells, the more likely it is to cause interference, and the earlier the inflection point occurs. The average reserve abundance of the Sulige gas field enrichment area is $1.5 \times 10^8 m^3/km^2$, which means that the well pattern in the Sulige gas field has the potential to be infilled to 4~5 wells/km^2.

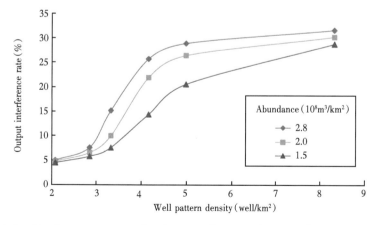

Fig. 4 Relationships between reserves abundance, well pattern density and output interference rate.

Based on the actual production data of 8 infilled well pattern test areas in the gas field, a diagram shows the relationship between the recovery, average final cumulative production of well, and gas production increment of the infilled wells, etc., with the density of the well pattern, has been compiled (Fig. 5). The gas production increment of the infilled well is defined as the increase in gas production per square kilometer of adding of 1well/km^2 than that of the original well density. The authors believes that "gas production increment of the infilled wells" is more scientific than "final cumulative output of the infilled wells": the final cumulative output of the infill wells is related to the infilling time. The later the infilling starts, the lower the final cumulative output; while the output increment of infilled well is produced from the inter-well non-connected effective reservoirs, which is independent of the infilling time, and is more closely related to the final recovery. As the density of well pattern increases, the interference between wells increases, the average cumulative output of single well continuously decreases, and the increase amplitude of recovery becomes smaller and smaller. This process can be divided into four stages (Fig. 5).: Stage Ⅰ, at well pattern density of 0~1.6 well/km^2, no inter-well interference, the increase in gas output of the infilled well pattern is equal to the cumulative output of the old wells, and the recovery increases linearly with the increase of the well pattern density; Stage Ⅱ, at the well pattern density of 1.6~4.5 wells/km^2, there is certain interwell interference, the gas output increment of the infilled well pattern is smaller than the cumulative output of the old wells, but the interference is not serious, and the recovery increases with the increase of the well pattern density; Stage Ⅲ, at the well pattern density of 4.5~8.4 wells/km^2, and the interwell interference gradually enhances, the gap between the gas output increment of the infilled well pattern and the cumulative gas output of the wells continuously expands, and the increase of the recovery drops with the increase of the well pattern density. Stage

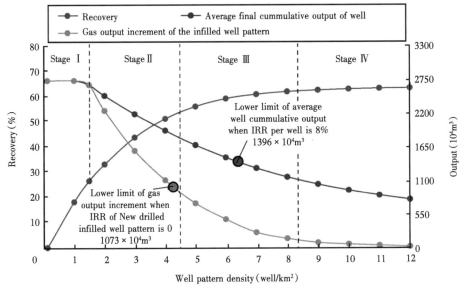

Fig. 5 Four stages of recovery changes with the well pattern density.

Ⅳ, at the well pattern density of more than 8.4 wells/km², the well pattern basically controls the reservoir completely, the infilled well can hardly drill new reservoirs anymore, and the output of the new well is low, the recovery has reached the limit by well pattern infilling. According to the percolation test simulation and modelling, numerical simulation, the technical limit recovery of Sulige tight gas field by well pattern infilling is 63%[15].

4.1.4 Economic-technical index evaluation

The tight gas reservoirs of low permeability are poor in physical properties, and the prediction of effective reservoir is difficult. The reservoir stimulation requires better technology and large investment, however the single well output is low, and the development profit gets poorer. Therefore, reducing the development cost and pursuing economic efficiency are very important in tight gas reservoir development. The 3 methods mentioned above are mainly cut in from the perspective of geology or gas reservoir, and don't consider the economic factors fully. The economic-technical index evaluation method is a comprehensive assessment method orienting around the development economic efficiency, and determining the well pattern density with the internal rate of return (IRR) as the core evaluation index. The IRR is a key indicator for evaluating the investment efficiency internationally, which refers to the discount rate when the total present value of capital inflows is equal to the total value of capital outflows and the net present value (NPV) is zero, and can be understood as the capability that the project investment income can withstand the currency devaluation and inflation. The IRR of 0 corresponds to the breakeven point. In recent years, natural gas is in a booming period. The domestic oil and gas industry lowered the IRR of tight gas development from the previous 12% to 8%. Presently, preferential policies are applied, and the IRR standard is expected to be lowered to 6% in the next three years. Calculated with the fixed cost of 8.0×10^6 CNY, bank loans 45%, interest rate 6%, operating costs 1.2×10^6 CNY, depreciation duration of 10 years, and considering urban construction, resource taxes and other related taxes and fees, the lowest ECP (estimated cumulative production) of gas well corresponding to IRR 8%, 6% and 0 were estimated at different gas prices. The higher the gas price, the lower the gas well ECP required to reach the IRR standard will be (Fig. 6). When the gas price is 1.15 CNY/m³, the lower limit of the gas well ECP required to reach the IRR standard 8%, 6%, 0 are 1396×10^4, 1289×10^4 and $1\,073 \times 10^4$ m³ respectively. In the future with the advancement of technology, rise of gas price or lower of IRR standards, the development economic efficiency of gas fields is expected to increase further.

Estimating the appropriate well pattern density requires balancing the recovery, single well output, and development economic efficiency. If the well pattern is too sparse, the reserves can't be effectively utilized, and the recovery would be low; if the well pattern is too dense, geological conditions and productivity interference would limit the development economic efficiency. In this paper, three basic principles of infilling adjustment are proposed, i.e., the recovery rate is significantly improved, wells are economically efficient overall, and the newly drilled infill wells can be kept above breakeven line. Hereunder are the details. (1) Great enhancement of gas recovery: to achieve this goal a certain degree of inter-well interference can be accepted. According to the previous analysis the well pattern density can be infilled to 1.6~8.4 wells/km². (2) The average IRR of all the wells in the block reaches 8%, and the corresponding average final cumulative output of the gas wells is not less than $1.396×10^7 m^3$, the density of the well pattern is no more than 6.3 wells/km². (3) No infilled well shall lose money and meet the requirement of IRR being 0. The infilled wells can improve gas recovery, although not necessarily reach 8% of IRR, they should not lose money, i.e. the output increment of the infilled well is not less than $1073×10^4 m^3$, and the rational well pattern density is no more than 4.2 wells/km².

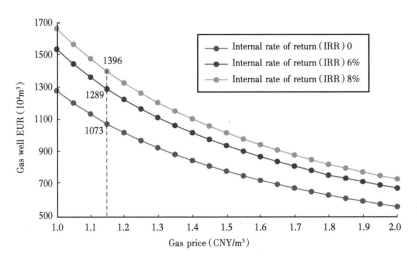

Fig. 6 Lower limit of ECP corresponding to gas well development economic efficiency at different gas prices

With the above analysis, also in consideration of the current economic and technical conditions, to meet the 3 basic principles of infilling adjustments, and at the same time in combination of the existing framewoek well pattern, the suitable density of well pattern is considered to be 4 wells/km². In the model, the average single well output of Sulige gas field simulated by using 600m×800m well pattern is $2420×10^4 m^3$, and the average gas well output after infilling is about $1920×10^4 m^3$, which still meets the economic efficiency of the development plan. The average output increment of infilled well is $1 110×10^4 m^3$, which is higher than the corresponding economic limit output when the IRR is 0. The recovery after infilling is about 50% (Fig. 5).

The appropriate well pattern density under different economic conditions were examined. From the perspective of geological conditions and gas reservoir dynamics, to meet the requirement of enhancing gas recovery significantly, the appropriate well pattern density should be maintained at 1.6~8.4 wells/km² and does not vary with gas price; from the perspective of economic evaluation, to meet the goals of keeping the average IRR at 8% and the infill well profitable, with the increase of gas price, the density of well pattern would increase gradually. When the gas price is 1.0~1.1 CNY/m³, the suitable well pattern density is 3 wells/km²; when the gas price is 1.1~1.5 CNY/m³, the suitable well pattern density is 4 wells/km²; when the gas price reaches 1.5~2.0 CNY/m³, the suitable well pattern density is 5 wells/km²(Table 3).

Table 3 Suitable well pattern density at different gas prices.

Gas price (CNY/m^3)	Average IRR of 8% per well		Breakeven of infilled well		Comprehensive judgement
	Ave. well EUR (10^4 m^3)	Well pattern density (well/km^2)	Output increment of infilled well (10^4 m^3)	Well pattern density (well/km^2)	Well pattern density (well/km^2)
1.00	⩾1 659	⩽5.1	⩾1278	⩽3.4	3
1.11	⩾1 396	⩽6.3	⩾1073	⩽4.2	4
1.30	⩾1 206	⩽7.5	⩾928	⩽4.4	4
1.50	⩾1 020	⩽9.0	⩾785	⩽4.8	4
1.80	⩾0 826	⩽11.5	⩾637	⩽5.3	5
2.00	⩾0 736	⩽12.8	⩾566	⩽5.6	5

4.2 Well pattern combining vertical wells and horizontal wells to improve the recovery

The combination well pattern of veritcle and horizontal wells is mainly suitable for the blocks with obvious major gas reservoirs (major gas reservoir profile reserve ratio is bigger than 60%), by which the breaking blocked zone and high in-layer recovery of horizontal well can be made use of fully, to save development investment and improve economic returns. In reality, the well pattern is in the form of grid. Firstly, the target area is divided into grid units according to the horizontal well spacing (600m×1600m in Sulige gas field), then the reservoir structure of each unit is defined through detailed reservoir description, and finally the corresponding well pattern is deployed according to the structural characteristics. From the analysis of typical blocks, in Sulige gas field there are 30% of grid units suitable to be developed with horizontal wells, and the remaining 70% area should be developed by vertical wells.

In the 150km^2 area of the SU 6 block, 3 designs, i.e. basic well pattern (600m×800m vertical well pattern), infilling well pattern (4 vertical wells in 1km^2), combined well pattern (1 horizontal well or 4 vertical wells in 1km^2) were simulated to compare. With the infilling well pattern (400m×600m infilled from the 600m×800m basic well pattern), the cumulative output of single well decreases by 21.9%, and the recovery increases by about 18% from 31.94% to 49.89%, and all the wells are economically efficient. With the combination of vertical well and horizontal well, the recovery is basically equal to that of the infilled well pattern. However, since the horizontal well investment in Sulige area is about three times that of the vertical well, the horizontal well pattern density is about one quarter of the veritcal well pattern, which means the investment of one vertical well per square kilometer is saved. Compared with the infilled well pattern, the well pattern combining horizontal and vertical wells increases the recovery from 49.89% to 50.70%, and decreases the investment from 49.14×10^8 CNY to 45.61×10^8 CNY, saving about 7% of the investment (Table 4). In recent years, in Changqing Oilfield the Second Accelerated Development Plan is proposed, according to which, within 3 years more than 5,000 vertical wells and 1,000 horizontal wells will be deployed. The above research results can provide strong support for the oilfield on-site capacity construction.

Table 4 Comparison of simulated indexes of the combined well pattern and the infilled well pattern.

Simulated plan	Number of vertical wells(well)	Average vertical well cumulative output($10^4 m^3$)	Number of horizontal wells(well)	Average horizontal well cumulative output($10^4 m^3$)	Recovery (%)	Investment (10^8CNY)
Basic well pattern	300	2306	0	0	31.94	24.57
Infilled well pattern	600	1801	0	0	49.89	49.14
Combined well pattern	432	1771	42	7 932	50.70	45.61

5 Matching technologies for improving recovery of tight gas reservoirs

For the four types of remaining gas in the tight gas reservoirs, the two main technologies, infilled well pattern and well pattern combining vertical well and the horizontal well are mainly used to produce the inter-well remaining gas, and can increase the recovery of the enrichment area from 32% to about 50%. The rest remaining gas needs to be recovered by relevant supporting technologies to improve the reservoir permeability and well drainage capacity to further enhance the recovery. According to the remaining gas genesis, in the developed enrichment area, 5 matching measures have been taken to enhance the recovery, namely tapping potential of old wells, technology optimization for new wells, optimization of production system, water drainage and gas recovery, and reduction of abandonment production rate. These measures are expected to increase the recovery by about 5% by increasing producing degree of minor gas layers.

5.1 Tapping potential of the old wells

The technical measures for tapping potentials of old wells mainly include producing new reservoirs in old wells, sidetracking in old vertical wells, and repeated fracturing of old wells. Among them, producing new reservoirs in old wells included reviewing gas-bearing layers in old wells, including up to HE 6 member and down to MA 5 member besides the current major reservoirs, HE 8 and Shan 1 members, and reforming the missed reservoirs to improve production. The sidetracking horizontal well from the old well is mainly for Class II and III gas wells in the favorable blocks. Firstly, the gas well condition was evaluated, the 3D inter-well reservoir was predicted for gas wells meeting the sidetracking requirements, to find out the connectivity between the production wells, and the cumulative output of the sidetracking horizontal well was predicted by numerical simulation, finally the remaining gas of wells that are economically efficient was tapped to improve the effective production of the inter-well residual reservoirs. The repeated fracturing is mainly for gas wells with large differences in dynamic and static evaluation. Firstly, the fracturing in original perforation layer and completion situation were analyzed, meanwhile the pressure relief of the well to be treated was compared with that of the surrounding wells to evaluate the feasibility of repeated fracturing, finally the remaining reserves due to engineering factors were produced, at the same time the missed reservoirs re-checked.

5.2 Technology optimization for new wells

After nearly 10 years of exploration and development, based on the continuous optimization and upgrading of the reservoir fracturing technology, the Sulige gas field has been developed effectively and become the largest natural gas field in China. At the same time, tight gas fields such as the Daniudi and the Denglouku are also successfully developed[16-18]. The fracturing in vertical well or directional well has evolved from mechanical packers to coiled tubing separated layer fracturing, which integrates precise positioning, sandblasting perforation, high flow rate fracturing, and zone isolation, to increase the layers fractured and improve significantly tight gas vertical reserve producing degree, and at the same time the wellbore conditions are more convenient for later operation. The technology solves the problems of limited flow rate, poor wellbore integrity, and low operation efficiency of the cluster well group, etc in the separate layer fracturing of vertical wells with multiple tight gas layers in the Sulige gas field. The multi-fracture fracturing technology in horizontal well section has made breakthrough. Degradable temperory plugging agents of different paritcle sizes and fiber compositions have been developed, which are similar in pressure bearing capacity and degrading time with overseas products. By using them, the effective fracturing volume of the tight gas horizontal well has been greatly increased. The problems of poor isolation in the external packer staged fracturing in horizontal wells and low fracturing degree of staged multi-cluster fracturing with bridge plugs.

5.3 Optimization of production system

The tight gas reservoirs have low porosity, serious heterogeneity, rich secondary pores and small throats, and complicated gas-water relationship, etc., which lead to the complexity of the percolation mechanism of underground fluid. This is usually manifested in production as small sweeping range of gas well pressure, rapid pressure drop, low natural productivity, and high decline rate[19-21]. To ensure long-term effective gas recovery, a rational production system is essential for increasing the cumulative production of the well and extending the relative stable production period. The dynamic physical simulation experiment of pressure release and pressure control exploitation in low permeability-tight sandstone gas reservoir shows that pressure release recovery has faster gas production rate, shorter production time, but lower cumulative gas production and recovery rate; pressure controlled recovery can use the formation pressure effectively and has higher gas recovery in unit pressure drop and final recovery. For gas-water producing wells, for example the gas wells in the west of Sulige gas field generally produce water, and the reservoir water body has a significant impact on the gas percolation capacity. The gas expands when pressure drops and pushes the water body to flow. Under the influence of the pressure gradient, the gas percolation capacity reduces and the water percolation capacity increases[22-25]. In this case, it is necessary to comprehensively consider the pressure control degree and liquid carrying capacity of the gas well, and set a rational output to achieve smooth production and higher recovery. The dynamic optimization method of production allocation proposed by Li Yingchuan etc.[26] considered the material balance principle, gas well productivity, wellbore temperature and pressure distribution and continuous liquid-carrying theory. According to this method, in the initial stage of production, the gas production is controlled slightly higher than the critical liquid flow at the wellhead, so as to make full use of the liquid carrying capacity of the gas well, reduce the workload of water drainage, save development cost and enhance the final recovery rate. This method has been applied to the production allocation of water-gas wells in the western area of Sulige gas field. As a result, the continuous liquid-carrying gas production wells account for about 90%, and the wells need to take water drainage measure are only about 10%, which has ensured the recovery rate and improved the development efficiency simultaneously.

5.4 Water drainage gas recovery

Tight gas wells usually have low output, weak liquid carrying capacity, relatively active formation water, and almost no pure gas enrichment zone[27]. Often from the beginning of production, they produce water and see rise of water cut. They are not able to drain the liquid loading by themselves. By the end of 2018, the wells with liquid loading exceeded 60%. In order to maximize production capacity, the effective production period, and the final cumulative gas production, many research and application tests have been carried out for the Sulige gas field, forming a series of water drainage technologies suitable for geological and technological characteristics of the gas field.

In the aspect of assisting the drainage of water in wells, a series water drainage technologies including the main measure, foam drainage, and acceleration string and plunger gas lift. In restoring production of liquid loading dead gas wells, compressor gas lift and high pressure nitrogen gas lift are employed[28-30]. Among them, the foam drainage is to convert the liquid at the well bottom into a low-density and easy-to-carry foamy fluid to improve the liquid-carrying capacity, to drain the water out of the wellbore. It is suitable for liquid loading wells with production of more than $0.5 \times 10^4 m^3/d$, and has the advantages of simple equipment, easy operation, and no influence on the normal production, etc. In water drainage and gas recovery with acceleration string, the coiled tubing of small diameter is suspended at the wellhead as production string to increase the gas flow rate so as to enhance the liquid carrying capacity, and drain the water out of the wellbore with the energy of the well itself. It is suitable for the liquid loading wells with production of more than $0.3 \times 10^4 m^3/d$, and has the advantage of one-time construction and no afterward maintenance needed. In water drainage gas recovery with plunger gas lift, the plunger is used as the mechanical interface between gas and liquid, and the energy of the well is used to push the plunger to carry out liquid periodically in the tubing, this way can effectively prevent the gas breakthrough and the liquid fall-back. It is suitable for the wells with gas production of more than $0.15 \times 10^4 m^3/d$, and has advantages of high degree automation, safety and environmental friendliness, etc. Compressor gas lift drains the water in the well with the pressure energy of the gas, during the lifting the natural gas produced from the tubing is injected into the gas well continuously by the compressor along the annulus between tubing and casing, and the injected natural gas is then taken up from the tubing, sent through the seperator and then is pressed into the wellbore again by the compressor, by such circulation the liquid loading is drained out. High pressure nitrogen gas lift is to inject high-pressure nitrogen from the tubing (or casing), to discharge the liquid loading through the casing (or oil pipe) to restore the gas well production.

5.5 Reduction of abandonment production rate

Abandonment production rate is an important economic and technical index for gas field development and the main basis for evaluating the final recovery of gas fields[31]. The determination of the abandonment output depends on the gas price and cost. Tight gas wells often enter into production decline shortly after being put into production, with output decreasing constantly. Finally, with comprohensive consideration of the pressure system matching of the formation, wellbore and export pipeline, the well is produced at fixed pressure at higher decline rate, until the annual cash inflow of the well is equal to the cash outflow, then the well is to be abandoned, and the gas output at this point is the abandonment production rate. The final abandonment production has strong impact on the recovery of the gas well and gas field. If the abandonment production rate of Sulige gas field reduces from $0.14\times10^4 m^3/d$ to $0.10\times10^4 m^3/d$, the cumulative gas production per well would increase by $150\times10^4 m^3$, the recovery increase by about 2%. At present, wellbore water drainage and wellhead pressurization are used to reduce the abandonment pressure and the abandonment production rate, so as to realize the goal of increasing the final cumulative production and recovery of gas wells.

6 Conclusions

Three major factors cause lower technical ultimate recovery of tight gas reservoirs (60%~70%) than conventional gas reservoirs (80%~90%), namely, serious heterogeneity, poor percolation capacity, and existence of gas-water two-phase flow. Based on the analysis of reservoir geology and gas reservoir development performance, the remaining reserves of the Sulige tight gas field are divided into four types: uncontrolled by well pattern, missed by horizontal wells, incomplete perforation, and left in blocked zone of complex sand body. Among them, the reserves uncontrolled by well pattern and left in blocked zone of complex sand body are inter-well unproduced reserves, accounting for 82% of the total remaining reserves, and are the major targets for tapping remaining gas. Well pattern adjustment is the major means to improve the production rate and recovery of such reserves.

In order to improve gas field recovery and gas well development economic efficiency, two kinds of well pattern adjustment, infilling of vertical well pattern and combined well pattern of vertical and horizontal wells have been proposed. The relationship diagram of the final cumulative production, gas production increment of infilled well, and degree of inter-well interference with the change of well pattern density has been established. Four stages of recovery variation with well pattern density have been identified, and the corresponding well pattern densities determined. In the Sulige tight gas field, the rational well pattern density is 4 wells/km² under the present gas price of 1.1~1.5 CNY/m³, and when the gas price is 1.5~2.0 CNY/m³, the rational well pattern density is 5 wells/km². The combined well pattern is similar in recovery with the infilled vertical well pattern, but can save about 7% of development investment. In general, the two main technologies of infilled vertical well pattern and the combined vertical well-horizontal well pattern can enhance the recovery from the current 32% to about 50% in the enrichment area.

For the unproduced residual gas missed by the horizontal well and incomplete perforation, a series of matching technologies to improve recovery have been established, including tapping potentials of old wells, technology optimization for new wells, production system optimization, water drainage gas recovery, and reduction of abandonment production rate, etc., which can increase the recovery rate by another 5% on the basis of well pattern optimization.

References

[1] GUO Zhi, JIA Ailin, JI Guang, et al. Reserve classification and well pattern infilling method of tight sandstone gasfield: A case study of Sulige gasfield. Acta Petrolei Sinica, 2017, 38(11): 1299-1309.

[2] MA Xinhua, JIA Ailin, TAN Jian, et al. Tight sand gas development technologies and practices in China. Petroleum Exploration and Development, 2012, 39(5): 572-579.

[3] HE Dongbo, WANG Lijuan, JI Guang, et al. Well spacing optimization for Sulige tight sand gas field, NW China. Petroleum Exploration and Development, 2012, 39(4): 458-464.

[4] HE Wenxiang, WU Shenghe, TANG Yijiang, et al. The architecture analysis of the underground point bar: Taking Gudao oilfield as an example. Journal of Mineralogy and Petrology, 2005, 25(2): 81-86.

[5] HE Dongbo, JIA Ailin, JI Guang, et al. Well type and pattern optimization technology for large scale tight sand gas, Sulige gas field. Petroleum Exploration and Development, 2013, 40(1): 79-89.

[6] QIU Yinan, JIA Ailin. Development of geological reservoir modeling in past decade. Acta Petrolei Sinica, 2000, 21(4): 101-104.

[7] LIU Xingjun, ZHANG Ji, YOU Shimei. Logging interpretation of sedimentary facies controlled reservoir in central Sulige. Well Logging Technology, 2008, 32(3): 228-232.

[8] WANG Feng, TIAN Jingchun, CHEN Rong, et al. Analysis on controlling factors and characteristics of sandstone reservoir of He 8 (Upper Paleozoic) in the northern Ordos Basin. Acta Sedimentologica Sinica, 2009, 27(2): 238-245.

[9] LI Hong, LIU Yiqun. Reservoir geology modeling of lithological reservoir with low permeability in Xifeng oilfield, Ordos Basin. Acta Sedimentologica Sinica, 2007, 25(6): 954-960.

[10] JIA Ailin. Research achievements on reservoir geological modeling of China in the past two decades. Acta Petrolei Sinica, 2011, 32(1): 181-188.

[11] YE Tairan, ZHENG Rongcai, WEN Huaguo. Application of high resolution sequence stratigraphy to the sand reservoir prediction for 8th Member of lower Shihezi Formation in Sulige gas field, Ordos Basin. Acta Sedimentologica Sinica, 2006, 24(2): 259-266.

[12] HU Xianli, XUE Dongjian. An application of sequential simulation to reservoir modeling. Journal of Chengdu University of Technology (Science & Technology Edition), 2007, 34(6): 609-613.

[13] JIA Ailin, WANG Guoting, MENG Dewei, et al. Well pattern infilling strategy to enhance oil recovery of giant low-permeability tight gasfield: A case study of Sulige gasfield, Ordos Basin. Acta Petrolei Sinica, 2018, 39(7): 802-813.

[14] WU Fan, SUN Lijuan, QIAO Guoan, et al. A research on gas flow property and starting pressure phenomenon. Natural Gas Industry, 2001, 21(1): 82-84.

[15] GUO Jianlin, GUO Zhi, CUI Yongping, et al. Recovery factor calculation method of giant tight sandstone gas field. Acta Petrolei Sinica, 2018, 39(12): 1389-1396.

[16] ZHANG Jinwu, WANG Guoyong, HE Kai, et al. Practice and understanding of sidetracking horizontal drilling technology in old wells in Sulige Gas Field. Petroleum Exploration and Development, 2019, 46(2): 370-377.

[17] LIU Naizhen, ZHANG Zhaopeng, ZOU Yushi, et al. Propagation law of hydraulic fractures during multi-staged horizontal well fracturing in a tight reservoir. Petroleum Exploration and Development, 2018, 45(6): 1059-1068.

[18] LI Jinbu, BAI Jianwen, ZHU Li'an. Volume fracturing and its practices in Sulige tight sandstone gas reservoirs, Ordos Basin. Natural Gas Industry, 2013, 33(9): 65-69.

[19] DAI Qiang, DUAN Yonggang, CHEN Wei, et al. Present state of low permeability reservoir percolation study. Special Oil and Gas Reservoirs, 2007, 14(1): 12-14.

[20] YANG Jian, KANG Yili, LI Qiangui, et al. Characters of micro-structure and percolation in tight sandstone gas reservoirs. Advances in Mechanics, 2008, 38(2): 229-234.

[21] LI Qi, GAO Shusheng, YE Liyou, et al. The research of percolation mechanism and technical countermeasures of tight sandstones gas reservoirs. Science Technology and Engineering, 2014, 14(34): 79-84.

[22] ZHOU Keming, LI Ning, ZHANG Qingxiu, et al. Experimental research on gas-water two phase flow and confined gas formation mechanism. Natural Gas Industry, 2002, 22(S1): 122-125.

[23] GAO Shusheng, YE Liyou, XIONG Wei, et al. Influence of the threshold pressure gradient on tight sandstone gas reservoir recovery. Natural Gas Geoscience, 2014, 25(9): 1444-1449.

[24] GAO Shusheng, YE Liyou, XIONG Wei, et al. Seepage mechanism and strategy for large and low permeability and tight sandstone gas reservoirs with water content. Journal of Oil and Gas Technology, 2013, 35(7): 93-99.

[25] YE Liyou, GAO Shusheng, YANG Hongzhi, et al. Water production mechanism and development strategy of tight sandstone gas reservoirs. Natural Gas Industry, 2015, 35(2): 41-46.

[26] LI Yingchuan, LI Kezhi, WANG Zhibin, et al. Study on dynamic production optimization for water produced gas wells in Daniudi low permeability gas reservoirs. Oil Drilling & Production Technology, 2013, 35(2): 71-74.

[27] MENG Dewei, JIA Ailin, JI Guang, et al. Water and gas distribution and its controlling factors for large scale tight sand gas fields: A case study of western Sulige gas field, Ordos Basin, NW China. Petroleum Exploration and Development, 2016, 43(4): 607-614.

[28] YU Shuming, TIAN Jianfeng. Study and application of drainage gas recovery technology in Sulige gas field. Drilling & Production Technology, 2012, 35(3): 40-43.

[29] ZHANG Chun, JIN Daquan, LI Shuanghui, et al. Progress and measurement of drainage gas recovery technology in Sulige gas field. Natural Gas Exploration and Development, 2016, 39(4): 48-52.

[30] ZHU Xun, ZHANG Yabin, FENG Pengxin, et al. Study and application of digitized drainage gas recovery system in Sulige gas field. Drilling & Production Technology, 2014, 37(1): 47-49.

[31] MAO Meili, LI Yuegang, WANG Hong, et al. Prediction of gas well abandonment production in the Sulige gas field. Natural Gas Industry, 2010, 30(4): 64-66.

原文刊于《Petroleum Exploration and Development》,2019,46(3):629-641.

低渗透—致密气藏开发动态物理模拟实验相似准则

焦春艳[1,2,3]　刘华勋[1,2]　刘鹏飞[4]　宫红方[5]

(1. 中国石油勘探开发研究院,北京 100083；
2. 中国石油集团科学技术研究院有限公司,北京 100083；
3. 中国石油天然气集团公司天然气成藏与开发重点实验室,廊坊 065000；
4. 中国石油长庆油田分公司第二采气厂,榆林 719000；
5. 中石化胜利油田石油开发中心有限公司,东营 257100)

摘要:低渗致密气藏地质条件复杂,普遍含有孔隙水,渗流规律不清,开发动态在早期难以预测。根据低渗致密气藏地质与开发特征,筛选出影响气藏开发动态的主要影响因素,并依据 π 定理确定了气藏开发物理模拟实验相似准数,阐述了其物理意义;以鄂尔多斯盆地低渗气藏 X 井为例,根据相似性准则,开展了气藏开发物理模拟实验研究,并与 X 井生产动态进行对比。结果表明:应用所建相似准则进行气藏开发物理模拟可以较好地预测气藏开发动态,得出的相似准则合理;本文得出的相似准则中:关键相似准数为动力相似、运动相似和含水饱和度相似。研究成果对于低渗致密气藏有效开发具有重要的理论指导意义。

关键词:低渗致密气藏；物理模拟；π 定理；相似准数；相似准则

物理模拟实验技术在油气田开发领域被广泛应用,尤其是低渗致密气藏,由于储层基质十分致密,非均质性强,普遍含水,渗流规律复杂,难以通过数值模拟和气藏工程方法进行气藏开发动态预测,多借助于物理模拟方法。模拟实验的操作是以相似理论为基础,当同一类物理现象的单值条件相似,并且对应的相似准则(由单值条件中的物理量组成)相等时,这些现象必定相似,这是判断两个物理现象是否相似的充分必要条件[1]。但是如何确定物理模拟中能反映矿场实际情况的各项参数,以及模拟结果如何在矿场应用,是目前所面临的重要问题。换句话说,需要建立物模实验参数与矿场参数的有效换算关系。目前,关于低渗气藏物理模拟的研究很多[2-15],但是针对低渗气藏物理模拟相似准则方面的研究则较少。因此,以低渗致密气藏为研究对象,首先根据气藏压裂后流动特征分析、气藏工程方法和相似性理论,确定气藏动态相似准数;然后根据相似准则,建立模型参数与原型参数之间换算关系及气藏开发动态物理模拟方法;最后选择鄂尔多斯盆地低渗致密岩样,开展了不同条件下气藏开发动态物理模拟实验,验证相似准则的合理性,并预测了气藏开发动态。

1 低渗致密气藏开发相似性实验相似准数

1.1 相似准则建立

低渗致密气藏储层致密,渗透率小于 1mD,其中致密储层渗透率小于 0.1mD,储层流动性差,自然产能低,普遍采取压裂增产措施后再投产[12],裂缝半长50~100m,裂缝面为主要泄流面,储层流动以垂直裂缝的直线流为主,因此,当以规则的矩形井网开发时,低渗致密气藏储层渗流问题可简化为若干个一维直线渗流问题,直线渗流宽度为b(排距),厚度为h,均匀布井时渗流长度压裂直井为1/2的井距或裂缝到流动单元边界距离,压裂水平井为1/2n的井距或1/2的裂缝间距,n为裂缝条数,压裂直井对应的n为1,井距等于裂缝间距,故在下文统一称为裂缝间距,图1为压裂直井近似流场示意图,压裂水平井可简化n个如图1所示流动单元。

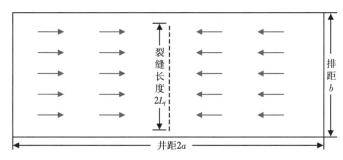

图1 压裂直井近似流场示意图

假定储层为均质储层,以图1中1/2的直线流为研究对象,根据渗流力学理论和油气藏工程理论,气藏气井井底压力(对应着物模岩心出口压力)P_w主要影响因素为:

(1)渗透率K,量纲为$[L^2]$;
(2)孔隙度ϕ,量纲为1;
(3)含气饱和度S_g,无量纲;
(4)渗流长度a(物模实验为岩心长度,矿场压裂直井为气藏边界距裂缝距离,压裂水平井为相邻两条裂缝间距),量纲为$[L]$;
(5)渗流面宽度b,量纲为$[L]$;
(6)厚度h,量纲为$[L]$,厚度与渗流面宽度乘积对应着物模岩心渗流截面面积;
(7)原始地层压力p_i,量纲为$[M/T^2/L]$;
(8)产气速度q,量纲为$[L^3/T]$,需要注意的此处流量为气井流量的1/2n,其中,n为裂缝条数,压裂直井$n=1$;
(9)时间t,量纲为$[T]$;
(10)气体压缩因子z,无量纲;
(11)储层温度T,量纲$[K]$;
(12)标准温度T_{sc},量纲为$[K]$;
(13)标准大气压,P_{sc},量纲为$[M/T^2/L]$。

以上共13个自变量(当选用模拟气藏储层岩心时,气相相对渗透率K_{rg}只是含水饱和度S_w的函数,非独立变量,也不予考虑),加上因变量P_w,共14个变量。

存在四个基本量纲,分别为长度量纲$[L]$、质量量纲$[M]$、时间量纲$[T]$和温度量纲$[K]$,根据相似理论,有10个相似准数,任何一个相似准数π表达式为:

$$\pi = K^{x_1}\phi^{x_2}S_g^{x_3}a^{x_4}b^{x_5}h^{x_6}P_i^{x_7}q^{x_8}t^{x_9}z^{x_{10}}T^{x_{11}}T_{sc}^{x_{12}}P_{sc}^{x_{13}}P_w^{x_{14}} \quad (1)$$

根据齐次原理,对应的线性方程组如下:

长度量纲为1:

$$2x_1 + x_4 + x_5 + x_6 - x_7 + 3x_8 - x_{13} - x_{14} = 0 \quad (2)$$

质量量纲为1:

$$x_7 + x_{13} + x_{14} = 0 \quad (3)$$

时间量纲为1:

$$-2x_7 - x_8 + x_9 - 2x_{13} - x_{14} = 0 \quad (4)$$

温度量纲为1:

$$x_{11} + x_{12} = 0 \quad (5)$$

式(2)—式(5)分别为长度量纲、质量量纲、时间量纲和温度量纲的齐次方程,对应的方程组为齐次线性方程组,根据矩阵论,有10个基础解系,即存在10个独立的相似准数,解方程得低渗致密气藏开发相似性物理模拟实验相似准数,见表1。

表1 气藏物理模拟相似准数

序号	相似准数	相似属性	用途	物模取值	矿场取值
1	$\pi_1 = \phi$	孔隙度相似	确定模型孔隙度	0.02~0.2	0.02~0.2
2	$\pi_2 = S_g$	含气饱和度相似	确定模型饱和度	0.4~0.8	0.4~0.8
3	$\pi_3 = z$	气体压缩性相似	确定模型气体	0.9~1.2	0.9~1.2
4	$\pi_4 = T/T_{sc}$	温度相似	确定模型温度	1~1.1	1.1~1.3
5	$\pi_5 = b/h$	几何相似	确定模型尺寸	1	10~50
6	$\pi_7 = b/a$	几何相似	确定模型尺寸	0.3~1	0.3~1
7	$\pi_7 = P_{sc}/P_i$	动力相似	确定模型原始地层压力	0.002~0.01	0.002~0.005
8	$\pi_8 = P_w/P_i$	动力相似	建立井底压力换算关系	0~1.0	0.1~1.0
9	$\pi_9 = \dfrac{q}{\dfrac{bhKK_{rg}T_{sc}P_i^2}{a\mu ZTP_{sc}}}$	运动相似	确定模型采气速度	0~0.5	0.1~0.3
10	$\pi_{10} = \dfrac{qt}{\dfrac{abh\phi S_g T_{sc}P_i}{zTP_{sc}}}$	采出程度相似	建立时间换算关系	0~1.0	0~0.95

1.2 相似准数物理意义

由相似准则表可以看出:根据定义,相似准数 π_1 为孔隙度相似,取模拟气藏主力层位岩心可实现物模模型与气藏原型相似准数一致。

相似准数 π_2 为含气饱和度相似,通过物模岩心抽真空饱和地层水,再气驱水实现物模岩心含气饱和度与矿场气藏储层含气饱和度一致,由于含气饱和度对低渗致密气藏储层气体渗流及开发影响显著,因此,低渗致密气藏开发相似性物理模拟实验中应做到含气饱和度一致,

即相似准数 π_2 一致。

相似准数 π_3 为气体偏差因子，即气体偏离理想气体程度，根据油层物理，在一定温压条件下，氮气和地层天然气偏差因子在 1 附近变化，选择 N_2 基本可实现模型与原型相似准数 π_3 一致，避免使用易燃易爆的天然气，提高气藏开发相似性物理模拟实验安全性。

相似准数 π_4 为温度与标准温度比值，相似性物理模拟实验温度一般在 50℃（323.15K）左右，气藏储层温度一般为 100℃（373.15K），物模模型与气藏原型储层温度差 50K，相对于分母标准温度 T_{sc} = 293.15K 来说相对较小，模型与原型相似准数 π_4 也基本一致。

相似准数 π_5 为渗流面宽度 b 与厚度 h 比值，根据压裂后流场分析（图 1），低渗致密气藏渗流面宽度 b 为排距，一般在 500m 左右，厚度一般在 10m 左右，相应的气藏原型相似准数 π_5 取值在 50 左右，而实验多采用柱状岩心，渗流截面为圆形，宽度与厚度比为 1，模型相似准数 π_5 为 1，模型相似准数与原型差异较大，但相似准数 π_5 主要反映垂直流动方向的几何相似，主要用于气藏三维空间流动规律研究，而对低渗致密气藏近似一维流动影响较小，相似性可以放宽要求。

相似准数 π_6 为渗流面宽度 b 与渗流长度 a 比值，反映的是流动平面的几何相似性，气藏原型渗流面宽度为排距，渗流长度为 1/2 井距，根据低渗致密气藏开发实践，气藏排距与井距 2 比值一般在 1:2，相似准数 π_6 取值在 1 左右，岩心模型泄流面宽度为岩心直径、渗流长度为岩心长度。因此，根据相似性准则，岩心长度与岩心直径相当即可，较为容易实现。

相似准数 π_7 和 π_8 为动力相似，其中 π_7 为标准压力与原始压力之比，反映气体被压缩程度，由于气藏原型和物模模型标准压力 P_{sc} 都一样，均为 0.101MPa，因此，根据相似准数 π_7，要求物模实验原始地层压力与气藏原始地层压力一致或基本一致，而一般低渗致密砂岩气藏原始压力 30MPa 左右，实验室能满足要求，实现相似准数 π_7 一致；相似准数 π_8 为井底压力 P_w 与原始地层压力 P_i 的比值，气藏生产过程中井底压力一般介于 $0.10 \sim 1.0 P_i$，相应的相似准数 π_8 为 $0.1 \sim 1.0$；物模实验井底压力（出口压力）介于 $P_{sc} \sim P_i$，相应的物模模型相似准数 π_8 介于 $0 \sim 1.0$，物模模型与气藏原型相似准数 π_7、π_8 也基本一致。另外，动力相似准数 π_8 也建立了物模岩心出口压力与气藏井底压力换算关系，为低渗致密气藏开发相似性物模实验关键相似准数，而且，通过相似准数 π_8、气藏原始地层压力和废弃井底压力以及物模实验原始地层压力可确定物模实验出口废弃压力。

当储层渗流为达西渗流时，相似准数 π_9 表达式分母为气井无阻流量表达式，即 π_9 为产气速度与无阻流量比值，这与依据无阻流量 $1/3 \sim 1/6$ 配产观点一致，反映的是运动相似，后续实验也将证明相似准数 π_9 是相似性实验一个重要的相似准数。根据矿场生产统计，原型取值一般为 $1/3 \sim 1/10$，低渗致密岩心由于渗透率低，无阻流量相对较小，流压 30MPa 时全直径岩心无阻流量 10000mL/min 左右，物模实验流量 $1000 \sim 3000$mL/min 即可满足模型与原型相似准数一致，实验上可行。

相似准数 π_{10} 分母为天然气地质储量，分子为累计采气量，即相似准数 π_{10} 为累计采气量与储量比值，反映气藏采出程度，根据统计，低渗致密气藏原型相似准数 π_{10} 介于 $0 \sim 0.6$，物模模型相似准数 π_{10} 介于 $0 \sim 0.95$，模型可做到与原型相似准数一致。

从表 1 中矿场原型相似准数和物模模型相似准数对比可以看出，两者基本一致，即低渗致密气藏开发动态物理模拟实验基本可以满足相似准则。

2 低渗致密气藏开发实验方法

根据低渗致密气藏开发动态相似性物理模拟实验相似准数,建立相应的相似性物理模拟实验流程。

(1)根据气藏开发与地质参数和表1中相似准数计算公式,计算气藏原型上述10个相似准数;

(2)根据气藏原型相似准数 π_1、π_2、π_4、π_5、π_6、π_7、π_8 确定物模全直径岩心长度、孔隙度、含气饱和度、初始饱和流体压力 P_i、温度 T 和出口压力取值范围;

(3)根据物模实验温度和初始饱和流体压力 P_i 计算实验气体偏差因子 Z、物模相似准数 π_3 和黏度 μ;

(4)根据气藏原型相似准数 π_9、π_{10} 和步骤(2)确定的物模实验参数,计算确定物模实验流量 q_m 和物模实验产气时间 t_m;

(5)将物模实验全直径岩心放置如图2所示物模实验装置的全直径岩心加持器中,并按照示意图连接好储气罐、压力传感器、质量控制流量计和阀门,并依次加围压、打开入口阀门、关闭出口阀门,注气直至初始饱和流体压力达到 P_i;再关闭入口阀门,打开出口阀门,以流量 q_m 恒定生产,记录岩心两端压力,直至生产至废弃时。

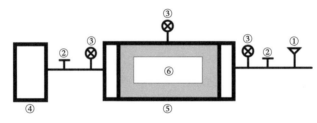

图2 低渗致密气藏开发动态物模实验装置示意图
①质量控制流量计;②阀门;③压力传感器;④储气罐;⑤岩心夹持器;⑥物模岩心

物模实验可获取物模岩心入口压力 P_1、出口压力 P_2、累计产气量和采出程度等关键开发动态数据,其中,物模岩心出口压力 P_2 对应着气藏井底压力 P_w,物模岩心入口压力 P_1 对应着气藏边界压力 P_e,因此,可根据物模实验结果和相似准数相似性换算获取矿场气井生产动态数据,其中根据相似准数 π_9 和物模岩心流量 q_m 计算矿场气井日产气量 q,公式为:

$$q = \frac{bhK_{rg}KT_{sc}P_i^2}{a\mu ZTP_{sc}} \left(\frac{a\mu ZTP_{sc}}{bhK_{rg}KT_{sc}P_i^2} q \right)_m \tag{6}$$

式中 m——表示括号里面参数为物模模型参数。

根据相似准数 π_{10} 和物模实验产气时间 t_m 计算矿场生产时间 t:

$$t = \frac{abh\phi S_g T_{sc} P_i}{qzTP_{sc}} \left(\frac{qzTP_{sc}}{abh\phi S_g T_{sc} P_i} t \right)_m \tag{7}$$

根据相似准数 π_8 和物模岩心出口压力 P_2 计算矿场气井井底压力 P_w:

$$P_w = P_i \left(\frac{P_w}{P_i} \right)_m \tag{8}$$

3 低渗致密气藏开发实验及应用

为了验证低渗致密气藏开发相似准则的合理性，以鄂尔多斯盆地低渗致密砂岩气藏某压裂直井(X井)为模拟对象,储层原始地层压力30MPa,厚度11.2m,含水饱和度为40%,孔隙度8%,根据试井解释成果,储层渗透率0.2mD,裂缝半长125m,无阻流量$18\times10^4 m^3/d$;产气量由初始$4\times10^4 m^3/d$降到$2\times10^4 m^3/d$,少量产水,生产比较平稳,累计生产3879天,累计采气$8769\times10^4 m^3$,平均日产气量$2.26\times10^4 m^3$,图3为X井生产动态曲线。

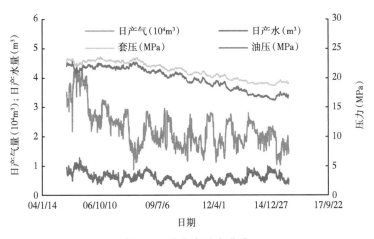

图3 X井生产动态曲线

选取X井主力储层全直径岩心开展相似性物理模拟实验,岩心覆压渗透率0.18mD,孔隙度7.8%,岩心长度15cm,实验前先采用气驱水的方法建立40%含水饱和度,物模岩心孔渗饱基本与储层平均孔渗饱相当;实验初始流压30MPa,与气藏原始地层压力一致,饱和进气量15798mL,无阻流量7600mL/min,物模实验流量400mL/min、954mL/min、1500mL/min、2000mL/min、2500mL/min,共五组;根据相似准数π_9和X井储层物性参数计算,物模实验流量对应着矿场日产气量$0.9\times10^4 \sim 5.9\times10^4 m^3$,其中,X井平均日产气量$2.26\times10^4 m^3$,对应着物模实验流量954mL/min。

图4为物模实验流量954mL/min实验曲线及相似性转换曲线,可以看出:依据物理模拟实验结果与相似准则计算得到的生产动态曲线与实际气井生产动态曲线基本

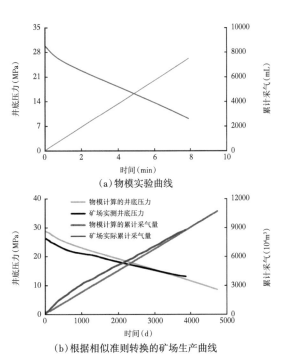

(a) 物模实验曲线

(b) 根据相似准则转换的矿场生产曲线

图4 X井物模实验及根据相似准则转换结果

吻合,早期预测的井底压力略微偏高,这是由于 X 井早期日产气量高于平均日产气量(约为平均日产气量的两倍),整体还是比较吻合,按当前采气速度继续生产的话,井底压力降到原始地层压力 1/3 时还可采出 $1934×10^4 m^3$,因此,本文中相似准则可靠,可依据本相似准则进行物理模拟实验,进而进行气藏(井)开发动态预测。

通过对比不同物模流量相似性转换曲线还可用于分析采气速度对气井开发的影响,确定合理采气速度,图 5 为 X 井不同流量时物模实验曲线及根据相似准则转换的矿场曲线,可以看出:不同采气速度下物模模型与气藏原型压降曲线差异较大,具体表现为采气速度越高,X 井井底压力速率越大,相同井底压降条件下采出程度越低,以井底压力降到原始地层压力 1/3 为例,气井日产气 $0.9×10^4 \sim 5.9×10^4 m^3$ 对应的采出程度为 30.3%～61.5%,相差一倍,采气速度对 X 井生产影响明显,从另一侧面反映运动相似准数 π_9 是相似准则中的关键参数,是建立物模实验流量与矿场气井日产气量关系的关键参数;当前采气速度下井底压降相对平缓,采气速度比较合理,可取得一个相对较高的采出程度,约为 50% 左右。

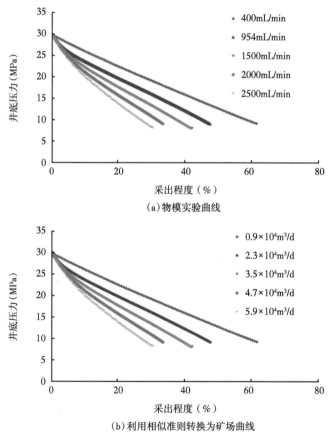

图 5 X 井不同流量物模实验及利用相似准则转换结果

4 小结

(1)气藏主要利用气体膨胀能量开发,油藏多依靠外界补充能量开发,开发方式不同是导致气藏开发相似准数与油藏开发相似准数不同的根本原因,气藏开发主要相似准数为动力相似性(π_8),含水饱和度相似(π_2)和运动相似(π_9)。

(2)根据相似性理论,建立了低渗致密气藏开发物理模拟的相似准则,利用该准则可进行物模实验流量、压力和时间与矿场气井日产气量、压力和生产时间的相互换算,实现了物模实验结果到矿场生产的直接应用,利用相似准则得到的转化曲线可用于指导气井生产,确定气井合理产能和采出程度等关键开发指标。

(3)本文针对均质程度好的低渗致密气藏建立了相似性物理模拟技术,也没有考虑井筒流动对气井生产的影响,而实际很大一部分低渗致密气藏储层非均质性强,气井普遍产水,井筒流动对气井生产影响大,尤其低产井,井筒积液易造成停产或间歇性生产,气藏开发物理模拟相似准则有待于进一步完善。

参 考 文 献

[1] 徐挺. 相似理论与模型实验[M]. 北京:中国农业机械出版社,1982.
[2] 胡勇,李熙喆,李跃刚,等. 低渗致密砂岩气藏提高采收率实验研究[J]. 天然气地球科学,2015,26(11):2142-2148.
[3] 胡勇,郭长敏,徐轩,等. 砂岩气藏岩石孔喉结构及渗流特征[J]. 石油实验地质,2015,37(3):390-393.
[4] 胡勇,李熙喆,卢祥国,等. 砂岩气藏衰竭开采过程中含水饱和度变化规律[J]. 石油勘探与开发,2014,41(6):723-726.
[5] 胡勇,李熙喆,卢祥国,等. 高含水致密砂岩气藏储层与水作用机理[J]. 天然气地球科学,2014,25(7):1072-1076.
[6] 胡勇,李熙喆,万玉金,等. 致密砂岩气渗流特征物理模拟[J]. 石油勘探与开发,2013,40(5):580-584.
[7] 朱华银,朱维耀,罗瑞兰. 低渗透气藏开发机理研究进展[J]. 天然气工业,2010,30(11):44-47,118-119.
[8] 朱华银,胡勇,朱维耀,等. 气藏开发动态物理模拟技术[J]. 石油钻采工艺,2010,32(S1):54-57.
[9] 朱华银,徐轩,安来志,等. 致密气藏孔隙水赋存状态与流动性实验[J]. 石油学报,2016,37(2):230-236.
[10] 朱华银,徐轩,高岩,等. 致密砂岩孔隙内水的赋存特征及其对气体渗流的影响——以松辽盆地长岭气田登娄库组气藏为例[J]. 天然气工业,2014,34(10):54-58.
[11] 朱华银,付大其,卓兴家,等. 低渗气藏特殊渗流机理实验研究[J]. 天然气勘探与开发,2009,32(3):39-41+74.
[12] 刘华勋,任东,高树生,等. 边、底水气藏水侵机理与开发对策[J]. 天然气工业,2015,35(2):47-53.
[13] 叶礼友,高树生,杨洪志,等. 致密砂岩气藏产水机理与开发对策[J]. 天然气工业,2015,35(2):41-46.
[14] 高树生,叶礼友,熊伟,等. 大型低渗致密含水气藏渗流机理及开发对策[J]. 石油天然气学报,2013,35(7):93-99.
[15] 李熙喆,万玉金,陆家亮,等. 复杂气藏开发技术[M]. 北京:石油工业出版社,2010.

原文刊于《大庆石油地质与开发》,2019,38(1):155-161.

神木气田低渗致密储层特征与水平井开发评价

王国亭[1]　孙建伟[2]　黄锦袖[2]　韩江晨[1]　朱玉杰[2]

(1. 中国石油勘探开发研究院,北京 100083；2. 长庆油田分公司第二采气厂,榆林 719000)

摘要：鄂尔多斯盆地东部神木气田逐渐成为长庆气区增储上产的重要组成部分,系统研究气田主力层位的储层特征、有效储层空间叠置类型对气田的高效开发有重要意义。气田山西、太原组储层岩石类型主要为岩屑石英砂岩、岩屑砂岩及石英砂岩,孔隙类型以溶蚀孔、晶间孔及粒间孔为主。储层孔隙度分布于2%～10%,平均为6.6%,渗透率分布于0.1～1.0mD,平均为0.83mD,评价结果表明神木气田整体属低孔隙度、低渗透率致密砂岩气藏。明确了孔隙度5%、渗透率0.1mD、含气饱和度45%为有效储层物性下限标准。将有效砂体空间结构类型划分为多层孤立分散型、垂向多期叠加型、侧向多期叠置型三种。建立了神木气田水平井开发地质目标优选标准,提出局部式水平井开发方式,并建立孤立式、复合式及丛组式三种水平井部署类型。

关键词：鄂尔多斯盆地；神木气田；储层特征；物性下限；有效储层结构；水平井开发

鄂尔多斯盆地低渗透率、致密砂岩气资源丰富,先后发现了榆林、子洲、苏里格、大牛地等多个探明储量超千亿方的气田,目前探明(含基本探明)储量超过 $5 \times 10^{12} m^3$。近年来,盆地东部神木地区逐渐成为长庆气区增储上产的重要组成部分。与盆地中部苏里格气田相比,东部气田储层发育特征、主力含气层系等都有明显差异,开展神木气田主力层位储层特征研究,确定有效储层物性下限、规模及叠置类型,对盆地东部地区储量评价、开发方式确定等具有重要意义,也可为国内外其他类似气田的开发提供借鉴。此外,水平井已成为长庆气区低渗透率、致密气藏开发的重要手段,开展盆地东部水平井适用性评价,明确合理的水平井开发方式,对神木气田的高效开发具有重要意义。

1 气田概况

神木气田位于陕西省榆林市榆阳区和神木市境内,西邻榆林气田、南抵米脂气田、西北接大牛地气田,勘探开发面积约 $2.5 \times 10^4 km^2$（图1）。构造位置处于鄂尔多斯盆地次级构造单元伊陕斜坡东北部,为宽缓西倾斜坡,倾角小于1°。神木地区内上古生界石炭、二叠系主要发育一套海陆交互相含煤地层,自下而上为本溪组($C_2 b$)、太原组($P_1 t$)、山西组($P_1 s$)、石盒子组($P_1 sh$)及石千峰组($P_3 q$),2007年双3井区的勘探突破标志着神木气田的发现[1]。截至目前,气田探明储量规模达 $3334 \times 10^8 m^3$,叠合含气面积 $4069 km^2$,属千亿级特大型气田。气田纵向发育太原组、山西组、石盒子组等多套含气层系,其中山西组、太原组为神木地区主力产层,平均产量贡献率分别为53%、33%,是研究的重点目标(图2)。气田于2009年开展前期评价,历经试采、规模建产后,2014年底井口建成 $20 \times 10^8 m^3/a$ 产能,初步实现规模开发。

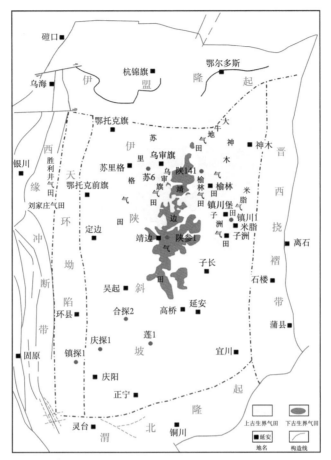

图 1 鄂尔多斯盆地神木气田位置图

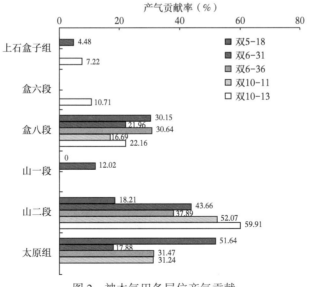

图 2 神木气田各层位产气贡献

2 储层地质特征

晚古生代受沉积物供给、构造沉降及海平面升降等多因素影响,鄂尔多斯盆地发育多种沉积体系,经历了由海相潟湖、潮坪、三角洲沉积体系到陆相河流、三角洲体系的演变,形成了大面积分布的河流—三角洲储集砂体[1-4]。神木地区太原期主要发育海相潮控三角洲沉积,山二段沉积期演变为海相河控三角洲沉积,山一段沉积期海水退出鄂尔多斯盆地,研究区以湖相三角洲沉积为主。

2.1 储层基本特征

2.1.1 岩石学特征

神木地区太原、山西组储层岩性主要为岩屑石英砂岩、岩屑砂岩及石英砂岩(图3)。碎屑颗粒总量平均为83.0%,以石英颗粒为主、平均为64.4%,岩屑次之、平均为18.4%,长石较少、平均为0.3%,从山一段到太原组石英含量逐渐增大,岩屑、长石含量依次降低(图4)。岩屑类型包括岩浆岩岩屑、变质岩岩屑及沉积岩岩屑,其中变质岩岩屑含量最高,平均为10.7%,

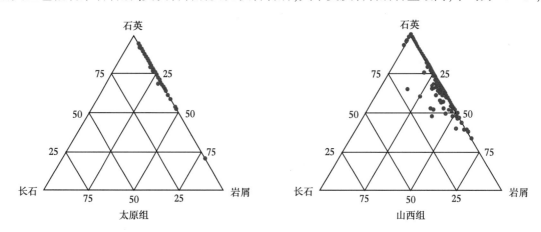

图3 神木气田太原组—山西组储层岩石分类

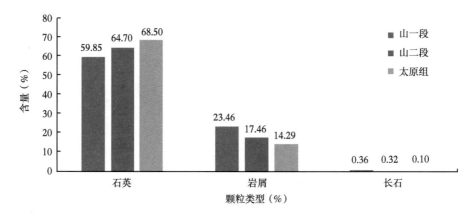

图4 神木气田太原组—山西组储层碎屑颗粒特征

以片岩、千枚岩、变质砂岩及板岩为主,岩浆岩岩屑含量次之,平均为5.5%,以喷发岩和隐晶岩为主,沉积岩岩屑含量最低,平均为0.6%,以粉砂岩和泥岩为主。填隙物总量平均为17.0%,以水云母为主,其次为硅质、高岭石、铁方解石及铁白云石。碎屑颗粒以中粒、粗粒结构为主,分选中等,磨圆以次棱角状—次圆状为主,胶结方式以孔隙式为主。太原组砂岩颗粒接触方式以点接触、点—线接触为主,山西组以点接触为主。

2.1.2 孔隙结构特征

薄片鉴定及扫描电镜分析表明,神木气田山西组、太原组储层孔隙类型由溶蚀孔、晶间孔及粒间孔组成,其中溶蚀孔包括岩屑溶蚀孔、长石溶蚀孔、杂基溶蚀孔及粒间溶孔,并以岩屑溶蚀孔为主体,晶间孔以高岭石、伊利石等黏土矿物晶间孔为主。不同层位孔隙类型不同,其中山一段以岩屑溶孔、晶间孔为主,分别占总孔隙比例的62%、26%,山二段以岩屑溶孔、晶间孔、粒间孔为主,分别占总孔隙比例的58%、15%、8%,太原组则以岩屑溶孔为主,占总孔隙比例的84%,其他类型孔隙较少。从山一段到太原组岩屑溶孔的比例逐渐增加、晶间孔比例逐渐降低。压汞实验分析表明,神木气田山西组、太原组储层中值半径为0.15μm,分选系数为2.13,变异系数为0.23,排驱压力为1.17MPa,最大进汞饱和度为66.38%(表1)。总体而言,神木气田山西组、太原组储层孔隙中值半径较小,最大进汞饱和度较低,无效孔隙占比相对较高。

表1 神木气田碎屑岩储层孔隙结构参数表

层位	孔隙度(%)	渗透率(mD)	中值半径(μm)	分选系数	变异系数	排驱压力(MPa)	最大汞饱和度(%)
山一段	6.77	0.74	0.07	2.44	0.29	0.97	60.60
山二段	6.65	0.68	0.13	1.94	0.20	1.69	63.01
太原组	8.10	0.55	0.24	2.01	0.20	0.87	75.52
平均	7.17	0.65	0.15	2.13	0.23	1.17	66.38

2.1.3 有效储层物性下限评价

神木地区山西组、太原组储层物性统计分析表明,孔隙度分布于2%~10%,平均为6.6%,渗透率分布于0.1~1.0mD,平均为0.83mD。依据砂岩储层划分标准,神木气田整体属于低孔隙度、低渗透率致密砂岩气藏。根据产气效果的差异,将储层划分为气层、差气层及干层,气层物性及含气性最好、单位厚度产气量高,差气层次之,干层则不具备产气能力。气层、差气层统称为有效储层,是气井产量的主要贡献者,其对应的孔隙度、渗透率、含气饱和度最低界限值为有效储层物性下限。明确有效储层物性下限对于储层类型划分、储量评价具有重要意义。结合神木气田储层物性、含气性数据和试气效果资料,开展有效储层物性下限评价。统计分析表明,当储层孔隙度低于5%、渗透率小于0.1mD、含气饱和度低于45%时,储层基本已不具备产气能力,已属无效干层。故将孔隙度5%、渗透率0.1mD、含气饱和度45%确定为有效储层物性下限(图5)。

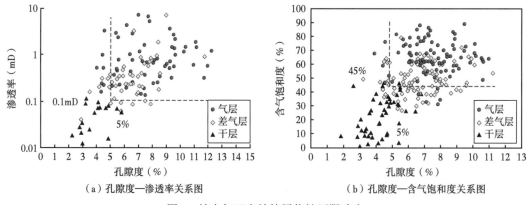

(a)孔隙度—渗透率关系图　　　　　(b)孔隙度—含气饱和度关系图

图 5　神木气田有效储层物性下限确定

2.2　有效储层规模

鄂尔多斯盆地苏里格气田开发不断深入,井网加密试验、井间干扰试验及野外露头描述等动静态资料逐渐丰富,为有效储层规模精细描述创造了条件,支撑了气田合理开发技术对策的制定[5-6]。神木气田处于早期开发阶段,储层地质条件与盆地中部地区具有较大差异,开展神木地区有效储层规模分析并建立空间发育模式,可为气田合理开发方式的确定奠定基础。

2.2.1　有效储层规模

有效储层规模分析包括厚度、宽度、长度等定量参数的评价。结合岩心、测井、局部密井网解剖等手段开展神木气田山西组、太原组有效储层规模评价。分析结果表明:神木气田山一段有效单砂体厚度范围为 0.5~4.5m,平均约为 2.5m,宽度范围为 300~500m,平均约为 400m,长度范围为 400~700m,平均约为 600m;山二段有效单砂体厚度范围为 0.8~6.5m,平均约为 3.5m,宽度范围为 400~800m,平均约为 600m,长度范围为 600~1000m,平均约为 800m;太原组有效单砂体厚度范围为 0.6~5.0m,平均约为 2.8m,宽度范围为 300~600m,平均约为 450m,长度范围为 500~800m,平均约为 600m(表2)。比较而言,山二段有效单砂体规模较大,太原组次之,山一段最小。总体而言,神木气田山西组、太原组有效单砂体规模有限,但多期单砂体复合叠置可形成相对较大规模复合有效砂体。

表 2　神木气田有效储层规模解剖参数表

层位	有效砂体类型	厚度(m)	宽度(m)	长度(m)	展布面积(km²)
山一段	单期	0.5~4.5	300~500	400~700	0.20~0.40
	复合	1.0~7.0	600~1000	900~1500	0.50~1.50
山二段	单期	1.0~6.5	400~800	600~1000	0.20~0.80
	复合	2.5~12.5	800~1600	1200~2000	0.80~3.00
太原组	单期	0.6~5.0	300~600	500~800	0.20~0.50
	复合	1.5~8.5	600~1200	1000~1600	0.60~2.00

2.2.2 有效储层空间叠置类型

受沉积环境、古地貌、可容纳空间等因素综合影响,神木地区上古生界发育多种形态砂体,精细表征砂体空间叠置类型可有效指导气田合理开发技术对策的制定。自河道砂体构型提出概念以来,国内外学者通过沉积露头和现代沉积特征的研究,依据垂向叠置和侧向叠置作用程度的强弱将河道砂岩的叠置模式划分成孤立式、多层式、多边式[8-10]。河道下切作用利于孤立式、多层式砂体形成,河道侧积作用利于多边式砂体形成。

神木气田储层表现出"二元"结构特征,有效储层呈透镜状包裹于背景砂体之中。结合砂体空间叠置类型,开展神木地区各层有效储层空间叠置方式分析。将神木地区有效储层空间结构类型划分为多层孤立分散型、垂向多期叠加型、侧向多期叠置型三种主要类型(图6)。多层孤立分散型有效储层呈多层系分散分布,以彼此孤立、不接触为主要特征;垂向复合叠加型有效储层多期垂向切割连通,多层系可复合叠加发育;侧向复合叠置型多期有效储层侧向切割连通,多层系可复合叠置发育。解剖分析表明,神木气田不同层系发育的有效储层空间结构类型具有一定差异,总体而言多层系孤立分散型是主要的有效储层结构类型,侧向叠置及垂向叠加型在山二段、太原组局部发育,山一段发育相对有限。多层孤立分散型有效单砂体通常规模较小、呈透镜状,多层系叠置后平面上表现为大面积连片分布的特征,适宜采用直井/定向井开发。垂向多期叠加式有效储层顺物源方向延伸较远,表现出"带状"分布特征,可采用水平井开发,水平段应以平行于物源方向为主。侧向多期叠置式有效储层垂直于物源方向延伸较远,表现出"片状"分布特征,可采用水平井开发,水平段应以垂直于物源方向为主。

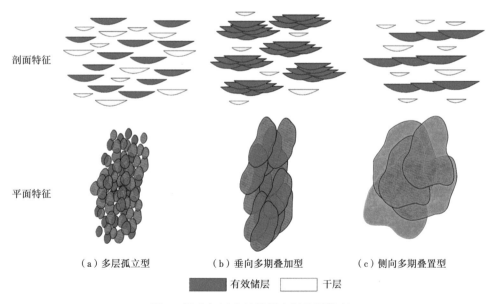

图6 神木气田有效储层空间叠置类型

3 水平井开发适用性评价

近年来,长庆气区水平井获得大规模推广应用,助推了气区天然气产量的快速上升,在产能建设中发挥了重要作用[10-12]。与直/定向井相比,水平井具有单井产量高、产能建设工作量低的特点。同时也面临着一定的开发风险,若开发地质目标选择不当,会造成气井产量低、经济效益差、储量大量遗留等问题。因此,结合实际开发区储层地质特征,开展水平井开发适用性评价对降低开发风险具有重要意义。

3.1 水平井地质目标筛选标准

长庆气区苏里格气田水平井推广深度最大,针对水平井开发地质目标筛选的研究也最深入,筛选参数主要包括主力气层厚度、含气面积、物性、储层连续性、夹层厚度及储量集中度等指标[13-14]。借鉴苏里格气田水平井开发实践,同时紧密结合神木气田有效储层结构特征,建立了水平井地质目标筛选标准,明确了适合水平井开发的有效储层叠置类型(表3)。统计分析表明,水平井产量与主力气层连续有效厚度、物性、有效水平段长度密切相关。为保证开发效益,开发地质目标应具备连续较大的有效厚度、较好的物性及含气性,同时兼具较长的有效延伸范围。综合分析认为,神木气田水平井开发地质目标至少应具备 8m 的连续有效厚度,孔隙度大于 6.5%、渗透率大于 0.5mD、含气饱和度大于 60%、气层有效延伸范围大于 1km。为保证水平井产量的稳定性,开发地质目标内部应相对均质,泥质夹层厚度以小于 2m 为宜。有效储层空间叠置类型应主要以侧向多期叠置和垂向多期叠加型为主。

表3 神木气田低渗致密砂岩气藏水平井地质目标筛选标准

类型	参数		指标
主力气层	连续有效厚度(m)		>8.0
	物性	孔隙度(%)	>6.5
		渗透率(mD)	>0.5
		含气饱和度(%)	>60.0
	有效延伸长度(km^2)		>1.0
	泥质夹层厚度(m)		<2.0
	叠置类型		侧向多期叠置、垂向多期叠加
储量	储量集中度(%)		>75.0
产量	初期日产(m^3)		>5.0

苏里格气田长期水平井开发实践表明,水平井虽可有效提高主力层段储量的动用程度,但同时也会造成非主力层储量的大量遗留。在目前经济技术条件下,缺乏动用此类遗留储量的有效技术手段,最终导致水平井开发方式下储量总体动用程度相对偏低。因此,尽可能提高储量总体动用程度、减少非主力层段储量遗留是水平井开发地质目标筛选需要考虑的重要问题。综合分析认为,水平井主力开发层段储量集中度应不低于75%(表2)。

3.2 水平井开发适用性评价

盆地中部苏里格气田有效储层主要发育于下石盒组盒八段和山西组山一段,单井最多可钻遇气层 6 层,大部分气井钻遇 2~4 层,气层发育数量相对较少[15-17]。盆地东部神木地区气层发育状况发生较大变化,石千峰组至马家沟组等都有气层发育,单井最多可钻遇23层,大部分井钻遇 7~13 层,多层系含气特征明显。结合纵向含气层的叠置模式,以多层系兼顾、提高储量动用程度和单井产量为原则,目前以多井型大井组立体式开发作为神木气田主体开发方式。该方式节约大量井场、集输、道路等费用,有效推进产建进程,实现了神木气田的低成本开发,有效缓解了矿区叠置复杂的难题[9]。

水平井部署技术包括整体式部署、局部式部署两种方式,整体部署适合于储层整体地质条件满足水平井地质目标筛选标准的面积较大的新建产区,局部式部署则适合的满足筛选标准的小面积局部井区[10]。结合神木气田水平井地质目标筛选标准,按重点层位开展气田山西组、太原组水平井开发区优选。结果表明,神木气田适合水平井开发的地质目标规模较小、呈小面积零星分布,且储量占比、面积占比都较低,总体不适合大面积整体式部署,可进行局部式甜点部署(表4)。

表4 神木气田水平井开发地质目标筛选

层位	面积(km^2)		储量占比(%)	面积占比(%)
	范围	平均值		
山一段	2.0~3.4	2.7	2.6	1.7
山二段	6.9~13.9	8.4	3.9	2.2
太原组	7.8~14.7	11.3	4.5	2.7

基于神木气田多井型大井组立体式开发方式,以多层兼顾、提高储量动用程度为核心,将局部式水平井部署划分为三种类型:孤立水平井、复合水平井及丛式水平井组。对于垂向主力层系单独发育的井区而言,适合采用孤立式水平井开发,对于垂向多套主力层发育井区而言,采用复合式水平井开发,而对于空间上多方向都发育主力层系的井区而言,则适合采用丛式水平井组开发(图7)。三种水平井局部式部署方式与井区直/定向井部署有机结合,可有助于实现神木地区多层系低渗透率、致密气藏的高效开发。

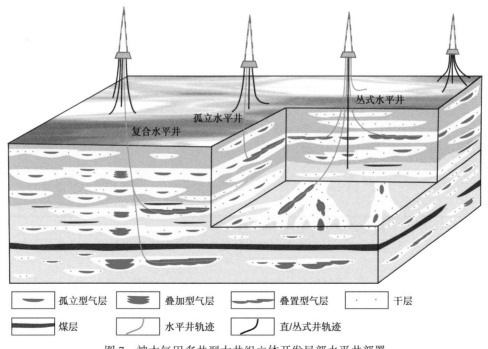

图7 神木气田多井型大井组立体开发局部水平井部署

4 结论

神木气田是鄂尔多斯盆地增储上产的重要组成部分,具有多层系含气特征。储集岩性主要为岩屑石英砂岩、岩屑砂岩及石英砂岩,孔隙类型主要为岩屑溶蚀孔及晶间孔,整体属于低孔隙度、低渗透率致密砂岩气藏。分析表明,孔隙度5%、渗透率0.1mD、含气饱和度45%为神木气田有效储层物性下限。

山西、太原组有效单砂体规模较小,多期叠置可形成较大规模复合有效储层。将神木地区有效储层空间结构类型划分为多层孤立分散型、垂向多期叠加型、侧向多期叠置型三种主要类型。不同类型有效储层适宜采用的开发方式具有一定差异。

建立了神木地区低渗砂岩气藏水平井地质目标筛选标准。分析认为,避免储量大量遗留,水平井主力开发层段储量集中度应不低于75%。评价表明,神木气田总体不适合整体式水平井部署,应以局部式部署为主。结合主力层系发育特征,提出了孤立式、复合式及丛式井组式三种水平井部署方式。

参 考 文 献

[1] 杨华,刘新社,闫小雄,等. 鄂尔多斯盆地神木气田的发现与天然气成藏地质特征[J]. 天然气工业,2015,35(6):1-13.

[2] 沈玉林,郭英海,李壮福,等. 鄂尔多斯地区石炭纪—二叠纪三角洲的沉积机理[J]. 中国矿业大学学报,2012,41(6):936-942.

[3] 席胜利,李文厚,刘新社,等. 鄂尔多斯盆地神木地区下二叠统太原组浅水三角洲沉积特征[J]. 古地理学报,2009,11(2):187-194.

[4] 刘锐娥,黄月明,卫孝锋,等. 鄂尔多斯盆地北晚古生代物源区分析及其地质意义[J]. 矿物岩石,2003,23(3):82-86.

[5] 何东博,贾爱林,冀光,等. 苏里格大型致密砂岩气田开发井型井网技术[J]. 石油勘探与开发,2013,40(1):79-89.

[6] 何东博,王丽娟,冀光,等. 苏里格致密砂岩气田开发井距优化[J]. 石油勘探与开发,2012,39(4):458-464.

[7] Miall A D. Architectural-element analysis: A new method of facies analysis applied to fluvial deposits[J]. Earth-Science Reviews,1985,22(4):261-308.

[8] Miall A D. Reconstructing the architecture and sequence stratigraphy of the preserved fluvial record as a tool for reservoir development: A reality check[J]. AAPG Bulletin,2006,90(7):989-1002.

[9] 武力超,朱玉双,刘艳侠,等. 矿权叠置区内多层系致密气藏开发技术探讨——以鄂尔多斯盆地神木气田为例[J]. 石油勘探与开发,2015,42(6):826-832.

[10] 卢涛,张吉,李跃刚,等. 苏里格气田致密砂岩气藏水平井开发技术及展望[J]. 天然气工业,2013,33(8):6.

[11] 位云生,贾爱林,何东博,等. 苏里格气田致密气藏水平井指标分类评价及思考[J]. 天然气工业,2013,33(7):47-51.

[12] 刘群明,唐海发,吕志凯,等. 鄂东致密气水下分流河道复合体储层构型布井技术[J]. 中国矿业大学学报,2017,46(5):1144-1151.

[13] 郝骞,卢涛,李先锋,等. 苏里格气田国际合作区河流相储层井位部署关键技术[J]. 天然气工业,2017,

37(9):39-47.

[14] 费世详,王东旭,林刚,等. 致密砂岩气藏水平井整体开发关键地质技术——以苏里格气田苏东南区为例[J]. 天然气地球科学,2014,25(10):1620-1628.

[15] 王国亭,何东博,王少飞,等. 苏里格致密砂岩气田储层岩石孔隙结构及储集性能特征[J]. 石油学报,2013,34(4):7.

[16] 郭智,孙龙德,贾爱林,等. 辫状河相致密砂岩气藏三维地质建模[J]. 石油勘探与开发,2015,42(1):76-83.

[17] 卢涛,刘艳侠,武力超,等. 鄂尔多斯盆地苏里格气田致密砂岩气藏稳产难点与对策[J]. 天然气工业,2015,35(6):43-52.

[18] 谭中国,卢涛,刘艳侠,等. 苏里格气田"十三五"期间提高采收率技术思路[J]. 天然气工业,2015,36(3):30-40.

[19] 李建奇,杨志伦,陈启文,等. 苏里格气田水平井开发技术[J]. 天然气工业,2011,31(8):60-64.

[20] 侯加根,唐颖,刘钰铭,等. 鄂尔多斯盆地苏里格气田东区致密储层分布模式[J]. 岩性油气藏,2014,26(3):1-6.

原文刊于《西南石油大学学报:自然科学版》,2019,41(5):1-9.

New method in predicting productivity of multi-stage fractured horizontal well in tight gas reservoirs

Wei Yunsheng Jia Ailin He Dongbo Wang Junlei

(Research Institute of Petroleum Exploration & Development, PetroChina, Beijing 100083, China)

Abstract: The generally accomplished technique for horizontal wells in tight gas reservoirs is by multi-stage hydraulic fracturing, not to mention, the flow characteristics of a horizontal well with multiple transverse fractures are very intricate. Conventional methods, well as an evaluation unit, are difficult to accurately predict production capacity of each fracture and productivity differences between wells with a different number of fractures. Thus, a single fracture sets the minimum evaluation unit, matrix, fractures, and lateral wellbore model that are then combined integrally to approximate horizontal well with multiple transverse hydraulic fractures in tight gas reservoirs. This paper presents a new semi-analytical methodology for predicting the production capacity of a horizontal well with multiple transverse hydraulic fractures in tight gas reservoirs. Firstly, a mathematical flow model used as a medium, which is disturbed by finite-conductivity vertical fractures and rectangular shaped boundaries, is established and explained by the Fourier integral transform. Then the idea of a single stage fracture analysis is incorporated to establish linear flow model within a single fracture with a variable rate. The Fredholm integral numerical solution is applicable for the fracture conductivity function. Finally, the pipe flow model along the lateral wellbore is adapted to couple multi-stages fracture mathematical models, and the equation group of predicting productivity of a multi-stage fractured horizontal well. The whole flow process from the matrix to bottom-hole and production interference between adjacent fractures is also established. Meanwhile, the corresponding iterative algorithm of the equations is given. In this case analysis, the productions of each well and fracture are calculated under the different bottom-hole flowing pressure, and this method also contributes to obtaining the distribution of pressure drop and production for every horizontal segment and its changes with effective fracture half-length and conductivity. Application of this technology will provide gas reservoir engineers a better tool to predict well and fracture productivity, besides optimizing transverse hydraulic fractures configuration and conductivity along the lateral wellbore.

Keywords: Tight gas; Multi-stage hydraulically fractured horizontal well; Single fracture; Production interference between adjacent fractures; Productivity prediction

1 Introduction

There are many transverse fractures with different forms around a horizontal well after being fractured multi-stage in tight gas reservoirs, this greatly increases the contact area a between gas well and the formation. Simultaneously, the flow conditions around the bottom-hole are improved. The states of gas flow include Darcy flow in the formation pores, variable Darcy flow in the hydraulic fractures, and variable pipe flow in the horizontal wellbore. The interferences and the coupling happen between the three flow patterns by means of the boundary conditions.

In terms of research on finite conductivity fractures, parts[1] unraveled the flow relationship between the elliptic fractures and matrix by conformal transformation; this presented the function relationship between finite conductivity and effective wellbore diameter. Cinco-Ley[2] evaluated the flow capacity of fractures with finite conductivity by the numerical discrete method. Liao[3] researched variable flow within fractures through transforming elliptic coordinate. On these bases, the analysis on unstable flow of fractured horizontal well within closed formation, Zerzar[4] obtained the characteristics of a linear (double) flow in the early stages, and a quasi-steady flow in the late stages done by the gradual approximation method; the parameters of the fractured horizontal well were then analyzed. Brown[5] used three linear flow models to reflect the flow law within the hydraulic fracture, the inner reservoir between hydraulic fractures, and the outer reservoir away from the tips of the fracture system. Regarding the productivity evaluation of fractured horizontal wells, based on the Joshi[6-7] productivity formula, Raghavan[8] introduced a method of predicting productivity of a multi-stage fractured horizontal well, which substitutes the equivalent wellbore radius (radial flow) for hydraulic fractures to simulate fluid flow. Wang[9] corrected the equivalent wellbore diameter of the horizontal wells with vertical rectangular fractures through introducing the influence function of finite conductivity fractures. Meanwhile, in combination with pressure superposition principle, the influence factors of fractured horizontal well productivity were evaluated. Wang[10] established the mathematical model flow of a multi-stage fractured horizontal well; the characteristics include the rectangular closed boundary by means of the two variables considering gas slippage, the pseudo-pressure, and the pseudo-time. Based on the wellbore with infinite conductivity, the change on horizontal well productivity is analyzed with the various fracture length, fracture conductivity, the number of fractures, fractured horizontal well length, and so on. Li[11] established the empirical plate to evaluate the open flow rate and cumulative production based on demonstrating the scale of the effective sand. Li[12] quickly evaluated the horizontal well productivity in low-permeability and tight gas reservoirs through combining ideal model with numerical simulation. In this paper, with the aid of previous research study results, the idea[13] of a single fracture is introduced to accurately evaluate the productivity of fractured horizontal wells in tight gas reservoirs. By a single fracture serving as a unit, the principle of mass conservation is applied to couple the elliptic flow in a typical reservoir, as well as variable flow within fractures and variable pipe flow in the horizontal wellbore. Meanwhile, considering the interference between the adjacent fractures, the theoretical formula, and the corresponding algorithm are established for predicting productivity of fractured horizontal wells; the practical examples are analyzed for model verification. Thus, a new method for predicting productivity of multi-stage fractured horizontal wells in a tight gas reservoir is formed.

2 Mathematical model

2.1 Flow model in formation with finite conductivity fracture

There are complex flow types in the reservoir with fractures. The gas in pores flow linearly into fractures across the surface, the streamline form within limit scope around fractures and is similar to an elliptic flow, and pseudo-radial flow is usually expressed out of the elliptic flow (Fig. 1).

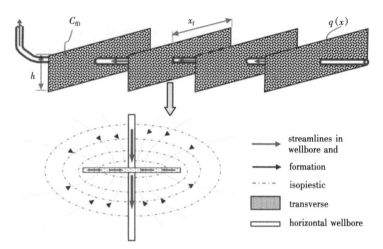

Fig. 1 Multi-stage fractured horizontal well and single fracture flow

Fracture conductivity has a great impact on gas well productivity in tight gas reservoirs. For that reason, the flow in the fracture is not negligible. In order to quantitatively describe the flow characteristics of fractures, some idealizations and simplifying were made assuming the following:

(1) The formation is homogeneous, and there's uniform thickness at the top and bottom.

(2) The fractures do not drain beyond the boundaries of this rectangular geometry ($x_e \times y_e$) with constant pressure.

(3) The perforated thickness of the fractures, h, is the same as the thickness of the reservoir. Furthermore, $q(x)$ is the variable for the length of the fracture.

Definitions of the dimensionless variables are as follows:

$$P_D = \frac{0.0786Kh(P_i^2 - P^2)}{\mu ZTQ_{sc}}, \quad q_D = \frac{2x_f q(x)}{Q_{sc}}, \quad J_D = \frac{j}{x_f}(j = x,y), \quad C_{fD} = \frac{K_f w_f}{k x_f}$$

The dimensionless control equation and the associated boundary conditions for the formation around fractures are given:

$$\frac{\partial P_D}{\partial x_D^2} + \frac{\partial^2 P_D}{\partial y_D^2} + q_D(x_D)\delta(y_D - y_{wD}) = 0 \qquad (1)$$

$$P_D(x_D, 0) = P_D(x_D, x_{eD}) \qquad (2)$$

and

$$P_D(0, y_D) = P_D(y_{eD}, y_D) \qquad (3)$$

where x_w, y_w are the coordinates of the fracture center, δ is the Dirac function.

In Eq. 1, Fourier finite sine integral of P_D and q_D along the direction of x_D and y_D are given, respectively, by

$$\hat{P}_D = \int_0^{x_{eD}} P_D \sin(\beta_m x_D) \, dx_D \tag{4}$$

$$\overline{P}_D = \int_0^{x_{eD}} P_D \sin(\gamma_n y_D) \, dy_D \tag{5}$$

and

$$\tilde{q}_D \int_0^{x_{eD}} q_D(x_D) \sin(\gamma_n x_D) \, dx_D \tag{6}$$

where the over-bar symbol indicates various integral transform direction.

The relationship between the dimensionless pressure through the double Fourier integral transform and dimensionless fracture production is given, depending on the boundary conditions Eq. 2 and Eq. 3, by means of

$$-\pi^2 \left(\frac{m^2}{x_{eD}^2} + \frac{n^2}{y_{eD}^2} \right) \hat{\overline{P}}_D + \tilde{q}_D \sin\gamma_n y_{wD} = 0 \tag{7}$$

The dimensionless pressure is obtained from Eq. 7 by two inverse transformations as follows:

$$P_D = \sum_{n=1}^{\infty} \frac{\sin(\gamma_n y_D)}{N(\beta_n)} \left(\sum_{m=1}^{\infty} \frac{\sin(\beta_m x_D)}{N(\beta_m)} \hat{\overline{P}}_D \right) \tag{8}$$

where the eigenvalues are given by

$$\beta_m = m\pi/x_{eD}; \quad \gamma_n = n\pi/y_{eD} \tag{9}$$

and the bottom of the norms is given by

$$N(\beta_m) = \int_0^{x_{eD}} \sin^2(\beta_m x_D) \, dx_D = \frac{x_{eD}}{2}; \quad N(\gamma_n) = \int_0^{y_{eD}} \sin^2(\gamma_n y_D) \, dy_D = \frac{y_{eD}}{2} \tag{10}$$

To substitute Eq. 7, Eq. 9 and Eq. 10 into Eq. 8, the formula for pressure is obtained as follows:

$$P_D = \sum_{m=1}^{\infty} \frac{2\tilde{q}_D}{x_{eD} y_{eD}} \sin\frac{m\pi x_D}{x_{eD}} \left[\sum_{n=1}^{\infty} \frac{\cos n\pi(y_D - y_{wD})/y_{eD} - \cos n\pi(y_D + y_{wD})/y_{eD}}{\pi^2(m^2/x_{eD}^2 + n^2/y_{eD}^2)} \right] \tag{11}$$

Assuming that there's only flux distribution along the fractures, Eq. 6 becomes

$$\tilde{q}_D = \int_{x_{wD}-1}^{x_{wD}+1} q_D(\alpha) \sin\left(\frac{m\pi\alpha}{x_{eD}}\right) d\alpha \tag{12}$$

and the transformation is as follows:

$$\sum_{k=1}^{\infty} \frac{\cos K\pi x}{K^2 + a^2} = \frac{\pi}{2a} \frac{\cosh[a\pi(1-x)]}{\sinh(a\pi)} - \frac{1}{2a^2}; \quad [0 \leqslant x \leqslant 2\pi] \tag{13}$$

The solution of the pressure at the intersection point (x_{wD}, y_{wD}) of the horizontal wellbore and finite conductivity fractures can be obtained after commencing Eq. 11-13 as follows:

$$P_D(x_D, y_D; x_{wD}, y_{wD}) = 2\int_{x_{wD}-1}^{x_{wD}+1} \left\{ \sum_{m=1}^{\infty} \frac{q_D(\alpha)}{m\pi} \sin\frac{m\pi x_D}{x_{eD}} \sin\frac{m\pi\alpha}{x_{eD}} \frac{\cosh\frac{m\pi(y_{eD}-|y_D-y_{wD}|)}{x_{eD}} - \cosh\frac{m\pi(y_{eD}-|y_D+y_{wD}|)}{x_{eD}}}{\sinh\frac{m\pi y_{eD}}{x_{eD}}} \right\} d\alpha \quad (14)$$

2.2 Flow model in finite conductivity fracture

Flow within fracture is assumed to be one-dimensional linear flow with a variable rate in the x direction (Fig. 2)

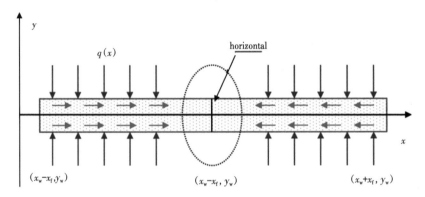

Fig. 2 Variable rate of linear flow in fractures

The elastic energy in the fracture is ignored because the volume of the fracture is minute. The dimensionless flow equation within the hydraulic fracture can be simplified to stabilize the state equation as follows:

$$\frac{d^2 P_{fD}}{dx_D^2} + \frac{2}{C_{fD}} q_D(x_D) = 0, \quad [-1 \leq x_D \leq 1] \quad (15)$$

The inner boundary condition is given by

$$\frac{dP_{fD}(x_{wD})}{dx_D} = -\frac{\pi}{C_{fD}} \quad (16)$$

After the second integral to x_D, Eq. 15 becomes

$$P_{wD} - P_{fD}(x_D) = \frac{\pi}{C_{fD}} \left[|x_D - x_{wD}| - \int_{x_{wD}}^{x_D} dv \int_{x_{wD}}^{v} q_D(u) du \right] \quad (17)$$

The coupling conditions between the fracture and the formation are given by

$$P_{fD}(x_D) = P_D(x_D, y_{wD}; x_{wD}, y_{wD}), \quad [-1 \leq x_D \leq 1] \quad (18)$$

The Fredholm integral equation is obtained by substituting Eq. 14 with Eq. 18. Then again this equation can't be solved by means of an analytical method. The numerical method is given as follows: the fracture is divided evenly into n sections and the flux and pressure of each section are uniform. Thus, we can obtain n+1 order equations (Eq. 19) about the flux at each section, $q_{Dj}(j=1, 2, 3\cdots, n)$, and the bottom-hole pressure, P_{wD}.

$$P_{wD} + 2\sum_{i=1}^{n} q_{Di} \sum_{m=1}^{\infty} \left\{ \begin{array}{l} \dfrac{x_{eD}}{m^2\pi^2}\sin m\pi \dfrac{x_{wD}+(j-0.5)\Delta x}{x_{eD}} \left[\cos m\pi \dfrac{x_{wD}+i\Delta x_D}{x_{eD}} - \cos m\pi \dfrac{x_{wD}+(i-1)\Delta x_D}{x_{eD}}\right] \\ \times \dfrac{\cosh[m\pi y_{eD}/x_{eD}] - \cosh[m\pi(y_{eD}-|2y_{wD}|)/x_{eD}]}{\sinh(m\pi y_{eD}/x_{eD})} \end{array} \right\}$$

$$= \dfrac{\pi}{C_{fD}}\left\{(x_{wD}+(j-0.5)\Delta x)\left(1 - \sum_{i=1}^{j-1} q_{Di}\Delta x_D - \dfrac{\Delta x_D}{2}q_{Di}\right) + \sum_{i=1}^{j-1} q_{Di}\Delta x_D[x_{wD}+(i-0.5)\Delta x_D] \right.$$

$$\left. + q_{Dj}\Delta x_D \dfrac{x_{wD}+(j-0.75)\Delta x_D}{2}\right\} \quad (19)$$

where the flux constraint equation is given as follows:

$$\sum_{i=1}^{n} q_{Di} = 1 \quad (20)$$

The Newton iteration is used to solve Eq. (19). Meanwhile, the relationship between the dimensionless conductivity C_{fD} and the dimensionless bottom-hole pressure P_{wD} are analyzed under diverse outer boundary conditions. The results indicate that the dimensionless bottom-hole pressure P_{wD} decreases with the increase of the C_{fD} in the finite conductivity fracture. Whenever $C_{fD}>1000$, P_{wD} tends to be constant, hence, P_{infwD}, which is the dimensionless bottom-hole pressure in the infinite conductivity fracture; this trend is only related to C_{fD}. Through data regression, the differential function, $f(C_{fD})$, that is, the influence function of fracture conductivity is obtained through Eq. 21. Kindly refer to Fig. 3.

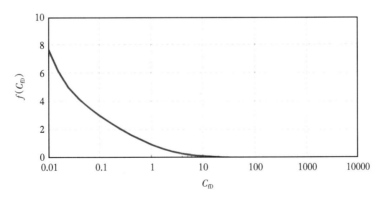

Fig. 3　Relationships between finite conductivity fracture influence function and dimensionless conductivity

$$f(C_{fD}) = \dfrac{0.95 - 0.56u + 0.16u^2 - 0.028u^3 + 0.0028u^4 - 0.0001u^5}{1.0 + 0.094u + 0.093u^2 + 0.0084u^3 + 0.001u^4 + 0.00036u^5}, \quad u = \ln C_{fD} \quad (21)$$

where the dimensionless bottom-hole pressure in the infinite conductivity fracture is given by

$$p_{infwD} = 2\left\{\sum_{n=1}^{\infty} \frac{x_{eD}}{n^2\pi}\sin n\pi \frac{x_D}{x_{eD}}\sin n\pi \frac{1}{x_{eD}}\sin n\pi \frac{x_{wD}}{x_{eD}} \cdot \frac{\cosh\frac{n\pi y_{eD}}{x_{eD}} - \cosh\frac{n\pi(y_{eD}-2y_{wD})}{x_{eD}}}{\sinh\frac{n\pi y_{eD}}{x_{eD}}}\right\} \quad (22)$$

In the subsequent derivations, the one-dimensional (linear) flow has been assumed within the hydraulic fracture; that is, the radial convergence of flow towards the wellbore has been ignored within the hydraulic fracture. Nonetheless, the radial flow near the horizontal wellbore exists objectively. Therefore, the skin factor is introduced to calculate the flow resistance. The formula[9] is given:

$$\text{skin} = \frac{Kh}{K_f w_f}\left[\ln\frac{h}{2r_w} - \frac{\pi}{2}\right] \quad (23)$$

Adding the choking skin to Eq. 22, we obtain the following solution. This is a good approximation for dimensionless wellbore pressure after the end of radial flow in the finite conductivity fracture:

$$P_{finwD} = P_{infwD} + f(C_{fD}) + \text{skin} \quad (24)$$

That is:

$$\frac{P_i^2 - P_{wf}^2}{Q_{sc}} = \frac{\mu ZT}{0.0786Kh}\left\{\frac{1}{C_{fD}}\frac{h}{x_f}\left[\ln\frac{h}{2r_w} - \frac{\pi}{2}\right] + f(C_{fD}) + 2\sum_{m=1}^{\infty}\left[\frac{x_e}{m^2\pi x_f}\sin\frac{m\pi x_f}{x_e}\sin^2\frac{m\pi x_w}{x_e} \cdot \frac{\cosh\frac{m\pi y_e}{x_e} - \cosh\frac{m\pi(y_e-2y_w)}{x_e}}{\sinh\frac{m\pi y_e}{x_e}}\right]\right\} \quad (25)$$

Using Eq. 25, the productivity of a single transverse fracture with finite conductivity can be quickly calculated. The calculation is the basis for predicting productivity of a multi-stage fractured horizontal well.

2.3 Horizontal wellbore model

Multiple transverse hydraulic fractures are coupled with each other through the pipe flow in the horizontal wellbore.

The single-phase pipe flow is considered because the diameter of horizontal wellbore is far greater than the size of the flow channel in the formation and fracture. Flow rate in the horizontal wellbore is changing, hence, the flow pressure is calculated piecewise. The pressure gradient is given by

$$\frac{dP}{dy} = f\frac{\rho v^2}{2r_w} \quad (26)$$

where the pressure drop of the kinetic energy is caused by the increase in velocity is ignored.

Through separating variables and definite integral based on Eq. 26, the relationship between the gas flow rate and pressure square difference is derived as follows:

$$P_{wfi}^2 - P_{wfi-1}^2 = 9 \times 10^{-12} \frac{ZT\gamma_g}{r_w^5} f_i d_i (\sum_{j=1}^{n} Q_{scj})^2 \quad (i = 1,2,\Lambda,n; P_{wf0} = P_{wf}) \quad (27)$$

where f is calculated by Eq. 28, in which the Reynolds number considering the turbulent condition is calculated by means of Eq. 29.

$$f_i = \left[1.14 - 2\lg(\frac{e}{1000D} + \frac{21.25}{R_{ei}^{0.9}})\right]^{-2} \quad (28)$$

and

$$R_{ei} = \frac{177.1\gamma_g \sum_{j=1}^{n} Q_{scj}}{2\mu r_w} \quad (29)$$

2.4 Productivity model with the interference between adjacent fractures

The interference[15-16] between adjacent fractures happens very often when the steady state flow or pseudo-steady state flow appears. For the quantitative evaluation to be easy, the interference degree, the outer flow boundary for every fracture is described as an elliptic boundary.

The flow resistance from outer boundary to horizontal wellbore is given by:

$$R = \frac{\mu ZT}{0.0786Kh} \ln \frac{\sqrt{ab}}{r_{we}} \quad (30)$$

where a, b is the semi-major axis and semi-minor axis of outer boundary elliptic, respectively, m; r_{we} is equivalent radius and the formula[17-19] is given by:

$$r_{we} = 2x_f \exp\left\{-\left[\frac{3}{2} + f(C_{fD}) + \text{skin}\right]\right\} \quad (31)$$

In combining Eq. 25 with Eq. 30, Eq. 32 is obtained by:

$$\ln \frac{\sqrt{ab}}{r_{we}} = \frac{1}{C_{fD}} \frac{h}{x_f} \left[\ln \frac{h}{2r_w} - \frac{\pi}{2}\right] + f(C_{fD}) \\ + 2\sum_{m=1}^{\infty} \left[\frac{x_e}{m^2 \pi x_f} \sin \frac{m\pi x_f}{x_e} \sin^2 \frac{m\pi x_w}{x_e} \frac{\cosh \frac{m\pi y_e}{x_e} - \cosh \frac{m\pi(y_e - 2y_w)}{x_e}}{\sinh \frac{m\pi y_e}{x_e}}\right] \quad (32)$$

where

$$a^2 - b^2 = x_f^2 \quad (33)$$

The elliptic boundary size and shape of every fracture can be calculated by means of Eq. 32 and

Eq. 33 where every fracture spatial location is combined. The interference of every fracture outer flow boundary is then determined (Fig. 4).

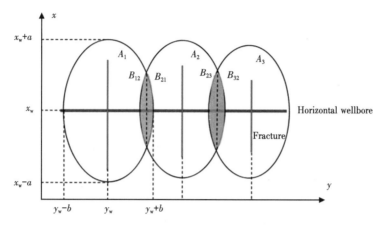

Fig. 4　Scheme with interference between fractures in fractured horizontal well

Assuming the flow area of every fracture is respectively A_1, A_2, $\cdots$, A_n, the intersecting area of adjacent elliptic boundary is respectively B_{12}, B_{21}, B_{23}, B_{32}, $\cdots$, $B_{(n-1)n}$, $B_{n(n-1)}$, and the production rate of every fracture is Q_{sc1}, Q_{sc1}, $\cdots$, Q_{scn} when the interference between the adjacent fracture is not considered. According to the principle of flow area[20-21], every fractures' contribution to the horizontal well production is proportional to its area of flow boundary; then the actual production rate of every fracture is acquired:

$$Q_{scr1} = \frac{A_1 - B_{21}}{A_1} Q_{sc1}$$
$$Q_{scri} = \frac{A_i - B_{(i-1)i} - B_{(i+1)i}}{A_i} Q_{sci} \quad (1 < i < n) \quad (34)$$
$$Q_{scrn} = \frac{A_n - B_{(n-1)n}}{A_n} Q_{scn}$$

then the production rate of the fractured horizontal well is given by:

$$Q_{scrt} = \sum_{i=1}^{n} Q_{scri} \quad (35)$$

By means of Eq. 25 and Eq. 34, the bottom-hole pressure of every fracture and the distribution of pressure in a horizontal wellbore are obtained.

3　Model Solution

Assuming the number of hydraulic fractures, n, Eq. 27 is a nonlinear equation which can be used to quickly and accurately solve the numerical iteration. The steps are as follows:

Step 1: The flow rates of the nth fracture, Q_{scnmax} and Q_{scnmin}, are assumed, therefore, $Q_{scnmid} =$

$(Q_{scnmax}+Q_{scnmin})/2$.

Prior Step 2 and 3, the pressure points of the nth fracture across the horizontal wellbore, $P_{wfnmax(0)}$, $P_{wfnmin(0)}$, $P_{wfnmid(0)}$, are calculated.

Step 2: The flow rate of the $(n-1)$th fracture, Q_{scn-1} is calculated on the basis of Eq. 25 and Eq. 27.

Step 3: According to Step 2, the flow rates of the i_{th} fracture are calculated in turn. When that of the first fracture is calculated, the pressure values of the first fracture across the horizontal wellbore, $P_{wf1max(0)}$, $P_{wf1min(0)}$, $P_{wf1mid(0)}$, can be calculated directly.

Step 4: If $(P_{wf1max(0)}-P_{wf})\times(P_{wf1mid(0)}-P_{wf})<0$, then $Q_{scnmin}=Q_{scnmid}$, or $Q_{scnmax}=Q_{scnmid}$.

Step 5: Step 2 to Step 4 are not looped until the precision of $P_{wfmid(j)}$ is fulfilled.

Through the preceding steps and the corresponding computer program, the flow rate of every fracture not considering the interference between adjacent fractures is obtained.

Step 6: Combining Eq. 32 with Eq. 33, the flow area of every fracture, $A_i(i=1,2,\cdots,n)$, their shapes, and locations are then determined.

Step 7: Through definite integral, the intersecting areas are obtained. Thereafter, the actual production rates of every fracture and horizontal well are obtained.

Step 8: Substituting the actual production rates of every fracture with Eq. 25, the bottom-hole pressure of every fracture and the distribution of pressure in a horizontal wellbore are also acquired.

4 Field example

In order for it to be easy to verify the results, the three horizontal wells, namely, the Well WH1, the Well WH2, and the Well WH3 have been selected as well examples because there are exists data on the interpretation results and monitored fracture data in these particular wells. Their fracture stages are 3, 4 and 5, respectively. The Well WH1 had three stages and a 1103m-horizontal well length that was analyzed in detail. The control range of the Well WH1 in the well pattern is 1600m×600m, and the other data are shown in Table 1 and Table 2.

Table 1 Basic parameters of the Well WH1

Temperature (K)	Original pressure (MPa)	Formation permeability (mD)	Gas viscosity (mPa·s)	Gas deviation factor	Gas relative density	Radius of horizontal wellbore(m)	Roughness of wellbore wall(mm)
385	31.7	0.74	0.023	0.98	0.598	0.076	3

Table 2 Fractures parameters of the Well WH1

No.	Half length (m)	Permeability (mD)	Width (m)	Height (m)	Spacing (m)	Semi-major axis (m)	Semi-minor axis (m)
1	70	5000	0.01	12	143	220	208
2	40	4500	0.01	10	310	133	127
3	30	3500	0.01	9	38	100	96

Whenever P_{wf} = 0.1MPa, the interference between adjacent fractures and horizontal wellbore friction (HWF) are considered, the production rates of the three fractures are Q_{sc1} = 19.33×10^{4}m^3/d, Q_{sc2} = 18.99×10^{4}m^3/d, and Q_{sc3} = 1.39×10^{4}m^3/d, respectively; the open-flow capacity of this well is Q_{AOF}, is 39.71×10^{4}m^3/d. Whenever HWF is not considered, the open-flow capacity of this well is 42.34×10^{4}m^3/d. Meanwhile, the open-flow capacity of this well evaluated by pressure buildup testing data and Topaze well test analysis software is 40.72×10^{4}m^3/d, which is closer to the result that considers friction; in addition, the relative error is −2.49%.

The flow rates of the well, every fracture, and the distribution of pressure in the horizontal wellbore are calculated under different bottom-hole pressures (0.1MPa, 1MPa, 5MPa, 10MPa, 15MPa and 20MPa) (Table 3 and Fig. 5).

Table 3 Production of each fracture in the WH1 under various bottom-hole pressures

P_{wf} (MPa)	Q_{sc1} (10^{4}m^3/d)	Q_{sc2} (10^{4}m^3/d)	Q_{sc3} (10^{4}m^3/d)	Q_{avg} (10^{4}m^3/d)	Variance
0.1	19.33	18.99	1.39	13.24	70.19
1	19.31	18.97	1.39	13.22	70.03
5	18.87	18.55	1.36	12.93	66.91
10	17.49	17.22	1.26	11.99	57.58
15	15.17	14.97	1.09	10.41	43.44
20	11.87	11.75	0.86	8.16	26.65

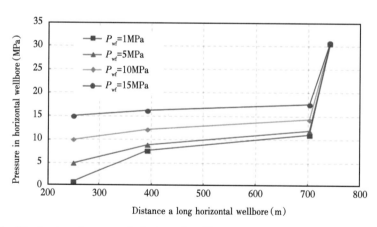

Fig. 5 Distribution of pressure in horizontal wellbore under various bottom-hole pressures

Table 3 shows that the production rates of fractures increase from toe to heel of the horizontal well whenever the HWF is considered. Fig. 6 indicates that the flow area and production rate of fracture No. 3 decreases evidently because of the spacing between fracture No. 2 and No. 3. The reservoir near the toe of the well is invalid, and there is no fracturing. Therefore, there is no production contribution, and the bottom-hole pressure in fracture No. 3 is higher in Fig. 6.

New method in predicting productivity of multi-stage fractured horizontal well in tight gas reservoirs

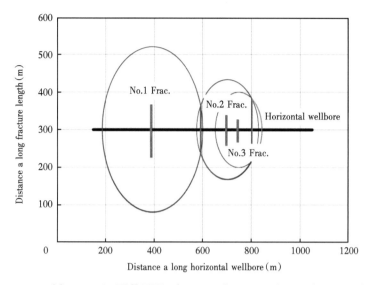

Fig. 6 Discharge area of fractures in Well WH1 when considering interference between adjacent fractures

In addition, the primary basic and fractures parameters of the Well WH2 and the Well WH3 is shown in Table 4 and Table 5. Their other parameters are the same to that of the Well WH1. It is consistent with calculated open flow rates with that of well-testing interpretation (Table 6).

Table 4 Basic parameters of the Well WH2 and the Well WH3

Well WH2				Well WH3			
Temperature (K)	Original pressure (MPa)	Formation permeability (mD)	Horizontal well length (m)	Temperature (K)	Original pressure (MPa)	Formation permeability (mD)	Horizontal well length (m)
387.8	30.5	0.51	505	389	31.3	0.35	969

Table 5 Fractures parameters of the Well WH2 and the Well WH3

Well Name	No.	Half length (m)	Permeability (mD)	Height (m)	Spacing (m)	Semi-major axis (m)	Semi-minor axis (m)
WH2	1	30	3500	11	65	100.2	95.7
	2	30	3500	10	135	100.4	95.8
	3	30	3500	9	100	100.3	95.7
	4	30	3500	7	105	100.0	95.5
WH3	1	30	1000	8	50	97.6	92.9
	2	30	1000	8	120	99.6	95.0
	3	55	1500	8	255	180.6	172.0
	4	50	1500	8	185	164.7	156.9
	5	50	1500	8	210	161.7	153.8

Table 6 Q_{AOF} of the Well WH2 and the Well WH3 as well as production of every fracture where $P_{wf}=0.1$MPa

Well Name	Production Rate of every fracture($10^4 m^3/d$)					Open flow Rate of well($10^4 m^3/d$)		
	No. 1	No. 2	No. 3	No. 4	No. 5	this results	Well testing results	Relative error (%)
WH2	11.48	9.09	8.10	10.37		39.04	36.91	5.8
WH3	5.83	5.74	5.85	4.85	6.17	28.44	25.92	9.7

5 Conclusions

Based on the basic flow principle within porous media, the whole flow process undergoes three stages according to the gas flow path from the matrix to hydraulic fracture, and lastly to the horizontal wellbore. The laws of variable flow in every part of the three stages are analyzed. Finally, we have presented a practical analytical model and analyzed three well examples of a tight gas reservoir. The following conclusions are warranted from the work presented in this paper:

(1) A single fracture segment is a unit, and all fracture sections are coupled by variable flow in a horizontal wellbore. Moreover, the effect of the interference between adjacent fractures on production and pressure of every fracture is evaluated quantitatively according to the principle of flow area. Thus, the new equations of predicting productivity are established and the method of solving the equations and its process are given.

(2) The open flow rates of three actual gas wells were calculated through the application of the equations, which conforms to the result of the well test evaluation. Additionally, we also quantitatively evaluated the contribution and effect of every fracture, thus, providing the theoretical basis for further optimization of the fracturing design.

(3) The method is suitable for predicting productivity of horizontal wells with non-uniform fracture system. That is unequal fracture spacing and fracture layout form, which is in line with the reality of complicated fracture system present in tight gas reservoirs. Meanwhile, it can also analyze the main factors influencing well productivity. Hence, the new method in this paper has broader prospects on predicting productivity of a multi-stage fractured horizontal well in the gas reservoir.

Nomenclature

P_D——dimensionless pressure

q_D——dimensionless fracture production

j_D——dimensionless length

C_{fD}——dimensionless fracture conductivity

P_i——original reservoir pressure, MPa

P——reservoir pressure, MPa

T——reservoir temperature, K

K —— reservoir permeability, mD

h —— effective reservoir thickness, m

μ —— gas viscosity, mPa · s

Z —— gas deviation factor

Q_{sc} —— fracture production under standard conditions, $10^4 m^3/d$

$q(x)$ —— production of per unit length fracture under standard conditions, $10^4 m^3/d$

x_f —— fracture half-length, m

k_f —— fracture permeability, mD

w_f —— fracture width, m

d —— fracture spacing, m

r_w —— horizontal wellbore radius, m

e —— roughness on horizontal wellbore wall, mm

γ_g —— gas relative density

f —— friction resistance coefficient

R_{ei} —— Reynolds number

References

[1] M Parts. Effect of vertical fractures on reservoir behavior-incompressible fluid case[J]. Society of Petroleum Engineers Journal, 1961,1(2),105-118.

[2] H Cinco-Ley, N Dominguez. Transient pressure behavior for a well with a finite-conductivity vertical fracture [J]. Society of Petroleum Engineers Journal, 1978,18(4),253-264.

[3] Y Z Liao, W J Lee. New Solutions for Wells with Finite-conductivity Fractures Including Wellbore Storage and Fracture-face Skin[C]. SPE East Regional Meeting, 2-4 November, Pittsburgh, Pennsylvania, SPE 26912. 1993.

[4] A Zerzar, Y Bettam. Interpretation of Multiple Hydraulically Fractured Horizontal Wells in Closed Systems[C]. SPE International Improved Oil Recovery Conference in Asia Pacific, 20-21 October, Kuala Lumpur, Malaysia. SPE 84888. 2003.

[5] M Brown, E Ozkan, R Raghavan, H Kazemi. Practical Solutions for Pressure Transient Responses of Fractured Horizontal Wells in Unconventional Reservoirs[C]. SPE Annual Technical Conference and Exhibition, 4-7 October, New Orleans, Louisiana. SPE 125043. 2009.

[6] S D Joshi. Argumentation of well productivity with slant and horizontal wells[J]. Journal of Petroleum Technology, 1988, 8(6): 729-739.

[7] S D Joshi. Production Forecasting Methods for Horizontal Wells[C]. International Meeting on Petroleum Engineering, 1-4 November, Tianjin, China. SPE 17580. 1988.

[8] R Raghavan, S D, Joshi. Productivity of multiple drainholes or fractured horizontal wells[J]. SPE Formation Evaluation, 1993,8(1),1-16.

[9] X D Wang, G H Li, F Wang. Productivity analysis of horizontal wells intercepted by multiple finite-conductivity fractures[J]. Petroleum Science, 2010,(7),367-371.

[10] Junlei Wang, Ailin Jia, Dongbo He, et al. Rate decline of multiple fractured horizontal well and influence factors on productivity in tight gas reservoirs[J]. Natural Gas Geoscience, 2014,25(2),278-285.

[11] Li Bo, Jia Ailin, He Dongbo, et al. Productivity evaluation of horizontal wells in Sulige tight gas reservoir with strong heterogeneity. Natural Gas Geoscience, 2015,26(3),539-549.

[12] Qin Li, Cheng Chen, Xiaoquan Xun. A new method of predicting gas wells productivity of fractured horizontal well of low-permeability tight gas reservoir[J]. Natural Gas Geoscience, 2013,24(3),633-638.

[13] Yunsheng Wei, Ailin Jia, Dongbo He, Guang Ji. A new way of evaluation productivity of staged fracturing horizontal well in tight gas reservoir[J]. Drilling & Production Technology, 2012,35(1),32-34.

[14] Li Shilun, et al. Natural Gas Engineering[M]. Beijing: Petroleum Industry Press, 2008.

[15] Mingqiang Hao, Yongle Hu, Fanhua Li. Production decline laws of fractured horizontal wells in ultra-low permeability reservoirs[J]. Acta Petrolei Sinica, 2012,33(2),269-273.

[16] Yanbo Xu, Tao Qi, Fengbo Yang, et al. New model for productivity test of horizontal well after hydraulic fracturing[J]. Acta Petrolei Sinica, 2006,27(1),89-91.

[17] M F Riley, W E Brigham, R N Horne. Analytic solutions for elliptical finite-conductivity fractures[C]. SPE Annual Technical Conference and Exhibition, 6-9 October, Dallas, Texas, SPE 22656. 1991.

[18] M Prats, P Hazebroek, W R Strickler. Effect of vertical fractures on reservoir behavior: Compressible-fluid case[J]. Society of Petroleum Engineers Journal, 1962,2(2),87-94.

[19] Xiaodong Wang, Yitang Zhang, Ciqun Liu. Productivity evaluation and conductivity optimization for vertically fractured wells[J]. Petroleum Exploration and Development, 2004,31(6),78-81.

[20] Tongyu Yao, Weiyao Zhu, Jishan Li, et al. Fracture mutual interference and fracture propagation roles in production of horizontal gas wells in fractured reservoir[J]. Journal of Central South University: Science and Technology, 2013,44(4),1487-1492.

[21] Jia Deng, Weiyao Zhu, Qian Ma. A new seepage model for shale gas reservoir and productivity analysis of fractured well[J]. Fuel, 2014,124(3):232-240.

原文刊于《Journal of Natural Gas Geoscience》,2016,1(5):397-406.

鄂尔多斯盆地东部奥陶系古岩溶型
碳酸盐岩致密储层特征、形成与天然气富集

王国亭[1] 贾爱林[1] 孟德伟[1] 郭 智[1] 冀 光[1] 程立华[1] 彭艳霞[2]

(1. 中国石油勘探开发研究院,北京 100083;2. 中国地质大学,北京 100083)

摘要:目前对鄂尔多斯盆地东部地区古岩溶型碳酸盐岩储层的研究相对薄弱,系统研究奥陶系古岩溶型有效储层形成机理与天然气富集可为盆地东部下古生界古岩溶储层天然气储量规模增加和开发潜力评价奠定基础。通过对鄂尔多斯盆地东部奥陶系岩溶储层特征、有效储层形成控制因素、天然气富集主控因素与富集潜力的综合分析,盆地东部碳酸盐岩风化壳储层较为发育,气源供给充足,良好的源—运—储配置关系良好,具备天然气大规模富集的条件。盆地东部岩溶储层总体表现为低孔隙度、致密的特征,孔隙度3%、渗透率0.05mD 确定为有效储层物性下限标准,孔隙直径30μm、喉道直径5μm 界定为有效储层孔、喉尺度下限标准。有利沉积微相/岩相组合、高效岩溶作用、综合成岩作用等共同影响有效储层形成,半充填型硬石膏结核溶孔云岩为最重要的有效储层类型。总结了天然气富集主控因素:(1)有效储层发育是天然气富集的基础物质条件;(2)良好的源—运—储配置关系是天然气富集的关键;(3)岩溶储层的强非均质性影响着气、水分布格局。总体而言,盆地东部奥陶系古岩溶型碳酸盐岩岩溶储层具备较大的勘探开发潜力。

关键词:古岩溶储层;奥陶系;物性下限;天然气富集;鄂尔多斯盆地东部

中国古岩溶型碳酸盐岩储层广泛发育并取得了一系列重大突破,重点以塔里木盆地塔河油田及塔里木油田奥陶系石灰岩油气藏、鄂尔多斯盆地靖边下古生界奥陶系气藏、渤海湾盆地任丘奥陶系古岩溶油气藏等为代表[1-3]。鄂尔多斯盆地是中国重要的含油气盆地,蕴含丰富的天然气资源,目前天然气探明(含基本探明)储量已达 $5.7×10^{12}m^3$,发现了苏里格、靖边、榆林、大牛地、神木、子洲等多个探明储量超千亿立方米的气田。盆地目前绝大部分探明储量都集中于上古生界山西组至下石盒子组碎屑岩中,下古生界探明储量主要集中于古岩溶型碳酸盐岩储层发育的靖边气田,其储量规模仅占目前盆地总探明储量规模的10%。目前针对盆地中部靖边地区的岩溶型碳酸盐岩储层研究较为系统深入[4-9],而盆地东部地区相关研究比较薄弱且没有储量发现。本文以盆地东部神木地区为依托,深入开展盆地东部碳酸盐岩岩溶储层特征分析、有效储层形成机理及天然气富集潜力评价,以明确盆地东部下古生界的勘探开发前景,从而为盆地东部下古生界古岩溶储层天然气储量规模增加和开发潜力评价奠定基础。

1 奥陶系构造沉积与岩溶古地貌格局

古生代时期鄂尔多斯盆地属于华北克拉通盆地的一部分,南北两侧分别为古秦岭洋和古兴蒙洋。早古生代盆地演化主要受控于南侧古秦岭洋的演化,伴随古洋盆的形成、扩张、俯冲消减及最终闭合消亡,盆地内部经历早期陆表海盆地、后期陆缘海盆地、洋盆闭合并整体抬升

遭受剥蚀的演化过程[4-8]。鄂尔多斯盆地奥陶系沉积期表现为"两隆两鞍两坳陷"的古地貌特征，两隆指北部伊盟隆起和西南部中央古隆起，两鞍指两个隆起间的衔接部位，两坳陷指盆地东部米脂坳陷及盆地西部、南部的秦祁海槽[9]，上述古地貌控制着奥陶系沉积厚度的变化和相带的展布，决定了鄂尔多斯盆地奥陶纪岩相古地理格局。

盆地中东部奥陶系马家沟组沉积期经历三次海进、海退旋回，沉积了一套以碳酸盐岩为主夹蒸发岩的地层，自下而上可换分为马一段至马六段六个岩性段。目的层马五段自上而下细分为马五$_1$亚段至马五$_{10}$亚段，形成于海退期，主要为以膏岩、盐岩、白云岩为主[10]。中奥陶世末，华北地块因晚加里东运动整体抬升，经历了130—150Ma的沉积间断，盆地主体缺失晚奥陶—早石炭世沉积，中奥陶统马家沟组经历了长期岩溶作用。该期盆地总体表现为西高东低的岩溶古地貌格局，表现为岩溶高地、岩溶斜坡、岩溶盆地的岩溶古地貌格局(图1、图2)。岩溶储层主要发育于马五$_1$亚段至马五$_4$亚段，是盆地中部靖边地区下古生界的天然气主要的储层、产层。

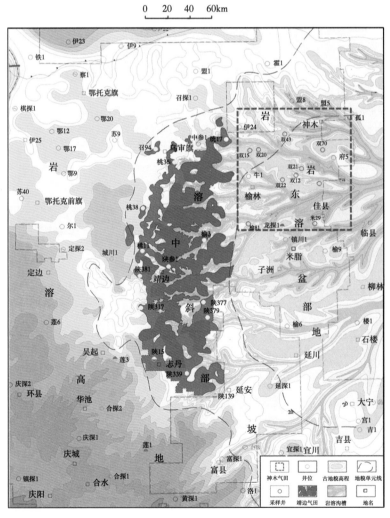

图1 鄂尔多斯盆地前石炭系岩溶古地貌格局

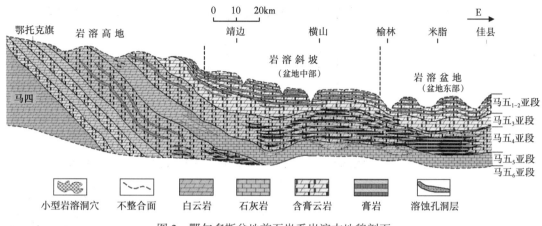

图 2　鄂尔多斯盆地前石炭系岩溶古地貌剖面

2　盆地东部古岩溶储层特征

2.1　古岩溶储层岩石学及物性特征

碳酸盐岩抬升裸露地表或近地表后受到各种复杂物理化学风化营力作用影响,并伴随各种机械、重力、化学等沉积作用,沉积期形成的原始地层结构发生改变,最终形成多种复杂类型岩溶岩,可细分为岩溶建造岩和岩溶改造岩两大类[2]。岩溶建造岩为岩溶溶洞中沉积并固化的机械沉积物、化学沉淀物及其他搬运至溶洞再堆积得物质,原始地层结构被彻底改变,可细分为残积岩、塌积岩、冲积岩、填积岩和淀积岩,此类岩石物性普遍较差,难以形成储层(图3a、b、c)。经历过岩溶作用而仍保持一定原始沉积结构的岩石称为岩溶改造岩,根据岩溶作用方式及引起的物理化学变化,可划分为岩溶溶蚀岩、岩溶变形岩及岩溶交代岩,前两种利于储层形成。盆地东部奥陶系顶部碳酸盐岩地层发育的岩石类型包括硬石膏结核云岩、白云岩、泥云岩、云膏岩、膏云岩、膏岩、云灰岩、灰云岩及石灰岩等多种类型,总体为潮坪沉积环境的产物。发育硬石膏结核及柱状晶体的白云岩在裸露风化期因大气淡水淋滤而形成溶模孔的岩溶溶蚀岩是研究区最重要的储集岩(图3d、f-i)。发生岩体张裂或假角砾化而形成的岩溶变形岩因发育溶滤缝和卸载缝也具有一定的储集性能,取决于裂缝系统的后期充填程度(图3e、j)。

盆地东部储集岩主要为硬石膏结核云岩和白云岩,并以前者为主,与盆地中部靖边气田储集岩性基本类似。盆地东部地区50余口探井密集取样的物性分析表明,下古生界岩溶储层总体表现出低孔隙度、致密的特征。孔隙度主要分布在1%~7%之间,在此范围的样品比例为76.9%,大于7%的比例为11.7%,平均为3.8%。渗透率主要分布在0.005~1mD,在该区间的样品比例为74.56%,大于1mD的比例为12.57%,平均为0.82mD(图4)。盆地中部靖边地区储层平均孔隙为6.70%,渗透率为3.80mD,与其相比,东部地区储层品质有所降低。

鄂尔多斯盆地上古生界碎屑岩储层孔隙度、渗透率表现出明显的线性相关性[11],受研究区碳酸岩储层孔喉结构、残余裂缝及孔洞的影响,下古生界储层物性线性相关性总体偏差,数据分布较为分散(图5a)。不发育裂缝、孔洞的基质储层孔隙度、渗透率正相关性明显,但线性

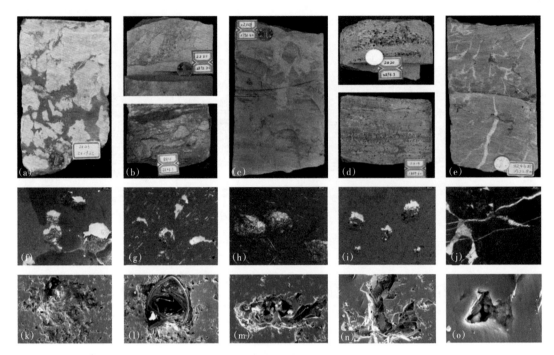

图 3　鄂尔多斯盆地东部奥陶系岩溶储层马五$_1$至马五$_4$段储层特征

(a) 双 43，马五$_2^1$，2329.6m，岩溶建造残积岩，可见溶解残余砾石组分，内部被细粒碎屑充填；(b) 双 20，马五$_1^4$，2872.2m；双 21，马五$_1^4$，2623.1m，岩溶建造塌积岩，可见塌积搬运过程中形成的磨蚀、圆化边界；(c) 双 22，马五$_1^4$，2794.4m 岩溶建造填积岩，可见近垂向充填的棱角状砾石结构；(d) 双 20，马五$_1^3$，2873.5m；双 15，马五$_1^4$，2859.6m，岩溶改造溶蚀岩，可见硬石膏结核溶蚀孔；(e) 双 43，马五$_1^4$，2522.8m，岩溶改造变形岩，可见大量岩溶裂缝；(f) 双 20，马五$_1^3$，硬石膏结核云岩，球状溶孔发育，白云石粉砂半充填；(g) 双 43，马五$_1^3$，硬石膏结核白云岩，球状溶孔、柱状溶孔发育，白云石粉砂半充填；(h) 陕 267，马五$_1^2$，硬石膏结核云岩，球状溶孔，方解石+白云石粉砂全充填；(i) 双 12，马五$_1^2$，硬石膏结核云岩，球状溶孔发育，白云石粉砂+高岭石全充填；(j) 榆 82，马五$_1^3$，白云岩，岩溶裂缝，方解石全充填；(k)-(o) 双 15，马五$_1^2$，2860.5m，泥云岩，石膏高结核溶蚀孔、白云石晶间孔发育，孔隙直径分布于几微米至几百微米

相关性一般，随着孔隙度增加渗透率逐渐增加，但数据点呈现分散、不集中的分布特征。部分样品表现出高孔隙度低渗透率特征，主要原因为结核溶蚀孔虽然发育，但孔隙呈孤立状分布，连通性差；部分样品表现出低孔隙度、高渗透率特征，主要受裂缝影响，渗透性虽好，但储集性差。

2.2　储集空间特征

孔隙、溶洞、裂缝三大类型储集空间在盆地东部碳酸盐岩风化壳岩溶储层中均有发育，以孔隙为主(图 3d、f-i)，偶见溶洞，且多被后期填充，受岩溶与后期构造作用影响，裂缝体系早期曾较发育，后期多被填充(图 3e、j)。孔隙类型主要为硬石膏结核球状溶孔、白云石晶间微孔、岩溶角砾间孔等类型，并以前两种为主。硬石膏结核球状溶孔呈圆形、椭圆形，直径范围 0.10~5.00mm，并以 1.5~3.0mm 为主，多被部分充填或全充填。白云石晶间微孔发育于细粉晶、泥晶白云岩中，也发育于硬石膏结核云岩基质中，直径范围几微米至近百微米(图 3n、o)。总体而言，盆地东部风化壳岩溶储层孔隙直径分布于微米级至毫米级范围，呈连续状分布特征。

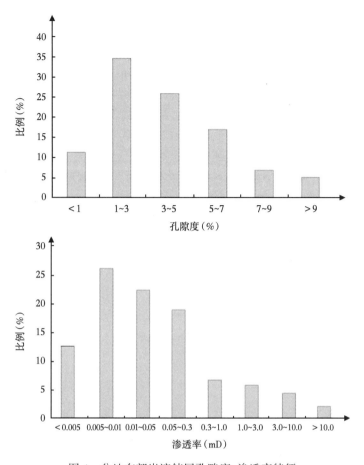

图 4 盆地东部岩溶储层孔隙度、渗透率特征

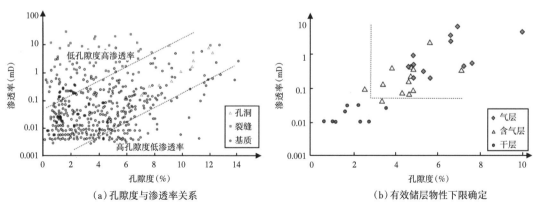

(a) 孔隙度与渗透率关系　　　　(b) 有效储层物性下限确定

图 5 鄂尔多斯盆地东部岩溶储层孔隙度渗透率关系及物性下限确定

铸体薄片孔喉图像分析表明,随着硬石膏结核球状溶孔填充程度的增加或发育程度的降低,对储集性能贡献作用较大的大孔隙发育比例逐渐降低,而微小孔隙发育比例逐渐升高。球状溶孔充填程度低的白云岩储层孔隙总体小于5mm,占总孔隙体积50%(即 $P_{50\%}$)以上的较大

孔隙对应的孔隙直径下限为 2mm,即直径为 2~5mm 的大孔隙占总孔隙的 50%,而球状溶孔充填程度高的白云岩、纯白云岩、泥云岩等 $P_{50\%}$ 对应的孔隙直径下限分别为 20μm、15μm、7μm。喉道也表现出类似特征,低球状溶孔充填程度的白云岩储层喉道总体小于 50μm,占总喉道 50%(即 $P_{50\%}$)以上的较大喉道对应的喉道下限直径为 10μm,即直径 10~50μm 的大喉道占总喉道的 50%,而球状溶孔充填程度高的白云岩、纯白云岩、泥云岩等 $P_{50\%}$ 对应的喉道直径下限分别为 4μm、3μm、2μm(图 6)。泥云岩为非储层类,分析表明其孔隙直径总体小于 30μm、喉道直径总体小于 5μm,可以此作为储层与非储层的孔—喉直径分界。球状溶孔充填程度低的白云岩、球状溶孔充填程度高的白云岩、纯白云岩等储层无效孔隙占的比例分别为 10%、65%、80%,无效喉道占的比例分别为 10%、55%、60%。总体而言,随着球状溶孔填充程度的增加或发育程度的降低,无效孔喉占的比例逐渐增加。

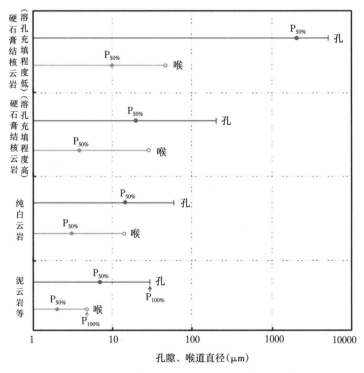

图 6　盆地东部岩溶储层孔隙、喉道直径分布

3　盆地东部古岩溶有效储层形成机理

结合大量试气、生产动态数据进行了储层物性下限分析,结果表明,孔隙度低于 3%、渗透率低于 0.05mD 的储层难以形成有效产层,为干层,孔隙度 3%、渗透率 0.05mD 可界定为有效储层的物性下限标准,高于物性下限标准的储层产气能力较强,为有效储层类型(图 5b)。有效储层形成过程极为复杂,综合受到有利沉积微相/岩相组合、高效岩溶作用、建设性及破坏性成岩作用共同影响。

3.1 有利沉积相/岩相组合

盆地中东部地区马家沟组为海平面周期性升降交替形成的碳酸盐岩与膏岩交互沉积。盆地东部马五$_{1-4}$亚段主体属于潮坪沉积环境的产物，主要亚相类型为潮上带、潮间带，岩相类型主要为含硬石膏结核云岩、白云岩、泥云岩、云灰岩、灰云岩、膏云岩及云膏岩等。盆地东部马五$_{1-4}$亚段在加里东期遭受较强烈的岩溶淋滤作用[2-15]，原始沉积产物受到岩溶作用改造。有利沉积相为储层形成提供了物质基础，有利岩相的发育是有效储层形成的关键。

潮间带及潮上带发育的含硬石膏结核云岩是盆地东部最有利的沉积相/岩相组合。硬石膏结核云岩由于含有大小适宜且易溶的硬石膏结核组分，便于在古岩溶作用中接受改造而形成百微米至毫米级溶孔，且此类溶孔又相对易于保存，因此硬石膏结核云岩为盆地东部地区最有利的沉积岩相(图3d)。白云岩、泥云岩、云灰岩、灰云岩、石灰岩等不含易溶组分，难以受到岩溶改造作用的影响，不利于有效储层的形成。膏云岩、云膏岩、膏岩等岩相易溶组分含量较高，在强岩溶淋滤作用下可能会因缺乏有效支撑而发生垮塌、填充，最终不利于储集空间的保存。

3.2 高效岩溶作用改造

有利岩相的存在为有效储层的形成创造了基础条件，但如果没有高效岩溶改造作用的存在，有利岩相难以转变为有效储层，因此高效岩溶改造是有效储层形成的关键。加里东末期，盆地整体抬升，马家沟组遭受长达1.3亿—1.5亿年的风化淋滤剥蚀。盆地前石炭纪古地貌分布以近南北向的中央古隆起为中心，古地势逐渐向东西两侧降低，经历剥蚀改造的奥陶系顶部界面呈现为以鄂托克旗—定边—庆阳一线为中心向东西倾伏的特点，并直接影响了盆地不同地区岩溶作用的发育强度(图1)。

根据盆地岩溶古地貌宏观格局，中东部岩溶古地貌可划分为岩溶高地、岩溶斜坡、岩溶盆地三种地貌单元(图1)[13-17]。岩溶高地古地势较高，侵蚀强度大、地层缺失严重，岩溶作用以垂向渗滤为主，形成垂向溶蚀带、落水洞等岩溶形态，非均质性较强。岩溶斜坡地带岩溶作用方式以水平状慢速扩散流溶蚀为主，有利于良好溶蚀性储层的形成，储层均质性较好，靖边气田即位于岩溶斜坡部位(图2)。岩溶盆地为岩溶斜坡以东大片地区古地势平坦开阔区，岩溶作用以沿地表侵蚀带的溶蚀及浅层地下径流带的岩石溶解为主，层状溶蚀作用偏弱，非均质性更强，此外，该区处于水流的汇水排泄区，充填、淀积作用强，岩溶空间充填作用高[13]。整体而言，盆地东部大部分地区都处于岩溶盆地范围，岩溶作用强度不如靖边气田所处的岩溶斜坡部位。

虽然盆地东部岩溶盆地区岩溶作用强度整体不如盆地中部区，但仍存在有效储层发育的条件。依据神木地区奥陶系风化壳上覆地层石炭系厚度及分布趋势、风化壳残余厚度及残余边界分布将岩溶盆地内古地貌刻画为丘台、坡地、沟槽—洼地等次级微地貌单元。丘台、坡地是古地形相对较高、马五段顶部(马五$_1^4$以上)保留相对完整的地带，其上发育有利沉积岩相的部位可受到较强岩溶作用的改造，是盆地东部地区有利于有效储层发育的沉积岩相—微地貌组合单元。有限的试气资料表明，产气井主要分布于丘台、斜坡部位，沟槽—洼地部位因岩溶作用过于强烈，导致地层缺失严重，为汇水排泄区，充填作用较强，不利于储层的保存发育(图7)，孔隙度平面等值线图也证明了这一点(图8)。同盆地中部靖边地区相比，神木地区岩溶作用总体偏弱且非均质性变强，有效储层连通性较差，西部、南部区比东部区有效储层发育。

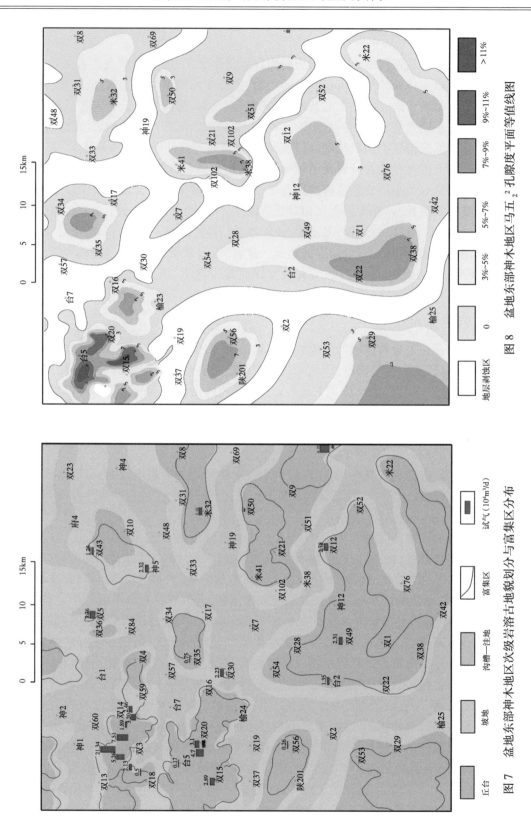

图8 盆地东部神木地区马五$_2^2$孔隙度平面等值线图

图7 盆地东部神木地区次级岩溶古地貌划分与富集区分布

3.3 建设性与破坏性成岩作用综合影响

鄂尔多斯盆地中东部下古生界碳酸盐岩地层经历了极其复杂的成岩作用过程,主要成岩作用类型包括溶蚀作用、白云石沉淀、干化脱水、机械压实、压溶、去白云石化、岩溶化和角砾化、胶结等众多类型[18]。沉积作用之后淋滤剥蚀期之前的浅埋藏期主要发生大气淡水溶蚀、白云石沉淀、干化脱水、机械压实、胶结等成岩作用,淋滤剥蚀期之后的埋藏期主要发生埋藏溶蚀、压溶、胶结、白云石化等成岩作用。

影响最大的成岩作用发生于在中石炭世之后的后期埋藏阶段,在淋滤剥蚀期岩溶作用改造了有利岩相储集物性的基础上,后期建设性与破坏性成岩作用的综合叠加决定有效储层的最终形成。埋藏溶蚀和白云石化对储层的改造是建设性的,在一定程度上提高了储层储集性能,而胶结充填作用是后期最主要的破坏性成岩作用,也是影响作用最大的成岩作用,此作用过程堵塞了岩溶改造阶段形成的溶蚀孔隙,使储层品质大幅降低。

盆地东部岩溶储层发育多种胶结充填类型,包括方解石(含铁方解石)、白云石(含铁白云石)、石英、高岭石、黄铁矿、萤石等。这些胶结充填物以单种、两种或多种组合的方式充填于岩溶孔隙空间中,其中白云石与其他类相组合的胶结充填方式最为普遍(图3f-i)。根据充填程度的强弱可分为半充填型和全充填型。全充填型的硬石膏结核溶孔虽然具备有利沉积岩相基础,也曾受到高效岩溶作用的改造,但因后期溶孔被完全充填堵死,储层品质变差,最终难以成为有效储层,而半充填型硬石膏结核溶孔云岩的溶蚀孔隙得以部分保存,储层品质较好,最终成为盆地东部地区最重要的有效储层类型(图3f-g)。

3.4 裂缝发育影响

盆地东部下古生界顶部风化壳地层受岩溶改造、构造运动、成岩收缩等多种因素影响,裂缝系统较为发育,主要为溶蚀缝、构造缝和成岩缝等三种类型。裂缝镜下宽0.01~1.00mm,岩心观察宽度一般在0.50~3.00mm,最宽可达1cm(图3m)。裂缝系统能够有效改善储层储集性能,尤其是渗流性能,物性分析也表明,残存裂缝系统的发育使部分储层具有低孔隙度、高渗透率的特征(图5a)。在后期成岩作用阶段裂缝网络大都被胶结物充填,因此对储层储集与渗流性能的改善作用有限,仅少量残存裂缝系统可局部改善储层物性。总体而言,裂缝系统对盆地东部地区下古生界碳酸盐岩风化壳型有效储层的形成影响有限。

4 盆地东部天然气富集潜力及与中部靖边气田对比

4.1 天然气富集潜力分析

4.1.1 有效储层发育与分布

盆地东部下古生界碳酸盐岩岩溶地层有效储层的形成受有利沉积相/岩相组合、高效岩溶改造、后期成岩作用与裂缝发育等因素的综合影响。有效储层发育是天然气能够富集的基本条件,是天然气富集的主要控制因素之一。半充填型硬石膏结核溶孔云岩是盆地东部地区最重要的有效储层类型,其分布受有利沉积相带、地层残存状况、次级古地貌单元的共同影响,主要出现在潮坪相硬石膏结核云岩较为发育、沉积地层保存相对完好、相对凸起的丘台、坡地等次级岩溶古地貌单元上。神木地区评价表明:马五$_1$亚段上部马五$_1^1$小层、马五$_1^2$小层严重缺失,有效储层不发育;马五$_1$亚段下部马五$_1^3$小层、马五$_1^4$小层保存较好,有利沉积/岩相发育,且主体位于丘台地貌背景,有效储层较为发育;马五$_2$亚段上部马五$_2^1$小层有利岩相不发育,有效储层不发育,下部马五$_2^2$小层有利岩相较为发育、地层保存较为完善,主体位于丘台地貌背景,有效储层非常发育;马五$_3$亚段整体以膏岩沉积为主,有利岩相不发育,虽地层保存良好,但有效储层不发育;马五$_4$亚段有利岩相发育,地层保存相当完整,但由于岩溶改造作用偏弱,有效储层发育较差。

据钻井资料揭示,神木气田厚度大于2m的有效储层钻遇率为51.81%,厚度主要分布于2～8m(图9)。整个盆地东部地区有效储层发育状况应该同神木地区基本类似,总体而言,盆地东部地区仍具备相对较好的有效储层发育基础。

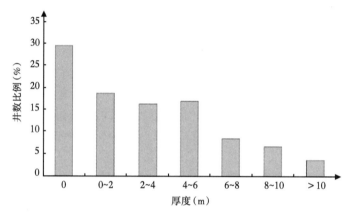

图9 盆地东部神木地区岩溶地层有效储层厚度分布(马五$_1$亚段—马五$_4$亚段)

4.1.2 源—运—储配置关系

鄂尔多斯盆地上古生界煤系烃源岩有机质丰度高,煤岩平均有机碳含量67.3%,碳质泥岩平均有机碳含量2.93%,具有较强的生气能力,是奥陶系顶部气藏的主要气源岩[19-24]。盆地东部地区上古生界源岩生烃强度大,生烃强度范围为$20 \times 10^8 \sim 40 \times 10^8 m^3/km^2$,平均为$36 \times 10^8 m^3/km^2$,高于盆地中西部地区,气源供给充分(图10)。在生排烃高峰期,天然气沿古沟槽与岩溶不整合面向下近距离运移,与风化壳岩溶储层构成良好上生下储组合关系[18-22]。

在气源岩、有效储层、运移通道都具备的条件下,三者良好的配置关系是盆地东部区域下古生界天然气能够有效富集的关键。盆地东部区域神木地区下古生界奥陶系顶部岩溶侵蚀差异明显,总体而言,神木地区的西部岩溶作用强于东部,西部地区岩溶沟槽多切割至马五$_3$亚段,局部甚至可达马五$_4$亚段,而东部普遍溶蚀切割至马五$_2$亚段(图11)。发育于丘台、斜坡部位的马家沟组上部马五$_{1-2}$亚段的有效储层在侧向或垂向可与岩溶不整合面及岩溶古沟槽等优势运移通道紧密相邻,并且临近上古生界气源岩,源—运—储配置关系最佳,有利于天然气富集。马五$_4$亚段总体发育完整,虽然其上部有效储层发育,但因被马五$_3$亚段厚层稳定泥云岩遮挡,优势运移通道难以与有效储层充分接触,源—运—储配置关系不佳,总体不利于天然气富集,仅在西部距离深切沟槽较近的区域有气层发育,东部区域则主要发育气水层、含气水层(图12)。此外,盆地东部的西部地区岩溶作用比东部地区相对强烈,有效储层发育程度高,

鄂尔多斯盆地东部奥陶系古岩溶型碳酸盐岩致密储层特征、形成与天然气富集

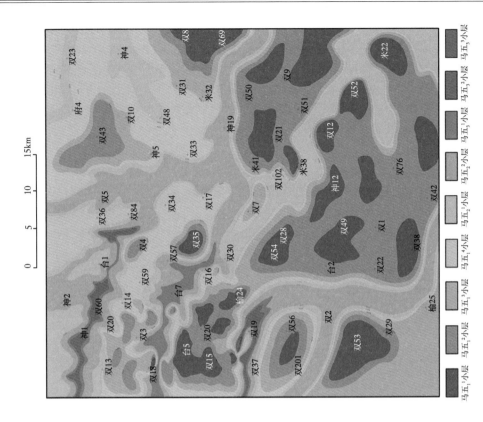

图11 盆地东部神木地区岩溶地层残存图

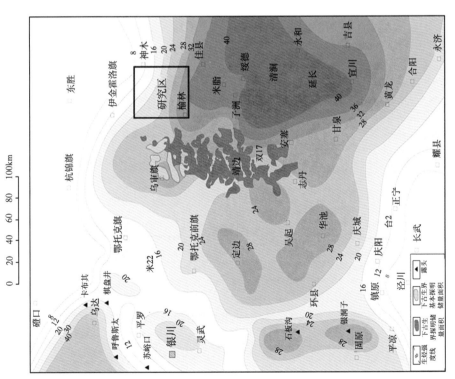

图10 鄂尔多斯盆地上古生界生烃强度等值线图

岩溶不整合面及古沟槽切割深度大，且上覆本溪组厚度薄，天然气运送距离短，源—运—储配置关系良好，因此气层发育程度比东部地区大。

总体而言，盆地东部区域上部层位比下部层位源—运—储配置关系好，西部地区比东部地区源—运—储配置关系好，即盆地东部区域发育于岩溶丘台、斜坡部位的马五$_{1-2}$亚段比保存相对较为完整马五$_4$亚段更有利于天然气富集，盆地东部区域靠近岩溶斜坡的西部地区比处于岩溶盆地的东部地区更利于天然气富集。

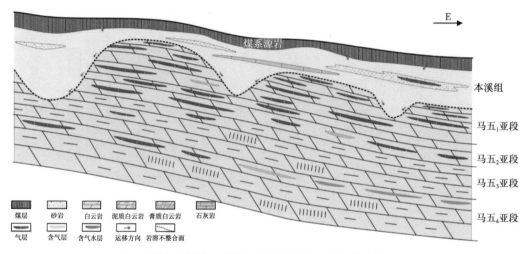

图 12 鄂尔多斯盆地东部下古生界顶部岩溶气藏富集模式

4.1.3 岩溶储层物性的强非均质性

室内模拟实验及相关分析表明，有效储层孔隙空间相对较大、物性较好，在天然气富集过程中起始充注压力低、运移阻力小，因此有利于天然气完全驱替可动地层水而高效富集，往往形成纯气层，而物性相对较差的储层因储层孔隙空间相对狭窄，天然气难以完全驱替地层可动水而不利于高效富集，往往形成气水层及含气水层[25-29]。

受沉积、岩溶、成岩等复杂多因素综合作用的影响，盆地东部下古生界岩溶储层非均质性要明显强于盆地中部靖边地区，储层的强非均质性影响着盆地东部下古生界储层气、水分布格局。从宏观来看，平面上西部储层物性整体要优于东部，纵向上上部地层储层物性要整体优于下部，因此，西部地区上部马五$_{1-2}$亚段纯气层发育比例较高，而东部地区下部马五$_4^1$小层气水层、含气水层比例较高。从局部来看，西部地区上部层位马五$_{1-2}$亚段物性普遍较好的储层中，也局部发育少量物性相对较差储层，因此表现为纯气层中夹有少量气水层的特征。

4.2 与盆地中部气田对比

靖边气田位于鄂尔多斯盆地中部，是盆地碳酸岩岩溶地层发现的唯一探明储量超千亿立方米的气田。对比评价表明，盆地东部有利沉积相/岩相组合不如中部靖边地区发育程度好，表生岩溶作用亦不如靖边地区高效，岩溶孔隙的填充程度也比靖边地区高，储层品质总体相对偏差（表 1）。但盆地东部地区下古生界碳酸盐岩岩溶储层平均有效厚度为 4.5m，局部地区可达 8m 以上（图 9），多层叠置也可大面积连片分布，总体仍较为发育，且盆地东部平均生烃强

度达 $36×10^8m^3/km^2$，远高于盆地中部地区，气源供给充分。盆地东部神木地区评价结果表明，马五$_{1-4}$亚段平均储量丰度为 $0.52×10^8m^3/km^2$，局部地区可与靖边地区平均值相当。鄂尔多斯盆地东部地区面积广阔，围绕神木—米脂—清涧—宜川一线，面积可达 $3×10^8km^2$，下古生界碳酸盐岩地层具备局部较大天然气勘探开发潜力。

表1 鄂尔多斯盆地中、东部下古生界碳酸盐岩岩溶储层参数对比表

对比指标	盆地中部	盆地东部
平均孔隙度(%)	6.00	3.80
平均渗透率(mD)	2.63	0.82
平均储层厚度(m)	6.6	4.5
平均生烃强度($10^8m^3/km^2$)	24	36
储量丰度($10^8m^3/km^2$)	0.72	0.52
孔喉特征	以溶蚀孔为主，晶间孔及膏模孔次之，属微米至毫米级孔隙，孔隙充填程度中等—偏高；可见微裂缝，可有效改善储层渗透性	以溶蚀孔为主，晶间孔及膏模孔次之，裂缝属微米至毫米级孔隙，孔隙充填程度普遍较高；微裂缝大多被充填，对储层渗透性的改善作用有限
有利沉积相/岩相组合	潮上含硬石膏结核云坪，分布广、连续性普遍较好，大面积连片分布	潮间含硬石膏结核云坪，分布局限、单层连续性较差，但多层叠合也可连片
岩溶古地貌背景	主体位于岩溶斜坡淋滤溶蚀区，存在垂直扩散渗滤、水平潜流及潜水面以下深部缓流等三种水流方式作用，表生岩溶作用较强，形成大量石膏结核溶蚀孔	主体位于岩溶盆地汇水区，表生岩溶作用偏弱，填充作用偏强，局部微构造高部位（丘台、坡地）存在强溶蚀区，有利于石膏结核溶蚀孔发育
储层分布特征	主力层位大面积连片分布，储层连续及连通性好；非主力层呈孤立井点状、孤岛状分布	主体以孤立井点状、孤岛状分布为主，多层叠置可连片，储层连通性总体偏差

5 结论

（1）鄂尔多斯盆地东部奥陶系顶部碳酸盐岩风化壳储层发育，以硬石膏结核云岩为主，储层孔隙直径分布于微米级至毫米级，喉道直径分布于几微米至几十微米，孔隙直径 $30\mu m$、喉道直径 $5\mu m$ 为储层与非储层的孔—喉界限标准。下古生界储层总体表现为低孔隙度—特低孔隙度、低渗透—致密的特征。

（2）有效储层的发育受有利沉积相/岩相组合、高效岩溶作用、成岩综合作用及裂缝的综合影响。有利岩相发育是有效储层得以形成的基本条件，高效岩溶改造是有效储层形成的关键，而成岩综合作用决定有效储层的最终形成，半充填型硬石膏结核溶孔云岩为盆地东部地区最重要的有效储层类型，由于大多被充填，裂缝系统对有效储层的形成影响有限。

（3）天然气富集受有效储层发育与分布、源—运—储配置关系、岩溶储层强非均质性的共同影响。有效储层发育是天然气能够富集的基本条件，源—运—储配置关系是盆地东部区域

下古生界天然气能够富集的关键,而储层物性强非均质性影响气、水分布格局。总体而言,盆地东部碳酸盐岩风化壳储层较为发育,气源供给充足,具备天然气大规模富集的条件,具有较大的天然气勘探开发潜力。

参 考 文 献

[1] 马永生,李启明,关德师,等. 鄂尔多斯盆地中部气田奥陶系马五碳酸盐岩微相特征与储层不均质性研究[J]. 沉积学报,1996,14(1):22-32.

[2] 何江,方少仙,侯方浩,等. 风化壳古岩溶垂向分带与储集层评价预测——以鄂尔多斯盆地中部气田区马家沟组马五$_5$—马五$_1$亚段为例[J]. 石油勘探与开发,2013,40(5):534-539.

[3] 杨华,付金华,魏新善,等. 鄂尔多斯盆地奥陶系海相碳酸盐岩天然气勘探领域[J]. 石油学报,2011,32(5):733-741.

[4] 杨华,刘新社,张道峰. 鄂尔多斯盆地奥陶系海相碳酸盐岩天然气成藏主控因素及勘探进展[J]. 天然气工业,2013,33(5):1-11.

[5] 韩品龙,张月巧,冯乔,等. 鄂尔多斯盆地祁连海域奥陶纪岩相古地理特征及演化[J]. 现代地质,2009,23(5):822-827.

[6] 杨华,付金华,包洪平. 鄂尔多斯地区西部和南部奥陶纪海槽边缘沉积特征与天然气成藏潜力分析[J]. 海相油气地质,2010,15(2):1-13.

[7] 冉新权,付金华,魏新善,等. 鄂尔多斯盆地奥陶系顶面形成演化与储集层发育[J]. 石油勘探与开发,2012,39(2):154-161.

[8] 黄正良,包洪平,任军峰,等. 鄂尔多斯盆地南部奥陶系马家沟组白云岩特征及成因机理分析[J]. 现代地质,2011,25(5):925-930.

[9] 付金华,白海峰,孙六一,等. 鄂尔多斯盆地奥陶系碳酸盐岩储集体类型及特征[J]. 石油学报,2012,32(2):110-117.

[10] 冯增昭,鲍志东. 鄂尔多斯奥陶纪马家沟期岩相古地理[J]. 沉积学报,1999,17(1):1-8.

[11] 王国亭,何东博,王少飞,等. 苏里格致密砂岩储层岩石孔隙结构及储集性能特征[J]. 石油学报,2013,34(4):660-666.

[12] 吴永平,王允诚. 鄂尔多斯盆地靖边气田高产富集因素[J]. 石油与天然气地质,2007,28(4):473-478.

[13] 夏日元,唐健生,关碧珠,等. 鄂尔多斯盆地奥陶系岩溶地貌及天然气富集特征[J]. 石油与天然气地质,1999,20(2):133-136.

[14] 顾岱鸿,代金友,兰朝利,等. 靖边气田沟槽高精度综合识别技术[J]. 石油勘探与开发,2007,34(1):60-64.

[15] 徐波,孙卫,宴宁平,等. 鄂尔多斯盆地靖边气田沟槽与裂缝的配置关系对天然气富集程度的影响[J]. 现代地质,2009,23(2):299-304.

[16] 苏中堂,陈洪德,林良彪,等. 靖边气田北部下奥陶统马五$_4^1$段古岩溶储层特征及其控制因素[J]. 矿物岩石,2011,31(1):89-96.

[17] 拜文华,吕锡敏,李小军,等. 古岩溶盆地岩溶作用模式及古地貌精细刻画——以鄂尔多斯盆地东部奥陶系风化壳为例[J]. 现代地质,2002,16(3):292-298.

[18] 杨华,王宝清,孙六一,等. 鄂尔多斯盆地南部奥陶系碳酸盐岩储层的胶结作用[J]. 沉积学报,2013,31(3):527-535.

[19] 夏新宇,赵林,李剑锋,等. 长庆气田天然气地球化学特征及奥陶系气藏成因[J]. 科学通报,1999,44(10):1116-1119.

[20] 戴金星,李剑,罗霞,等.鄂尔多斯盆地大气田的烷烃气碳同位素组成特征及其气源对比[J].石油学报,2005,26(1):18-26.
[21] 陈安定.陕甘宁盆地中部奥陶系天然气的成因与运移[J].石油学报,1994,15(2):1-9.
[22] 林家善,周文,张宗林,等.靖边气田下古气藏相对富水区控制因素及气水分布模式研究[J].大庆石油地质与开发,2007,26(5):72-74.
[23] 程付启,金强,刘文汇,等.鄂尔多斯盆地中部气田奥陶系风化壳混源气成藏分析[J].石油学报,2007,28(1):38-42.
[24] 肖晖,赵靖舟,王大兴,等.鄂尔多斯盆地奥陶系原生天然气地球化学特征及其对靖边气田气源的意义[J].石油与天然气地质,2013,34(5):601-609.
[25] 姜福杰,庞雄奇,姜振学,等.致密砂岩气藏成藏过程的物理模拟实验[J].地质评论,2007,53(6):844-849.
[26] 姜福杰,庞雄奇,武丽.致密砂岩气藏成藏过程的地质门限及其控气机理[J].石油学报,2010,31(1):49-54.
[27] 邹才能,陶士振,张响响,等.中国低孔渗大气区地质特征、控制因素和成藏机制[J].中国科学 D 辑,2009,39(11):1607-1624.
[28] 邹才能,陶士振,袁选俊,等.连续型油气藏形成条件与分布特征[J].石油学报,2009,30(3):324-331.
[29] 邹才能,陶士振,朱如凯,等."连续型"气藏及其大气区形成机制与分布——以四川盆地上三叠统须家河组煤系大气区为例[J].石油勘探与开发,2009,36(3):307-319.

原文刊于《石油与天然气地质》,2018,39(4):685-695.

致密砂岩气藏有效砂体规模及气井开发指标评价
——以鄂尔多斯盆地神木气田太原组气藏为例

孟德伟[1]　贾爱林[1]　郭　智[1]　靳锁宝[2]　王国亭[1]　冀　光[1]　程立华[1]

(1. 中国石油勘探开发研究院,北京 100083;2. 中国石油长庆油田分公司,西安 710018)

摘要:新层系有效砂体规模尺度不明确及作为新投产气田尚缺乏系统的开发指标评价体系是面临的两大问题。以储层地质特征为基础,从静、动态结合角度出发,综合利用野外露头测量、密井区地质解剖、不稳定试井边界探测及生产动态泄流半径分析等手段,系统论证神木气田有效砂体规模尺度,提出气田合理的开发井网;以压降法和产量不稳定分析法为基础评价气井动态控制储量并建立低渗透—致密藏气井动态储量变化趋势图版加以修正;综合静态储层物性参数、生产测试及产能试井解释交会分析,回归建立适合目标气藏的产能计算公式;指示曲线、合理生产压差及数值模拟相结合确定气井合理的生产制度。研究结果表明:神木气田太原组气藏有效砂体宽度范围 452~650 m,长度范围 678~1 000 m,合理的开发井网为 600 m×800 m;形成了以动态控制储量、无阻流量、稳产三年合理配产三项关键指标为基础的分类气井开发指标综合评价体系,明确了气井的开发潜力,该评价成果可为气田实现"规模建产、高效开发"提供技术支撑。

关键词:鄂尔多斯盆地;神木气田;太原组气藏;有效砂体规模;开发潜力;指标体系

　　致密砂岩气藏是中国天然气开发的主力气藏类型之一,目前产量占全国天然气总产量的比例超过 30%,其中以鄂尔多斯盆地苏里格气田最为典型[1],已成为中国最大的天然气田。致密砂岩气通常大面积含气,但储层薄、储量丰度低,储体内部存在较强的非均质性,具有主力含气砂体和基质储层的"二元"结构。受储层致密、非均质性强、主力含气砂体小而分散的影响,气井具有控制储量低、单井产量低且递减快的特征,只有通过井间接替、不断增加钻井数量的方式才能达到一定规模的生产能力并保持长期稳产。充分利用动、静态资料精细刻画、评价有效砂体发育规模、几何形态及展布方向,进而优化井网井距是提高致密气藏储量动用程度和采收率的关键手段。

　　勘探始于 20 世纪 90 年代初期的神木气田与苏里格气田、靖边气田、榆林气田、子洲气田等共同构成了保障长庆油田 5 000×10^4 t 油气当量长期稳产的半壁江山,从 1996 年陕 201 井在太原组钻遇 10 m 厚气层并获得 2.7×10^4 m^3/d 的无阻流量至今,经过 20 年持续不断的勘探开发,神木气田探明地质储量超过 3 000×10^8 m^3,且已经具备了 20×10^8 m^3/a 的生产能力,展示出了良好的发展前景。神木气田是鄂尔多斯盆地少数以太原组为主力产层的气田,投产时间短,最长气井仅有两年时间,研究论证太原组气藏有效砂体规模尺度并评价钻遇该储层气井的开发潜力,对完善整个鄂尔多斯盆地储层体系及指导神木气田今后的高效开发具有十分重要的意义,同时可以为进一步扩展深化盆地东部天然气勘探开发提供借鉴。文中主要从静态和动态两个角度论证太原组气藏有效砂体规模尺度,首先对气藏有效砂体开展精细解剖,结合露头观测地质统计数据库分析有效砂体形态及规模,进而采用生产动态泄流半径评价及不稳定试井边界探测辅以动态响应验证,确定有效砂体规模尺度,并与已成熟开发的苏里格气田有效砂体发育情况进行对比,以此为依据提出合理的开发井网。分别运用多种方法评价研究区气井的潜在产能、动态控制储量及合理生产制度,明确太原组主力砂体下气井的开发潜力,建立系统的开发指标评价体系,通过与苏里格气田盒八段、山一段主力储层气井开发指标对比分析,为神木气田逐步投产后合理高效开发提供技术支撑,为鄂尔多斯盆地东部后续的产能接替部署提供参考。

1 研究区概况

神木气田地处陕西省榆林市神木县境内,构造上位于鄂尔多斯盆地伊陕斜坡东北部,西接榆林气田,北与大牛地气田相邻,南抵子洲、米脂气田(图1)。上古生界以海陆过渡相—内陆湖盆沉积为主。自下而上发育石炭系本溪组、二叠系太原组、山西组、下石盒子组、上石盒子组及石千峰组,本溪组顶部的9号煤层在该地区普遍发育且分布稳定,形成了良好的地区性标志层[2]。其中二叠系太原组和山西组河流相砂岩储层为气田的主力含气层,厚度约100m。太原组自上而下划分为太一段、太二段两个小段,山西组自上而下划分为山一段、山二段两个小段[2-3](图1)。截至2016年底,神木气田投产气井近600口,年产量规模在$11×10^8m^3$左右。

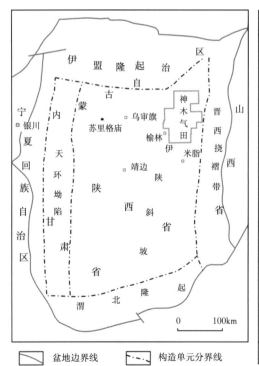

图1 鄂尔多斯盆地神木气田位置及地层发育简况

2 基本地质特征

2.1 沉积相特征

太原组气藏沉积模式为潮控三角洲—潮坪相交互沉积[4-8]。潮控三角洲主要发育三角洲前缘亚相,其中的水下分流河道、分流间湾微相主要见于太二段,远沙坝、席状砂微相主要见于太一段。潮坪相主要发育潮间带和潮下带亚相,包括灰坪、灰泥坪等微相,多见于太一段。

太二段沉积时期,低缓地形使得潮汐作用对三角洲沉积的影响加大,三角洲前缘水下分流河道垂直海岸线呈条带状延伸,横向上比邻三角洲前缘的海岸发育潮坪沉积,沉积灰质、砂、泥和泥炭,外围发育潮汐沙坝,反映三角洲砂体受到了波浪和潮汐的改造。太二段沉积早期发育多条三角洲平原分流河道且比较固定,分流河道间发育洼地和沼泽。太二段沉积晚期,分流河道发育程度减弱,到太一段沉积时期几乎不发育水下分流河道,以前缘泥和灰泥坪为主,总体上海水向北推进,砂体发育受到了抑制,南部逐渐发育潮坪相沉积(图2)。

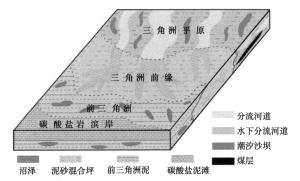

图2 神木气田太原组沉积微相划分及太二段沉积模式

2.2 储层特征

太原组砂岩类型有岩屑石英砂岩、石英砂岩和岩屑砂岩三类,从太二段到太一段,岩屑砂岩比例降低,砂岩成熟度升高。薄片资料显示碎屑成分主要为石英、岩屑,平均面积百分比分别为70.2%和12.3%,长石含量极少,平均面积百分比0.11%。岩屑成分主要包括岩浆岩岩屑、变质岩岩屑和沉积岩岩屑。岩浆岩岩屑以喷发岩和隐晶岩为主,平均4.8%;变质岩屑以片岩、千枚岩、变质砂岩和板岩为主,平均7.1%;沉积岩岩屑以粉砂岩和泥岩为主,平均0.7%。胶结物百分比为1.5%~45.0%,主要为水云母、绿泥石、铁白云石、高岭石、次生石英和铁方解石等。杂基主要是泥质、泥岩岩屑和云母类在压实作用过程中形成的假杂基。这些假杂基充填在颗粒之间,大部分水云母化,或被绿泥石和方解石交代。

太原组气藏总体表现出低孔隙度、低渗透率的储层特征[9]。岩心样品资料分析表明,孔隙度主要分布范围为4%~12%,平均7.1%,渗透率的分布范围为0.01~1.00mD,平均0.20mD。同相邻的苏里格气田储层相比,孔隙度、渗透率及含气饱和度均偏低,但单井钻遇砂体个数较多,平均钻遇储层厚度较大,达到苏里格气田的1.2倍。

孔隙结构方面,太原组气藏以发育小孔喉为主,具有排驱压力低、中值压力低和中值半径小的特点。排驱压力分布区间为 0.2~1.4MPa,中值压力为 1.0~10.0MPa,中值半径分布在 0.33~0.83μm。孔喉分选方面,太原组储层分选系数分布范围为 1.5~3.5,变异系数为 0.10~0.35。分析孔喉连通性,最大进汞饱和度范围 64%~85%,平均 78%,退汞饱和度范围 33%~52%,平均 44%,反映了太原组储层的孔喉连通性较差。

3 有效砂体规模分析

储层有效砂体规模分析对气田开发合理的井网部署至关重要,准确的有效砂体规模可指导开发井距优化,可以达到气井对储量的最大控制和动用程度。若井距设计过大,则井间会有部分有效储层未被钻遇而造成开发井网对储量的控制程度不足,影响采收率;若井距设计过小,则会增加两口井同时钻遇同一砂体的概率,增加井间干扰,导致单井累计产气量减少,影响经济效益。本文从静、动态结合角度考虑,综合运用野外露头测量对比、密井网精细地质解剖、生产动态泄流半径分析及不稳定试井砂体边界解释四种方法论证神木气田太原组气藏有效砂体规模尺度,为合理的井网部署提供建议和指导。

3.1 有效砂体精细解剖

确定有效砂体分布规模和几何形态的地质方法主要有两种[10]:一是露头测量对比法,即选取储层沉积露头,开展砂体二维或三维测量描述,建立露头地质知识库,预测气田有效砂体的规模尺度。如南 Piceance 盆地 Williams Fork 组发育透镜状致密砂体,应用露头资料建立了曲流河点沙坝单砂体的分布模型,为井距优化提供了依据[11];二是密井网解剖法,综合应用地质、地球物理和动态测试资料,开展井间储层精细对比,研究一定井距条件下砂体的连通关系,评价砂体规模的大小,对比过程中辅以地质统计分析,即在确定有效砂体厚度分布区间的基础上,根据定量地质学中同种沉积类型砂体的宽厚比和长宽比来估计有效砂体大小。通过对山西保德太原组露头观察测量,以及神木气田太原组气藏井网精细解剖,储层有效砂体主要呈孤立单砂体、垂向叠置型和侧向叠加型三种。有效单砂体厚度分布范围为 2~5m,宽厚比 80~150,长宽比 1.5~3.8,80% 以上的有效砂体宽度小于 650m,长度小于 1000m(图 3)。因此,从静态地质分析的角度来看,神木气田太原组气藏井距应首先小于 650m,排距小于 1000m。

3.2 生产动态泄流半径评价

生产动态泄气半径评价主要利用气井动态资料分析气井的控制储量和动用范围,进而为优化井网井距提供依据。在明确储层基本地质特征的基础上,综合考虑人工裂缝半长、表皮系数、渗流边界等工程相关参数建立解析模型,对单井生产动态数据(包括产量和井底流压)进行历史拟合,实现模型运行结果与气井实际生产数据在趋势和数值上的吻合,从而确定气井的泄流半径,为制定合理的井网井距提供参考和指导。致密气藏气井通常为压裂后投产,考虑裂缝的评价方法主要有 Blasingame、AG Rate vs Time、NPI、Transient 四种典型无因次产量分析图版及考虑压力历史的解析模型[12-15]。四种典型图版分析方法均以气井的实际产量数据对比拟合不同泄气半径与裂缝半长比值下的无因次产量、无因次产量积分及无因次产量导数等在无因次时间下的典型样板曲线,预测气井的裂缝半长和泄流半径。裂缝解析模型在给定气井产量数据条件下,拟合井底流压,最终确定裂缝半长和泄流半径。神木气田气井经过四种典型

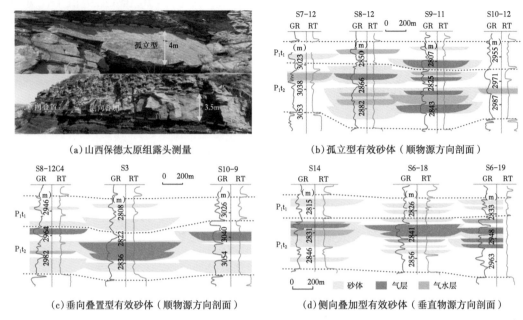

图 3　山西保德太原组露头测量及神木气田太原组气藏井网解剖

图版分析及裂缝解析模型拟合获得泄流半径范围为 130～390m，平均 242m，即得到砂体宽度在 484m 左右（图 4，表 1）。

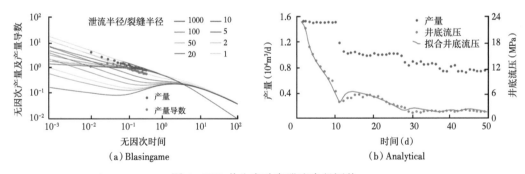

图 4　M38 井生产动态泄流半径评价

表 1　神木气田气井生产动态泄流半径评价

井名	泄流半径(m)	井名	泄流半径(m)
S8-21	380	S6-8	145
S6-12	384	S7-35	134
S7-14	362	S16	163
S14	318	S28	279
S5-18	256	M38	207
S9-12	183	S4-30	210
SA201	161	平均	242

3.3 不稳定试井砂体边界探测

不稳定试井是一种以渗流力学为基础,以高精度压力计等测试仪表为手段,开展油、气井生产动态及关井恢复数据测试,研究和确定油、气藏物性参数,井筒及工程压裂参数,判断油、气井泄压边界距离及几何形态的最为科学有效的方法[16-18]。通过优选储层模型、井筒及工程参数模型、控制边界模型对油、气井测试压力数据进行压力、压力导数双对数曲线,压力半对数曲线及压力变化历史进行综合分析与拟合,区分和识别油、气藏类型及包括井筒储集、近井人工裂缝、远井基质、边界控制在内的不同流动阶段,进而获得反映各流动阶段的特征参数值。神木气田11口气井经不稳定试井分析解释,显示气井钻遇的有效砂体主要具有单条、平行不渗透及条带型三种边界特征,边界距离范围为100~350m,平均226m,即得到砂体宽度约为452m(图5,表2)。

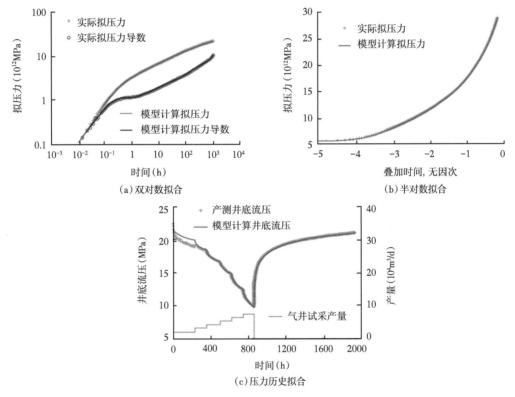

图5 神木气田S6-12井不稳定试井解释

表2 神木气田试采井不稳定试井边界解释

井名	解释模型	复合内区半径R及边界距离L_i(m)
S8-21	有限导流裂缝,径向复合气藏,不渗透边界	$R=136,L=188$
S5-18	有限导流裂缝,平行不渗透边界	$L_1=143,L_2=169$
S6-8	有限导流裂缝,条带型边界	$L_1=182,L_2=349,L_3=84,L_4=178$
S6-12	无限导流裂缝,不渗透边界	$L=313$

续表

井名	解释模型	复合内区半径 R 及边界距离 L_i(m)
S7-14	有限导流裂缝,平行不渗透边界	$L_1=221, L_2=186$
S7-35	无限导流裂缝,径向复合气藏,平行不渗透边界	$R=92, L_1=195, L_2=220$
S14	无限导流裂缝,径向复合气藏,平行不渗透边界	$R=58, L_1=L_2=171$
S16	有限导流裂缝,条带型边界	$L_1=L_3=328m, L_2=L_4=169m$
S28	有限导流裂缝,条带型边界	$L_1=151, L_2=229, L_3=60, L_4=228$
M38	有限导流裂缝,条带型边界	$L_1=168, L_2=209, L_3=344, L_4=214$
SP1	有限导流裂缝,条带型边界	$L_1=327, L_2=211, L_3=218, L_4=348$

3.4 合理开发井距

通过动态、静态两个角度,分别运用野外露头观察测量、密井网精细地质解剖、生产动态泄流半径分析及不稳定试井边界解释四种方法分析论证了神木气田太原组气藏的有效砂体规模,综合得出,宽度范围为452~650m,长度范围为678~1000m。与已成熟开发十余年并已成为国内最大天然气田的苏里格气田做对比[19],结果显示无论在静态有效砂体规模尺度还是动态泄气半径及气井控制边界,均非常接近,因此,提出神木气田适合的主体开发井网600m×800m,同苏里格气田一致(表3)。

表3 神木气田与苏里格气田有效砂体及合理井网对比

井名	地质精细解剖(m)	泄流半径评价(m)	试井边界解释(m)	长宽比	有效砂体规模(m)	主体井网(m×m)
神木气田	长<1000 宽<650	242	226	1.5~3.8	长:678~1000 宽:452~650	600×800
苏里格气田	长<1200 宽<600	246	220	1.2~4	长:720~1200 宽:440~600	600×800

4 开发指标评价

针对处于开发初期的气田,基于早期少量探井或评价井的短期测试资料,以单井为单元进行开发潜力评价,分析预测气井开发动态指标特征,明确气井、气藏的开发潜力,对气田制定合理的开发方案具有重要的指导作用和现实意义。本文研究对象神木气田尚未全面投入开发,自2016年开始仅部分产能建设投入运营,因此主要借助气井试气及短期试采数据分析气井无阻流量、动态控制储量、合理生产制度等生产指标,同时通过与苏里格气田对比开展分类气井评价,形成分类气井指标界限,建立针对鄂尔多斯盆地东部太原组气藏的开发指标评价体系。

4.1 动、静态交会产能评价

应用较广泛的气井无阻流量评价方法主要有三种:试气法、一点法及产能试井法。其中试气法普遍应用于现场投产前的多数新井,数值一般偏大;一点法为统计经验方法,多用于试采井和生产井,通过找到一个稳定的生产点计算无阻流量,根据地区差异经验公式将随之变化;产能试井是基于严格的渗流理论科学推导而来,以试采产量及对应压力数据为基础的最准确的无阻流量评价方法,但受限于气田开发成本和产建压力,往往只选择少数气井开展产能试井,这为气田数百口井实现准确的产能评价带来挑战。为此,通过采用静态地质参数与试气交会分析,并结合产能试井的方式,回归建立一种可以在不要求开展大规模产能试井测试的基础上,能够准确而便捷地对神木气田太原组气藏气井进行产能评价的方法。

以神木气田51口开发评价井测试成果为基础,交会分析产层物性参数与试气无阻流量的相关关系,结果显示地层系数(即有效储层厚度与渗透率的乘积)与试气无阻流量具有良好的正相关性(图6),趋势拟合相关系数高达0.8,说明基于气井测井解释获得的地层系数指标可以对气井试气无阻流量形成良好的预判。同时,经九口气井试气无阻流量与产能试井计算取得的无阻流量进行对比分析,两者具有较好的线性比例关系,即产能试井评价的无阻流量占试气无阻流量的比例均在38%左右,进而综合气井测井解释、试气成果及产能试井评价,回归建立基于静态地层系数的可快速、准确评价神木气田太原组气藏气井无阻流量的计算式(同理,对于其他类型气藏,可采用该思想建立反映目标气藏地质特征的,具有无阻流量计算针对性的评价方法):

$$q_{AOF} = 0.38 \times (1.6241 \times Kh + 2.1986) \tag{1}$$

式中 q_{AOF}——经过静态地层系数与试气无阻流量交会分析,并结合产能试井校正后的无阻流量,$10^4 m^3/d$;

Kh——地层系数,$mD \cdot m$。

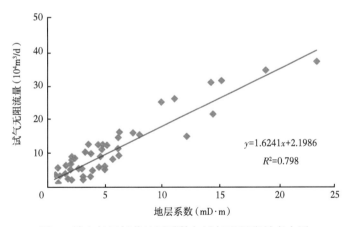

图6 神木气田气井地层系数与试气无阻流量交会图

4.2 动态控制储量校正

为获取准确的单井动态控制储量,多数评价方法要求气井生产到一定阶段,如压降法和流动物质平衡法要求气井采出程度大于10%,弹性二相法和产量不稳定分析法要求气井渗流到达边界控制的拟稳定状态[20-21]。针对投产时间较短的气井尚没有一种适合的动态控制储量评价方法,特别对于低渗透致密气藏,由于渗流速度缓慢,气井泄流到达边界的时间远长于常规气藏,往往计算结果偏小,随着气井生产时间的延长,泄流范围不断扩大,动态控制储量随之变大。为此,在采用产量不稳定分析和压降法两种方法评价神木气田气井动态控制储量的基础上(图4、图7),建立了气井动态控制储量变化趋势图版(图8),用以校正由于气井生产时间短所带来的计算误差。

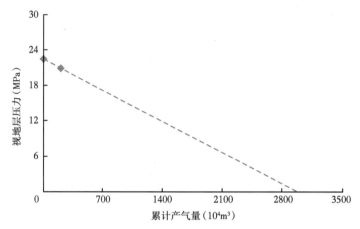

图7 神木气田 S8-21 井压降法动态控制储量评价

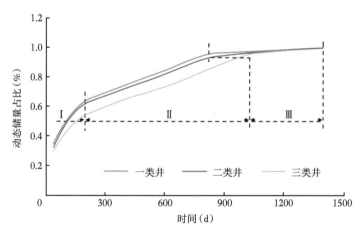

图8 气井动态控制储量变化趋势图版

动态控制储量变化趋势图版主要借助苏里格气田生产时间超过六年的 200 余口气井编制而成,苏里格气田与神木气田为储层物性相近的低渗透—致密砂岩气田,气体流动机理相似,因此该图版可有效对神木气田气井动态控制储量进行校正,提高准确度。图版呈现出三段式的变化特征:Ⅰ是快速上升段,Ⅱ是缓慢上升段,Ⅲ是稳定段,分别反映了气井近井人工裂缝、远井基质及边界波及三个流动控制阶段(图8)。反映储层物性相对较差的三类气井到达边界控制的拟稳态流动阶段时间(1000 天)要明显晚于一类、二类气井(800 天),很好地印证了流体在低渗透致密储层基质中的渗流特征。对于短时间生产的各类气井,计算当前生产时间的动态储量,应用图版折算校正即可获得最终所能控制的实际动态储量。

4.3 合理配产分析

综合运用指示曲线、合理生产压差及数值模拟三种方法评价神木气田气井的合理产量。从气井的二项式产能方程可知,在一定的气井产量范围内,压力平方差与气井产量可近似为线性关系,该范围内非达西流效应对气井生产的影响较小,消耗在非达西流的压降较低,更有效地利用了地层能量,因此采气指示曲线上直线切点位置对应产量即为气井的合理产量(图9a);合理生产压差法主要根据气田开发实践经验,通常取原始地层压力的10%~15%作为气井的合理生产压差,利用气井流入动态曲线获取气井合理产量(图9b);数值模拟方法则通过建立实际气藏井组的三维精细地质模型,经过气井产量及压力历史拟合的动态校正,预测气井的合理产量(图9c)。最终综合对比三种方法的评价成果,获得神木气田气井的合理产量。

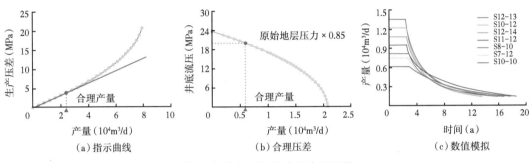

图9 神木气田气井合理产量评价

4.4 分类气井开发指标评价

重点对气井动态控制储量、无阻流量、合理生产制度等关键开发指标开展分析研究,形成了神木气田分类气井的开发指标综合评价体系,明确以太原组为主力产层的神木气田气井的开发潜力。同时,与已实现成功开发的低渗透—致密砂岩气藏典型代表——苏里格气田进行对比,在储层物性近似条件下,各项动态指标均低于苏里格气田的气井(表4)。结合苏里格气田开发经验,评价成果可为气田后续产能建设部署及稳定、高效开发提供依据。

表4 神木气田与苏里格气田开发指标对比

气田	有效厚度（m）	孔隙度（%）	渗透率（mD）	气井	无阻流量（$10^4 m^3/d$）	动态控制储量（$10^4 m^3$）	稳产三年合理产量（$10^4 m^3/d$）
苏里格气田	5~8	8.75	0.81	一类	10	3700	2.4
				二类	6.9	2580	1.2
				三类	2.8	1100	0.7
神木气田	6~10	7.8	0.72	一类	8.3	3300	2.2
				二类	4.6	1860	0.8
				三类	1.5	700	0.4

5 结论

(1)以鄂尔多斯盆地神木气田太原组气藏为例,系统论证致密砂岩气藏有效砂体规模及气井开发潜力。从动态、静态结合的角度出发,综合运用野外露头测量,密井网精细解剖,生产动态不稳定分析及试井边界探测解释方法,在静态地质特征认识的基础上加以动态响应佐证,明确了气藏有效砂体规模尺度,宽度为452~650m,长度为678~1000m,同成熟开发的苏里格气田对比,提出合理的开发井网600m×800m。

(2)针对致密砂岩储层流体渗流到达边界所需时间长、气井试气无阻流量误差大的问题,建立了动态储量校正曲线和动静交会产能计算式,同时应用指示曲线、生产压差和数值模拟等方法,系统评价了气井动态控制储量、无阻流量、合理生产制度等关键开发指标,确定了研究区气井的开发潜力。所形成的致密砂岩有效砂体规模及气井开发潜力分析方法,可为气田规模产能建设顺利推进、后期持续稳产及其他同类气田的有效开发提供指导和借鉴。

参 考 文 献

[1] 王猛,唐洪明,刘枢,等. 砂岩差异致密化成因及其对储层质量的影响:以鄂尔多斯盆地苏里格气田东区上古生界二叠系为例[J]. 中国矿业大学学报,2017,46(6):1282-1300.

[2] 周拓,周兆华,谷江瑞,等. 神木气田二叠系气藏有利区预测[J]. 东北石油大学学报,2013,37(1):38-44.

[3] 兰朝利,张永忠,张君峰,等. 神木气田太原组储层特征及其控制因素[J]. 西安石油大学学报(自然科学版),2010,25(1):7-12.

[4] 杨明慧,刘池洋,兰朝利,等. 鄂尔多斯盆地东北缘晚古生代陆表海含煤岩系层序地层研究[J]. 沉积学报,2008,26(6):1005-1013.

[5] 付金华,魏新善,任军峰. 伊陕斜坡上古生界大面积岩性气藏分布与成因[J]. 石油勘探与开发,2008,35(6):664-667.

[6] 席胜利,李文厚,刘新社,等. 鄂尔多斯盆地神木地区下二叠统太原组浅水三角洲沉积特征[J]. 古地理学报,2009,11(2):187-194.

[7] 刘群明,唐海发,吕志凯,等. 鄂东致密气水下分流河道复合体储层构型布井技术[J]. 中国矿业大学学报,2017,46(5):951-958.

[8] 兰朝利,张君峰,陶维祥,等. 鄂尔多斯盆地神木气田太原组沉积特征与演化[J]. 地质学报,2011,85(4):534-541.

[9] 蒙晓灵,张宏波,冯强汉,等. 鄂尔多斯盆地神木气田二叠系太原组天然气成藏条件[J]. 石油与天然气地质,2013,34(1):37-41.

[10] 贾爱林,程立华. 数字化精细油藏描述程序方法[J]. 石油勘探与开发,2010,37(6):709-715.

[11] Kuuskraa V A. Tight gas sands development: How to dramatically improve recovery efficiency[J]. Gas TIPS,2004,10(1):15-20.

[12] Fetkovich M J,Vienot M E,Bradley M D,et al. Decline-curve analysis using type curves case histories[J]. SPE Formation Evaluation,1987,2(4):637-656.

[13] Fraim M L,Wattenbarger R A. Gas reservoir decline—curve analysis using type curves with real gas pseudo pressure and normalized time[J]. SPE Formation Evaluation,1987,2(4):671-682.

[14] Blasingame T A,McCray T L,Lee W J,et al. Decline curve analysis for variable pressure drop/variable flow

rate systems[R]. SPE21513-MS,1991.
[15] Agarwal R G,Gardner D C,Kleinsteiber S W,et al. Analyzing well production data using combined type curve and decline curve analysis concepts[R]. SPE 49222-MS,1998.
[16] 庄惠农. 气藏动态描述和试井[M]. 北京：石油工业出版社,2004.
[17] 刘能强. 实用现代试井解释方法[M]. 北京：石油工业出版社,2008.
[18] 聂仁仕,贾永禄,朱水桥,等. 水平井不稳定试井与产量递减分析新方法[J]. 石油学报,2012,33(1)：123-127.
[19] 何东博,王丽娟,冀光,等. 苏里格致密砂岩气田开发井距优化[J]. 石油勘探与开发,2012,39(4)：458-464.
[20] 郭奇,陈开远,李祯. 低渗透气藏合理动态储量计算方法[J]. 特种油气藏,2016,23(1)：113-115.
[21] 刘琦,罗平亚,孙雷,等. 苏里格气田苏五区块天然气动态储量的计算[J]. 天然气工业,2012,32(6)：46-49.

原文刊于《中国矿业大学学报》,2018,47(5):1046-1054.

3D geological modeling for tight sand gas reservoir of braided river facies

Guo Zhi[1]   Sun Longde[2]   Jia Ailin[1]   Lu Tao[3]

(1. PetroChina Research Institute of Petroleum Exploration & Development, Beijing 100083, China;
2. PetroChina Company Limited, Beijing 100007, China; 3. Sulige Gas Field Research
Center of PetroChina Changqing Oilfield Company, Xi'an 710018, China)

Abstract: Considering the poor applicability of conventional geological modeling to tight sand gas reservoir in braided river facies, a modeling method of "multi-stage constraints, hierarchical facies control and multi-step modeling" was put forward taking Sulige gas field in Ordos Basin as the study object. The method obtains the GR field by seismic inversion constrained by logging data, and GR model is built under the control of the prior geological knowledge; the relation regression is realized between the GR model and the sandstone probability, sandstone probability model is built, and rock facies model is obtained by multi-point geostatistics theory; sedimentary microfacies model controlled by rock facies and braided-river-system is made; and eventually effective sand body model is built by integrating sedimentary microfacies, effective sand body scale and reservoir properties distribution. The research method discussed in this paper has put geological constraints into the model as far as possible, enhanced the inter-well sand body predictability and improved the precision rate, thus it can provide a more reliable geological basis for gas reservoir development.

Key words: Sulige gas field; tight sand gas; geological modeling; rock facies model; sedimentary microfacies model; multi-stage constraint; hierarchical facies

Introduction

Wide in distribution and huge in development potential[1], with recoverable resources of $913 \times 10^{12} m^3$, tight gas in China already accounts for more than 30% of the total gas production[2]. The Sulige gas field in Ordos Basin, the largest tight sandstone gas reservoir in China[3], has some unique sedimentary and reservoir characteristics different from other gas fields: firstly, the sedimentary environment is continental braided river in terrestrial lake basin background with fast facies change and poor reservoir stability[4]; secondly, the effective sands, mostly in lenticular shape, are "sweet spots" with relatively high permeability in a universally low permeability background, not all sands are effective ones, and they usually form a dual structure of "effective sand in sand".

For the tight sandstone reservoirs of braided river facies in Sulige gas field, conventional geological modeling methods have some shortcomings[5]; firstly, neither in "one step model" method (reservoir attribute model with no facies control) nor in "two-steps model" method (reservoir attribute model under the control of lithofacies or sedimentary facies)[6-8], geological constraints included are enough to describe the reservoir accurately; secondly, the combining effect of well logging and seismic data is not ideal, especially when the reservoir is buried deep and seismic data is poor in quality, the precision of conventional wave impedance inversion can not meet the demand of reservoir development[9-10]; thirdly, in the conventional sedimentary facies modeling, channel bars are simulated according to fixed proportion and nearly the same scale in the fluvial channel, which can't reflect the actual complex facies change, and is not consistent with the real sedimentary characteristics; finally, due to the complexity of tight sand reservoirs in Sulige and the limitations of previous modeling methods, it is difficult to identify and predict inter-well effective sands[11].

In view of the shortcomings of the existent geological modeling methods and the geologic features of Sulige gas field, a modeling method of "multi-stage constraints, hierarchical facies control and multi-step modeling" is put forward in this paper, in order to improve the accuracy of geological model gradually. The core of the modeling is to show the dual structure of "effective sands in ineffective sands" accurately in three-dimensional space, so this study stresses facies modeling and effective sand body modeling.

"Multi-stage constraints" refer to add constraints to the model in multiple stages[12] to reduce data interpretation uncertainty and to enhance geological meaning of the model. "Hierarchical facies control" means to build sedimentary microfacies model restricted by both rock facies and braided-river sedimentary system. "Multi-step model building" means breaking the geological model building into several steps: rock facies modeling, sedimentary facies modeling, reservoir property modeling and effective sand body modeling, in which each step is controlled by the previous step or several steps.

1 Basic geological setting

Located in north-central Yishaan slope in the Ordos Basin (Fig. 1), Sulige gas field is controlled by the gentle tectonic background of Ordos Basin. The structure is a large monocline dipping SW with few faults developed, just 4~8 m/km in sloping and only 0.2°~0.5° in slope angle. Main gas-bearing intervals in the Upper Palaeozoic gas reservoir include Member He8 of Permian Shihezi Formation and Member Shan1 of Shanxi Formation, which are divided into 7 sublayers. The overall sedimentary environment of Sulige gas field is terrestrial braided-river, which includes 3 microfacies: channel bar, channel filling and flood plain.

Block Su6, an infill drilling block which is representative in sedimentary and reservoir features, long in production time, relatively small in well spacing and abundant in geological data, is selected as the area for the modeling study. In the sedimentary background of frequent channel shifting, multistage sands in the study area are often stacking one over the other, forming developed reservoirs of large thickness, with a drilling rate of 60%~90%. Under the dual control of sedimentation and diagenesis[13], effective sands are mainly grit, which are concentrated in the center and bottom of channel bars or at the bottom of the river channel filling; scattered, thin, in lenticular shape, the effec-

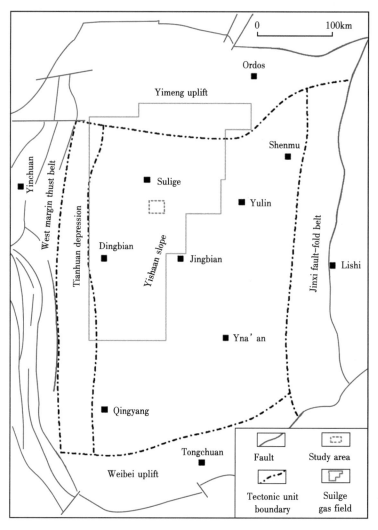

Fig. 1 Location of Sulige gas field

tive sands are only about 20%~70% in drilling rate, on average 44.7%. Effective sands often wrapped in between sands, forming a dual structure with the net/gross ratio (the ratio of effective sand thickness to sand thickness) of only 1/4~1/3 (Table 1).

Fine dissection of sands, interference well test and horizontal well drilling show that single effective sands are generally 1~5m thick, mostly 1.5~2.5m. The effective sands are 2~3m thick, 200~500m wide, and 400~700m long on average. According to outcrop observation, sedimentary physical simulation and previous research results, channel bars are 20 to 110 in width thickness ratio, and 2-6 in length width ratio; channel filling is 50~120 in width thickness ratio and 2~5 in length width ratio.

Table 1 Average thickness and drilling rate of sands in each layer of Sulige gas field

System	Member	Layer	Sandstone thickness(m)	Effective sand thickness(m)	Sand drilling rate(%)	Effective sand drilling rate(%)	Sand/formation ratio	Net/gross ratio
Permian	He 8[1]	1	6.89	1.05	87.50	29.17	0.38	0.15
		2	7.07	2.53	93.75	56.25	0.46	0.36
	He 8[2]	1	7.21	2.60	89.58	64.58	0.46	0.36
		2	7.98	2.76	97.92	70.83	0.51	0.35
	Shan 1	1	3.65	0.98	58.33	22.92	0.23	0.27
		2	4.42	1.53	70.83	50.12	0.3	0.35
		3	3.37	0.83	56.25	18.75	0.22	0.25
Average			5.80	1.75	79.17	44.66	0.37	0.30

2 Rock facies model

In view of well spacing and group interval, the modeling grid is designed as 25m×25m×1m. Based on the structural model established by drilling and seismic data, the rock facies distribution is delineated in 3D space. Here "rockfacies" refers to sandstone or mudstone, rather than specific lithology like grit sandstone and fine sandstone, thus it is different from the lithofacies in a broad sense, which characterizes the rock physical features.

The Member He 8 (abbreviated to "He 8" hereinafter) and the Member Shan1 (abbreviated to "Shan 1" hereinafter) are buried deep (3200~3600m on average), and the surface desert or half desert is weak in seismic reflection condition, together they result in poor seismic data quality and low signal-to-noise ratio and resolution, so it is necessary to calibrate seismic data with logging data to improve its vertical resolution. Although conventional wave impedance inversion can convert seismic wave records into lithology information, the method can only distinguish large sections of sandstone from mudstone, but can't tell single sandstone layer or mudstone layer precisely, because the sands and mudstones are similar in wave impedance in Sulige gas field.

Since some logging curves can show lithological changes more clearly, and are inherently connected with seismic data in reflecting the lithological interface, following the "multi-stage constraints" idea, well logging curves suitable are selected, and the function relationship between well logging and seismic data are analyzed by neural network recognition technology. Based on the relationship between log curve and seismic data, the stochastic field of geophysical characteristic curves is inverted, a geophysical characteristic curve model is built by the constraint of the stochastic field, sand probability volume is generated, and the rock facies model is established through multiple point geo-statistics method (Fig. 2).

2.1 Natural gamma field

Affected by fluvial sedimentary environment, reservoirs in Sulige gas field have multiple sand

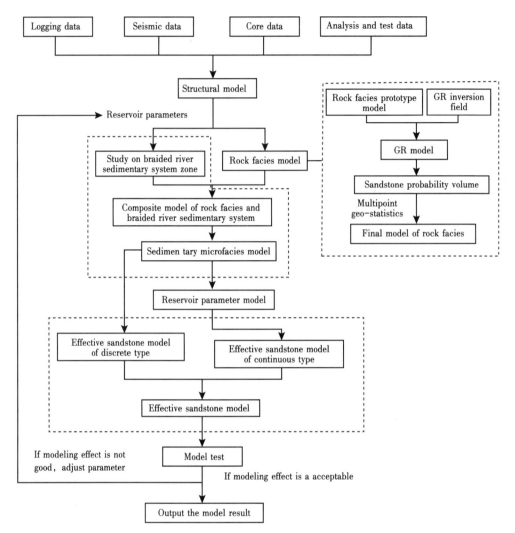

Fig.2　Geological modeling process

and mudstone intelayers in vertical direction. The analysis of several well logs including acoustic time, spontaneous potential (SP), natural gamma (GR), neutron log, and resistivity log[14] shows that the GR curve has the best match with the rock facies, and is most sensitive to lithological change. Seismic reflection wave also has a certain correlation with lithology, which is the very theoretical basis of the traditional wave impedance inversion. GR curves and seismic data were input into the neural network pattern recognition[15] to do match training to get the learning samples, and a series of seismic features close to the GR curves, and taking this as standard, GR field was inverted with seismic data by logging constraint.

Comparing the effect of wave impedance inversion and GR inversion (Fig. 3) shows that wave impedance values of sandstone and mudstone are both in the range of $10.0 \times 10^6 \sim 12.8 \times 10^6$ kg/(m^2·s), so wave impedance is poor in telling sandstone from mudstone in the study area, while inverted GR field can distinguish sand from mudstone because the inverted GR value of sandstone is generally low and that of mudstone is relatively high; the GR inversion and logging GR values match well with a correlation coefficient of 0.76, so prior geological information can be used to constrain the inverted GR field in order to reduce the multiplicity of interpretation.

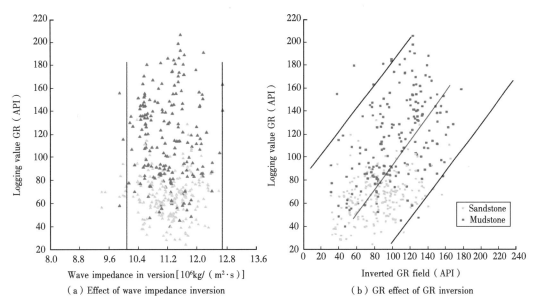

Fig. 3　Comparison of wave impedance inversion and GR inversion

By combining seismic data and logging data judiciously, GR field inversion ensures the quality and multiple sources of the analysis data, breaks the traditional seismic resolution limits, and in theory, can obtain the same resolution as the logging data. But the inverted GR field has an obvious defect, namely lack of definite geological meanings between wells and thus ambiguous interpretation. Under the present technical condition, it is difficult to raise the seismic data resolution, so the best way to reduce the ambiguity is obtaining more accurate geological information and adding it to the geological model.

2.2　Natural gamma model

Considering the inter-well limitations of inverted GR field based on seismic data, taking GR values at well points as hard data and inverted GR field at inter-well zones as soft data, choosing geological information obtained from fine geologic anatomy as constraints (provenance direction, major variation range, minor variation range and vertical variation range of sands, etc.), GR model was set up by using sequential Gaussian method. This way reduces the multiplicity of seismic data, clarifies the geological meaning of predicted sands, and guarantees the GR value continuity between the well points and inter-well places.

2.3　Sandstone probability volume

GR model is corresponding with sandstone probability to some extent. With the increase of GR value, the sandstone probability decreases and mudstone probability increases in general, but this does not mean if GR values of any interval is given, the lithology, (sandstone or mudstone) of the interval can be identified accurately. In view of that, the statistical relationship between GR value and sandstone probability was regressed, and the GR model was transformed into sandstone probability volume. When modeling software identified rock facies automatically according to GR value, on the basis of the calculated sandstone probability, multiple rock facies models were generated stochastically to be chosen to avoid the error caused by only one GR threshold.

2.4 Rock facies modeling method

At present, the two most commonly used facies modeling methods are sequential indicator simulation and object based simulation. Sequential indicator simulation, a method based on pixel, investigates the correlation of geological variables at any two points in underground space with variation function. Although faithful to hard data at well points, this method cannot simulate sophisticated relationships of multiple variables. The method could result in fluvial channel offset, sands in clumps and with jagged rim, which does not conform to the braided river sedimentary mode. With discrete objects as the simulated unit, object based simulation method can characterize channel morphology, but often fails to reconcile with the well point data when there is a large number of well points in the study area.

In view of the defects of traditional stochastic simulation methods based on variation function and based on object, the multipoint geo-statistics has emerged as the times require, and quickly become the research frontier and hot spot of stochastic modeling. Replacing variation function with training image, this method reveals the spatial structure of geological variables[16], thus overcomes the disadvantage that the target geometry can't be reconstructed; meanwhile, multipoint geo-statistics method, adopting sequential algorithm, is loyal to hard data[17], thus overcoming the disadvantages of object-based simulation algorithm.

2.5 Rock facies model

Training images acquirement is the key basis of multipoint geo-statistics method. Training image is a digital image which is able to describe geological information such as the actual sand structure, geometry and distribution patterns. Larger scale training image contains more geological information and is higher in simulation precision, but it is more time-consuming[18]. The training image is not necessarily loyal to the actual well point information, instead it only reflects a kind of prior geological concept and statistical characteristics[19]. The sources of training images include outcrop observation, modern sedimentary prototype model, unconditional simulation results based on object, sedimentation simulation parameters, and digital sketch.

Due to different reservoir characteristics of different layers, this study optimized training image simulation scale with unconditional simulation method based on object, and set up seven 3D training images respectively of 7 layers. In general, sands are more developed in Member He 8 than in Member Shan 1. With rock facies data at well points as hard data and sandstone probability volume in inter-well zones as soft data, on the basis of the training images, 3D rock facies model was established by multipoint geo-statistics method. Thanks to the fine geological anatomy of the dense well pattern area, accurate sandstone probability volume, and advanced algorithm of multipoint geo-statistics, the established 3D rock facies model is faithful to hard data at the well points and shows fluvial channel morphology well in inter-well zones.

After the rock facies model was built, seismic waveform was used to validate, revise and consummate it. Seismic waveform refers to the comprehensive changes of seismic wave amplitude, frequency and phase, which can describe the distribution of sands of a certain thickness on the plane, thus predicting sand distribution in interwell zones with seismic waveform can reach certain accuracy.

3 Sedimentary facies model

3.1 Study on braided river sedimentary system

Member He 8 and Shan 1 were deposited in position near provenance with gentle sloping, strong hydrodynamics, which resulted in frequent shift of channel, and thus a large scale of braided river sedimentary system (braided river complex) in sheets on the plane. Affected by factors such as provenance, hydrodynamics and paleotopography, braided river sedimentary system of Sulige gas field can be divided into 3 facies zones: braided river superimposed zone, transitional zone and intersystem zone. The braided river sedimentary system corresponds to formations of several kilometers on the plane, and sand groups vertically including 2~3 layers.

Studies show that braided river sedimentary system has strong influence on development types, frequency and scale of sedimentary microfacies. Sedimentary microfacies distribution of layer grade is under the control of braided river system of sand group grade generally, and both of them show great relevance in source direction and river trend, but sedimentary microfacies shows certain changes in local area (Fig. 4). The braided river superimposed zone was located in the lowest area of ancient

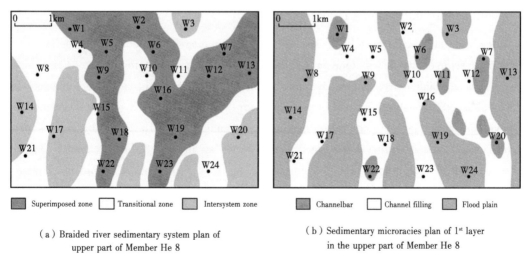

(a) Braided river sedimentary system plan of upper part of Member He 8

(b) Sedimentary microracies plan of 1st layer in the upper part of Member He 8

Fig. 4 Braided river sedimentary system and sedimentary mirofacies plan

terrain where ancient channels deposited continuously, resulting in high occurrence frequency and large scale of channel bars. Braided river transitional zone is the interactive area of high energy and low energy channels, where deposits channel filling primarily, and channel bar secondarily. Statistics shows the development frequency of channel bars in the superimposed zone is nearly as twice as that of the transitional zone (Table 2). The channel bars in superimposed zone are generally 0.3~0.5m thicker, 70~80m wider, and 100~200 m longer than those in the transitional zone.

Table 2 Size proportion of channel bar and channel filling in the superimposed zone and transitional zone

System	Member	Layer	Superimposed zone		Transitional zone	
			Channel bar(%)	Channel filling(%)	Channel bar(%)	Channel filling(%)
Permian	He 8^1	1	59.04	40.96	21.77	78.23
		2	59.56	40.44	36.02	63.98
	He 8^2	1	63.52	36.48	33.15	66.85
		2	72.10	27.90	28.12	71.88
	Shan 1	1	43.62	56.38	23.65	76.35
		2	62.73	37.27	34.59	65.41
		3	45.00	55.00	19.99	80.01
	Average		57.94	42.06	28.18	71.82

Channel bars don't occur at fixed proportion evenly in the river channel, in fact, they occur at high frequency and large scale in the superimposed zone, and relatively low frequency and small scale in the transitional zone. Through examining the braided river sedimentary system, the uneven spatial distribution of sedimentary microfacies can be described. Since channel bars correspond to the most favorable reservoirs in Sulige gas field, different distribution patterns of channel bars in the superimposed zone and transitional zone, result in obviously different quality of the reservoirs in them.

3.2 Sedimentary microfacies model

In the traditional facies modeling, "facies" means "lithology" or "sedimentary facies"[20], but lithofacies or sedimentary facies alone cannot characterize the strong heterogeneity of tight sandstone gas reservoirs in braided river facies. Therefore, under the premise of the highly reliable rock facies model established, this study first tries to set up, with the object-oriented simulation method, a facies model using rock facies to control sedimentary microfacies facies. The modeling process includes two steps: first, channel filling was merged with channel bar into the "fluvial channel" facies, as the simulation facies, corresponding to sandstone in the rock facies model, while flood plain was taken as the background facies, corresponding to the mudstone in the rock facies model; second, channel bars were simulated only in fluvial channel generated in the first step, while other grids were kept unchanged. The question is that channel bars simulated by this modelling method are distributed in the fluvial channel according to fixed proportion or nearly the same scale. In other words, the channel is roughly viewed as a homogeneous whole, which is inconsistent with the known sedimentary understanding.

In view of the strong control of braided river sedimentary system on sedimentary microfacies, we came up with the idea to constrain the sedimentary microfacies model with both the braided river sedimentary system and rock facies. But there are two problems need to be solved in order to do that: one, the braided river sedimentary system and rock facies are different in formation scale, a braided river sedimentary system is the comprehensive reflection of sedimentary environment, corresponding to the formation of sand group level, while 3D rock facies model, similar to a complex of isochronous formation slices, is much smaller in formation scale, the braided river superimposed system can not even correspond to sandstone in rock facies model sometimes; the other, in the process of establishing facies model, relevant software only allows to enter just one 3D model as constraint. Therefore, the braided river sedimentary system and 3D rock facies must be combined together, which was done with the method below: the grid which is sandstone, and located in the superimposed zone was defined as the superimposed zone; the grid belonged to sandstone, and located in the transitional zone or intersystem zone was defined as the transitional zone; the grid which is mudstone was defined as the intersystem zone. According to the statistic features of distribution frequency and scale of sedimentary microfacies such as channel bar and channel filling in different braided river sedimentary systems, the sedimentary microfacies model was set up under the control of both rock facies and braided river sedimentary system.

In sedimentary microfacies model built under the dual constraints of rock facies and braided river sedimentary system, channel bars are more concentrated and large in scale in local zones, which is consistent with sedimentary characteristics. By contrast, in sedimentary microfacies model only controlled by rock facies, channel bars are distributed in the fluvial channel in fixed probability and almost equal scale, which played down the inherent inhomogeneity of sedimentary facies, thus resulted in poor simulation effect. The sedimentary microfacies model not controlled by the rock facies is even poorer in simulation result.

4 Effective sand body model

Spatial distribution of effective sand body follows certain geological and statistical regularities, and also is affected and controlled by sedimentary microfacies and reservoir parameters. Controlled by deposition and diagenesis jointly, the effective sands have higher reservoir parameter values (porosity, permeability and gas saturation) than ineffective sands, and a good correspondence with such sedimentary microfacies as channel bars. Statistics shows that 86% effective sands are in channel bars in the study area.

On the basis of establishing rock facies model, sedimentary facies model and reservoir parameter model, the effective sand body model was built with two methods respectively. The first method was discrete modeling method-simulation based on object, in which the effective sand body confirmed by logging or well test was taken as hard data, according to the distribution and statistical characteristics of effective sand body in space (Table 3), the effective sands (gas layers and gas-bearing layers) were simulated as facies properties, while the non-effective sands were taken as the background facies. The second method was continuous modeling method-sequential Gaussian simulation, in which, through analysis of formation testing and production test data, the lower limit values of effective sand reservoir parameters were obtained (porosity ≥5% and gas saturation ≥45%), then grids with higher value above the lower limits in the porosity (see porosity variation function in Table 4), permeability and saturation models were picked out, and defined as effective sands.

Table 3　Effective sand body modeling parameters

System	Member	Layer	Thickness(m)			Width(m)			Length(m)		
			Min	Average	Max	Min	Average	Max	Min	Average	Max
Permian	He 8¹	1	1.1	2.6	5.2	158	210	368	316	547	921
		2	1.1	3.0	7.2	177	236	413	354	614	1 033
	He 8²	1	0.9	2.8	6.7	170	227	398	341	591	994
		2	1.3	3.0	7.5	182	243	426	365	632	1 064
	Shan1	1	0.8	2.4	4.3	142	190	332	284	493	830
		2	0.9	2.5	5.2	151	201	351	301	522	879
		3	0.7	2.2	4.1	130	173	302	259	449	756

Table 4　Variance function of porosity modeling

System	Member	Layer	Major variance range(m)	Minor variance Range(m)	Vertical variance Range(m)	Azimuth (%)
Permian	He 8¹	1	803.5	368.2	7.36	12.48
		2	901.2	413.0	8.26	9.15
	He 8²	1	867.6	397.6	7.95	14.98
		2	928.7	425.6	8.51	14.14
	Shan 1	1	724.0	331.8	6.64	14.14
		2	766.8	351.4	7.03	9.98
		3	659.9	302.4	6.05	11.65

5　Model test

The precision and accuracy of geological modeling depends largely on the understandings of underground geology condition, application effect of basic modeling data, and rationality of modeling method and algorithm. Modeling effect was testified by several ways including geological knowledge validation, well pattern coarsening verification, reservoir parameter comparison, reserve calculation and dynamic verification. If the model is high in accuracy, then the data is output; if not, parameters will be adjusted and the model reestablished until reaching the ideal effect. This modeling method has been used in a development block of Sulige gas field, and worked well in sand prediction and hydrocarbon detection.

5.1 Verification by geological understanding

The 2nd layer in lower part of Member He 8 has thick sand and good reservoir quality in Well-block W9 and W12 in the study area. At well points, the established geological model is consistent with the sand isopach map. In inter-well areas, by integration of seismic data, sand probability volume and modeling algorithm, 3D geological model provided accurate sand distribution prediction (Fig. 5).

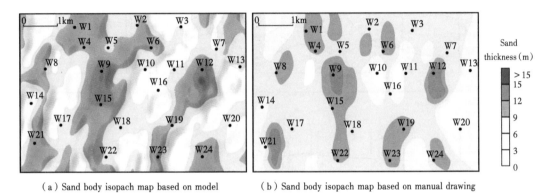

(a) Sand body isopach map based on model (b) Sand body isopach map based on manual drawing

Fig. 5 Comparison of sand isopach map based on model and sand isopach based on manual drawing

5.2 Well pattern coarsening verification

Well spacing of dense well pattern in Sulige gas field is 400 m by 600 m. Modeling pattern was coarsened step by step, and the removed wells were taken as testing wells and not used in the simulation, and the model was established with the rest well data to tell the accuracy of interwell sand identification and reliability of the model. The accuracy of inter-well sand identification was obtained by comparing the distribution of sand and mud in the places of the removed wells in the model with that from actual drilling profile. Statistics shows that with the increase of well spacing, the accuracy of inter-well sand identification decreases (Fig. 6). At the well pattern of 800m×1200m, 1200m×1800m and 1600m×2400m, the accuracy of sand identification were respectively 85.7%, 72.7% and 55.2%. When the well pattern was coarsened to 1600m×2400m, multiple sands were identified wrong, and the accuracy of inter-well sand identification decreased to about 50%, making the identification of sand and mud almost meaningless. The model can predict thick sands more accurately than thin sands. It is generally believed, a model with an identification accuracy of 70% is basically reliable. Comparison shows the rock facies modeling method discussed in this paper is applicable to 1200m×1800m well pattern, while the conventional rock facies modeling method is only suitable for 800m×1200m, and the model established in this study has much better precision.

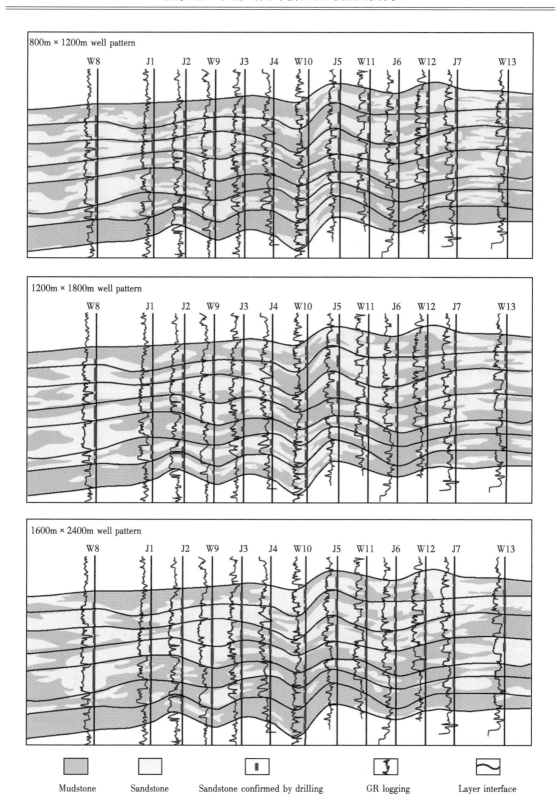

Fig. 6 Verification by modeling well pattern coarsening

5.3 Reservoir parameter comparison

By comparing property modeling parameter results, the discrete data from logging curves and the original logging curves, it is found the distribution range of the three is close, and the distribution ratio difference is small at the same range: the reservoir porosity ranging between 5% and 12%, the permeability $0.01\times10^{-3} \sim 10.00\times10^{-3}\mu m^2$, and gas saturation 20%~70%. The established reservoir property model under the control of rock facies and sedimentary facies can reflect the basic characteristics of the input properties parameters, thus the simulation of the model is relatively high in precision.

5.4 Reserves calculation

Since reserves concentration and scale is a comprehensive reflection of parameters such as porosity, gas saturation and net/gross ratio, reserves calculation can be used as a test standard of simulation result of reservoir parameters and effective sand quality. Dense well pattern zone in the study area is one of the most favorable development blocks in Sulige gas field, and exploration well data shows that the reserve abundance is $1.3\times10^8 \sim 1.5\times10^8 m^3/km^2$. Gas reserves calculated by the established geological model is $44.53\times10^8 m^3$, in which the 2nd layer in lower part of Member He 8 is largest in reserves, the 2nd layer in upper part of Member He 8 is the second largest among the 7 layers. In the model, reserves concentration degree of He 8 is generally higher than that of Shan 1, with an average reserves abundance of $1.362\times10^8 m^3/km^2$, which is in good agreement with known geological understandings.

5.5 Dynamic data validation

The accuracy of geological model was also testified by numerical simulation. Geological model grids were coarsened to 100m×100m×3m, production and wellhead pressure were matched historically, then the predicted performance was compared with actual production performance to find the difference, then corresponding adjustment were made to the model to analyze the matching effect. Statistics shows that 83.3% wells in the study area are less than 5% in fitting error, indicating the established geological model is high in precision.

6 Conclusions

In view of the characteristics of tight sandstone gas reservoirs of a braided river facies in Sulige gas field, making use of probability theory, integrating well logging, seismic and geological data, a modeling method of "multi-stage constraints, hierarchical facies control, and multi-step modeling" has been put forward. In this method, multiple geological constraints were introduced stage by stage (GR inversion field-GR model-sandstone probability volume-rock facies model) to reduce the ambiguity and uncertainty of the data interpretation; facies model was established by hierarchy, including rock facies model, braided river sedimentary system model and sedimentary microfacies model; eventually, a 3D geological model was set up by multiple steps, in which the results of the previous step or steps was the data base of the latter step.

Compared with conventional modeling method, the modeling method of "multi-stage constraints, hierarchical facies control, and multi-step modeling", puts prior geological information to the model as far as possible, enhancing the predictability of inter-well sands (GR inversion has better effect in distinguishing sandstone from mudstone than the conventional wave impedance inversion), and making the established model more consistent with sedimentary characteristics and known geological understandings. In summary, the new modeling method discussed in this paper can improve the model precision to a great extent.

References

[1] Sun Longde, Fang Chaoliang, Li Feng, et al. Petroleum exploration and development practices of sedimentary basins in China and research progress of sedimentology[J]. Petroleum Exploration and Development, 2010, 37(4): 385-396.

[2] Zhang Huanzhi, He Yanqing, Chen Wenzheng. Accelerate policy research on compact gas development[J]. Oil Forum, 2014, 33(4): 13-16.

[3] Sun Longde, Zou Caineng, Zhu Rukai, et al. Formation, distribution and potential of deep hydrocarbon resources in China[J]. Petroleum Exploration and development, 2013, 40(6): 641-649.

[4] Zhang Wencai, Gu Daihong, Zhao Ying, et al. Characteristics and generating mechanism of Permian relative low-density sandstone in Sulige Gas Field, Northwest China[J]. Petroleum Exploration and Development, 2004, 31(1): 57-59.

[5] Chen Fengxi, Lu Tao, Da Shipan, et al. Study on sedimentary facies of braided stream and its application in geological modeling in Sulige gas field[J]. Petroleum Geology and Engineering, 2008, 22(2): 21-24.

[6] Wu Shenghe, Li Yupeng. Reservoir modeling: Current situation and development prospect[J]. Marine Origin Petroleum Geology, 2007, 12(3): 53-60.

[7] Jia Ailin. Research achievements on reservoir geological modeling of China in the past two decades[J]. Acta Petrolei Sinica, 2011, 32(1): 181-188.

[8] Jia Ailin, Cheng Lihua. The technique of digital detailed reservoir characterization[J]. Petroleum Exploration and development, 2010, 37(6): 623-627.

[9] Wu Jian, Li Fanhua. Prediction of oil bearing single sand body by 3D geological modeling combined with seismic inversion[J]. Petroleum Exploration and development, 2009, 36(5): 623-627.

[10] Sun Longde, Sa Liming, Dong Shitai. New challenges for the future hydrocarbon in China and geophysical technology strategy[J]. Oil Geophysical Prospecting, 2013, 48(2): 317-324.

[11] Lu Tao, Li Wenhou, Yang Yong. The sandstone distribution characteristics of He-8 gas reservoir of Sulige gas field[J]. Mineral Petrol, 2006, 26(2): 100-105.

[12] Jia Ailin, Guo Jianlin, He Dongbo. Perspective of development in detailed reservoir description[J]. Petroleum Exploration and development, 2007, 34(6): 691-695.

[13] He Dongbo, Jia Ailin, Tian Changbing, et al. Diagenesis and genesis of effective sandstone reservoirs in the Sulige gas field[J]. Petroleum Exploration and development, 2004, 31(3): 69-71.

[14] Guo Zhi, Yang Shaochun, Jia Ailin, et al. Fine log interpretation for oilfields with thin sands and multiple developed layer systems[J]. Acta Petrolei Sinica, 2013, 34(6): 1137-1142.

[15] Ai Ning, Tang Yong, Yang Wenlong, et al. Fuzzy neutral network-based tight sandstone reservoir inversion: A

case study from the Denglouku Formation in Changling 1 gas field[J]. Oil & Gas Geology, 2013, 34(3): 413-420.

[16] Deutsch C V, Wang Libing. Hierarchical object-based geostatistical modeling of fluvial reservoirs[J]. Mathematical Geology, 1996, 28(7): 857-880.

[17] Strebelle S B, Journel A G. Reservoir modeling using multiple-point statistics[J]. SPE 71324, 2001.

[18] Shi Shuyuan, Yin Yanshu, Feng Wenjie. The development and prospect of multiple-point geostatistics modeling[J]. Geophysical & Geochemical Exploration, 2012, 36(4): 655-660.

[19] Wu Shenghe. Reservoir characterization & modeling[C]. Beijing: Petroleum Industry Press, 2010.

[20] Li Shaohua, Liu Xiantai, Wang Jun, et al. Improvement of the Alluvsim algorithm modeling based on depositional processes[J]. Acta Petrolei Sinica, 2013, 34(1): 140-144.

原文刊于《Petroleum Exploration & Development》,2015,42(1):83-91.

辫状河储层构型规模表征及心滩位置确定新方法
——以苏 6 区块密井网区盒八段为例

董 硕 郭建林 李易隆 郭 智

(中国石油勘探开发研究院)

摘要：针对辫状河储层构型研究中对构型规模表征及心滩位置不确定性较大的问题，以苏 6 区块为例，结合 Miall 河流相构型划分及鄂尔多斯盆地东部柳林地区盒八段露头解剖，将辫状河储层构型划分为五个级次：(1)复合河道带；(2)单一辫流带；(3)心滩与河道充填；(4)心滩内增生体及冲沟；(5)层系组。根据岩心及测井数据，划分单井构型级次，通过辫流带定界原则确定单一辫流带展布范围，提出"高程定中边，形态定前后，冲沟近中线，落淤层辅助"的心滩位置判别方法，干扰试验表明该划分方法具有较强的可靠性，适合大井距辫状河储层构型表征及心滩位置确定。采用经验公式约束、井间精细对比统计法，确定了研究区构型展布规模：心滩厚度主要分布于 4~12m，单一辫流带宽度 600~1400m，心滩长度 500~800m，心滩宽度 250~400m。

关键词：苏 6 区块；辫状河；储层构型；规模；心滩

储层构型的研究即对储层内部结构进行的精细划分，对不同级次储层构成单元的形态、规模、展布、叠置关系等进行精确描述。储层内部结构影响油气的分布，因而储层构型的研究对于有效储层的预测及开发后期剩余油气的挖潜都具有重要的意义。辫状河储层作为一种重要的油气储层，其内部构型的解剖成为研究重点。目前对于辫状河构型解剖的方法主要有露头解剖、现代沉积解剖、密井网解剖、基于 Google Earth 软件的测量研究、沉积模拟实验等。随着多种方法研究的逐渐深入，辫状河内部构型表征取得了丰硕的成果。于兴河[1]等通过对大同中侏罗统辫状河露头解剖，详细阐述了辫状河构型界面划分、沉积特征及沉积模式；Kelly[2]利用现代沉积和野外露头建立了心滩长度、宽度及心滩内增生体规模的经验公式；Bridge[3]通过大量数据统计，建立了辫流带最小宽度和最大宽度的经验公式；徐东齐等[4]通过 Google Earth 软件，测量了九个典型辫状河沉积区心滩坝及辫流带规模数据，建立了经验公式；孙天健等[5]在识别沉积微相的基础上，通过测井和岩心资料分级划分辫流带、单砂体和心滩内增生体三个构型单元，并利用经验公式得出各级构型的展布规模；邢宝荣等[6]利用密井网测井、岩心、卫星照片等资料，构建了大庆长垣油田喇萨区块葡一组辫状河储层地质知识库；牛博等[7]统计分析了辫状河储层中夹层的产状和平面几何参数，建立了落淤层地质知识库；孙天健等[8]采用灰色理论进行单井隔夹层识别，结合辫状河现代沉积测量，建立不同类型隔夹层规模的定量计算关系式。

虽然前人对于辫状河构型的研究取得了较多成果，但对于心滩的发育位置仍具有较强的推断性，且在不同沉积环境下，辫状河发育规模具有较大差异，经验公式可做参考，但直接套用经验公式推算研究区构型展布特征，稍欠考虑。因此，本文以苏 6 区块为例，研究辫状河内部构型划分方法,定量描述其展布特征。

1 地质背景

苏里格气田位于鄂尔多斯盆地西北部[9](图1),上古生界自下而上发育上石炭统本溪组、下二叠统太原组、山西组、中二叠统下石盒子组、上石盒子组及上二叠统石千峰组,沉积岩总厚度约为700m,主要含气层段位于山西组和下石盒子组[10]。其中,下石盒子组盒八段为典型的辫状河沉积,受北部物源控制,砂体呈南北展布,沉积物主要为岩屑石英砂岩、岩屑砂岩及石英砂岩。苏6区块位于苏里格气田中部,总面积约484km^2,探明地质储量1038.82×10^{8}m^3,密井网区面积24km^2,布井48口,平均井距543m。在较大的井距下识别河道、心滩的位置和规模,是本次研究的重点。

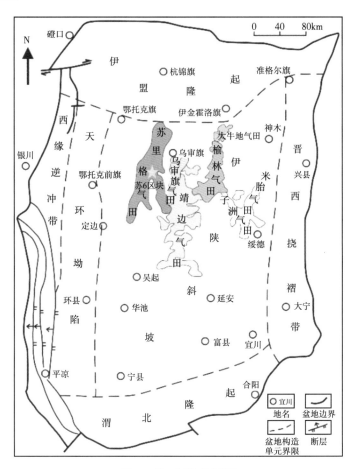

图1 苏6区块位置图

2 辫状河储层构型单元划分

储层构型研究以Miall提出的河流相构型分级为理论基础,该理论提出以来,国内外学者对不同沉积环境下的储层进行了构型解剖,尤其体现在曲流河构型表征中。对于辫状河构型表征,近年来也取得了一定的进展,但辫状河由于其频繁的改道,导致储层在纵向上多期叠置,沉积特征复杂,其构型级次划分尚未形成统一的理论。孙天建等[5]将砂质辫状河划分为辫流带、单砂体和心滩内增生体三个级次;李易隆等[11]根据成因机理与规模,将辫状河划分为五级构型单元;邱隆伟等[12]提出了辫状河储层六级构型单元划分方法。

2.1 构型单元划分

综合前人研究成果,辫状河储层构型的划分以 Miall 的河流相构型分级为指导,结合鄂尔多斯盆地东部柳林地区盒八段露头解剖及苏 6 区块密井网区资料分析,将辫状河储层构型划分为五个级次:(1)复合河道带;(2)单一辫流带;(3)心滩与河道充填;(4)心滩内增生体及冲沟;(5)层系组。其中,①级构型界面为复合河道带与泛滥平原泥岩的界面,构型单元由垂向上多期叠置、平面上交错变迁的单一辫流带组成;②级构型单元在垂向上表现为单砂体,构型界面即为单砂体之间的切割面或者单砂体与泛滥泥岩的接触面;③级构型界面为心滩、河道沉积之间的界面,常表现为心滩与河道沉积在侧向相邻的接触面;④级构型界面为心滩内部落淤层等夹层及心滩顶部小型冲刷面;⑤级构型界面为层系组的接触面,代表水流流向变化或流动条件变化,该级构型常由槽状交错层理、楔状交错层理组成(图2)。

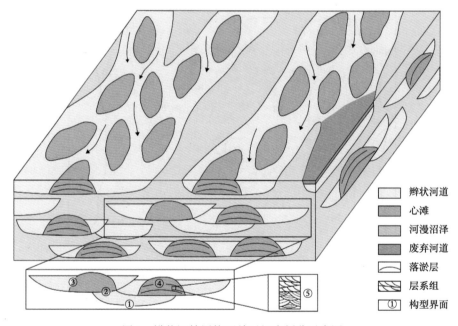

图 2 辫状河储层构型单元级次划分示意图

2.2 野外露头分析

鄂尔多斯盆地东部柳林地区盒八段露头位于鄂尔多斯盆地东缘晋西挠褶带,与苏里格地区盒八段沉积物物源相同,且均为辫状河沉积,因而选取该地区的露头进行解剖,用以指导苏里格盒八段辫状河构型划分。

图3(a)为一复合河道带露头,由两期单一辫流带叠置而成。可见单一辫流带界面(②级构型界面)出露,两期辫流带砂体皆呈正旋回沉积特征。上覆辫流带可见河道充填沉积,底部

(a)单河道剖面

(b)废弃河道剖面

(c)心滩剖面

河道　　心滩　　沟道　　废弃河道　　②级构型界面　　③级构型界面

④级构型界面　　槽状交错层理　　波纹层理　　泥岩

图3　野外露头构型解剖

发育中粒砂岩,可见槽状交错层理,向上层面减薄,粒度变细,为粉砂岩与粉砂质泥岩沉积,可见小型波纹层理,顶部发育泥岩。下伏辫流带左侧为心滩沉积,为典型的底平顶凸形态,粒度相较于河道充填沉积粗,为中粗粒砂岩,发育槽状交错层理;右侧为河道沉积,呈顶平底凸形态,河道充填沉积与心滩沉积体在侧向上相邻,接触面即为③级构型界面。因上覆河道的冲刷作用,上覆河道底部可见河道带底侵蚀面,下伏心滩披覆泥岩遭受冲刷侵蚀作用遗留下来,因而厚度较薄。

图3(b)左侧可见废弃河道,即河道逐渐因改道而被废弃时,水动力减弱、沉积泥质,形成的废弃河道,呈楔状,与右侧的心滩沉积体相邻,二者的接触面为③级构型界面。

图3(c)为心滩出露剖面,可见三个心滩内部增生体发育,内部多为槽状交错层理,呈叠瓦状排列。该心滩在露出水面后遭受水流冲刷形成冲沟[13],冲沟底部见冲刷面构造,冲沟内沉积物多为正韵律,底部可见小型槽状交错层理,顶部为泥质沉积。增生体之间可见薄层泥岩发育,为季节性洪水间歇期发育的泥质沉积,称为"落淤层"[14],该界面即为④级构型界面。

2.3 构型界面单井识别

在辫状河构型单元的研究中,核心问题是如何利用有限的资料识别地下各级构型界面,而第①至第④级次构型界面常在油气田开发过程中表现为隔层与夹层,因而是研究的重点。

复合河道带构型界面(①级构型界面)为多期叠置砂体与河漫泥岩的接触面,在GR测井曲线上表现为箱形—钟形复合形态的曲线底部(顶部)出现大段的泥岩回返(图4)。而箱形—钟形复合形态的GR曲线代表多期砂体的叠置,GR曲线形态、幅度出现突变的界面即为多期砂体间的接触面。在岩心观察中(图5a),可见砂体叠置于泛滥泥岩之上,由于砂体沉积时水流的冲刷作用,卷起了已固结的泥岩形成泥砾,包裹在了砂体底部,砂岩与底部泥岩的接触面即为冲刷面,也是①级构型界面。

单一辫流带构型界面(②级构型界面)为单砂体的顶底面(图4),或单砂体之间的切割面,在GR曲线上具有分离型、相邻型、切截型三种表现形式。分离型构型界面为河道与泥岩的接触面,河道发育完整,且两期河道之间夹有小段泥岩,GR曲线在泥岩段回返至泥岩基线附近;相邻型构型界面表示为两期完整的河道在垂向上相邻发育,无泥岩夹层,GR曲线在两期河道间出现回返,且回返幅度约为1/2;切截型构型界面为两期河道在垂向上切割叠置而成,上覆河道较完整,下伏河道的顶部被冲刷侵蚀,GR曲线在界面附近出现回返,且回返幅度小于1/2。在岩心观察中(图5b),可见两套颜色、结构均不同的砂体叠置而成,上覆砂体颜色较深,粒度较细,泥质含量较高,呈平行层理;下伏砂岩颜色较浅,粒度较粗,泥质含量少,发育交错层理。两套砂体物源不同,分属于两期河道,上覆河道对下伏河道具有冲刷作用,二者的接触面即②级构型界面,为切截型构型界面。

心滩与河道砂体的接触面(③级构型界面)常表现为砂体的侧向相邻,因而在单井上不易识别。

由测井及岩心资料分析发现,心滩在GR曲线上多为箱形、锯齿状箱形(图4),因而心滩内增生体构型界面(④级构型界面)常出现于箱形GR曲线段内,表现为GR曲线出现轻微的回返,回返程度由落淤层的厚度决定。对应测井曲线,可从岩心上识别出落淤层的位置,图5(c)中,底部为心滩内增生体沉积,为灰白色砂岩,顶部发育落淤层,为灰黑色泥岩,二者的接

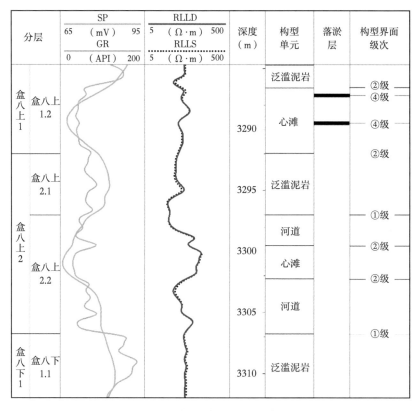

图 4 构型单元测井识别

触面即为4级构型界面。

图5(c)可见平行层理与交错层理,上下层系组的接触面即为⑤级构型界面,代表水流流向变化或流动条件变化。

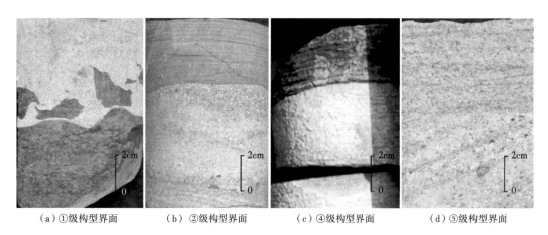

(a) ①级构型界面　　(b) ②级构型界面　　(c) ④级构型界面　　(d) ⑤级构型界面

图 5　构型界面岩心识别

3 构型单元展布特征

在识别各级构型单元界面的基础上,需进一步开展构型单元展布特征研究,本文采用单一辫流带边界识别法,确定单一辫流带的展布;利用测井曲线形态识别心滩、河道砂体沉积;采用"高程定中边,形态定前后,冲沟近中线,落淤层辅助"的判断方法,确定心滩的发育位置。

3.1 单一辫流带

在单井上识别单一辫流带的前提下,以盒八上亚段1.2小层为例,对研究区的辫流带平面展布进行分析。首先要根据测井资料确定河道的边界,辫流带边界通常位于井间,需要对比相邻井的测井曲线,来决定两口井在同一层内钻遇的砂体是否属于同一辫流带,继而判断两口井间是否发育辫流带边界。辫流带边界的划分应遵循以下原则:(1)砂体尖灭。井 a 钻遇砂体,临井 b 未钻遇砂体,砂体在 a、b 两井间发生尖灭;(2)高程差异大。井 a 钻遇的砂体顶面距小层顶面的高程,与井 b 钻遇的砂体顶面距小层顶面的高程,两者相差较大;(3)旋回性质差异。井 a 钻遇两套厚度较小的正旋回砂体,井 b 钻遇一套厚度大的正旋回砂体;(4)废弃河道阻隔。在单一河道的边部,常发育有废弃河道,代表该河道的消亡,在测井曲线上表现为 a 井钻遇砂体厚度大,b 井钻遇砂体顶部发育细粒沉积,导致钻遇砂岩厚度小。

运用以上原则,在平面上将辫流带边界绘出,继而将同一期辫流带侧向上相连,最后结合砂体厚度图,可将完整的辫流带绘制出来。由图6可知,研究区在盒八上亚段1.2小层可划分为上下两期砂体。

3.2 心滩及落淤层识别

前人研究表明,在单一辫流带内部,发育心滩和辫状河河道沉积,二者皆由砂质充填且发育规模相近,心滩呈底平顶凸形,河道沉积呈顶平底凸形[13]。于兴河等[1]认为,心滩内部发育多期垂向增生体,是洪水期作用的产物,每个增生体皆为正韵律段,洪水退却后,心滩露出水面,其后部形成一个受心滩保护的静水区,细粒沉积物沉积,导致增生体之间由细粒沉积物分隔,界面近水平展布,顺水流方向向下倾斜,称为"落淤层";吴胜和等[15]认为,心滩的发育演化可分为心滩坝的形成、生长及向下游方向迁移,心滩坝的侧向迁移,"坝尾沉积"及复合心滩坝的形成三个阶段;牛博等[7]认为,心滩坝具有"纵向平缓前积、横向多期增生体加积"的特点,落淤层一般发育在心滩坝的后部。

心滩在 GR 曲线上多为箱形、锯齿状箱形(图4),发育大型槽状交错层理、板状交错层理等。因形成时水动力较强,多发育粗砂岩、中粗砂岩,且物性较好,常形成有效砂体。心滩的识别相对较容易,但如何根据测井曲线识别心滩的位置,一直是研究的重点,尤其是在本区井距较大的情况下,因此,本文利用 GR 测井曲线,提出"高程定中边,形态定前后,冲沟近中线,落淤层辅助"的判断方法(图7)。

3.2.1 高程判别法

心滩在沉积末期常露出水面,因而沉积厚度稍大于同期河道充填沉积。将盒八上亚段1.2小层沿顶部拉平,心滩顶部将稍高于同期河道充填沉积。而由于钻遇心滩的井可能钻在心滩边部的位置,因心滩底平顶凸的形态,边部较薄,使得心滩砂体顶部略低于同期河道充填

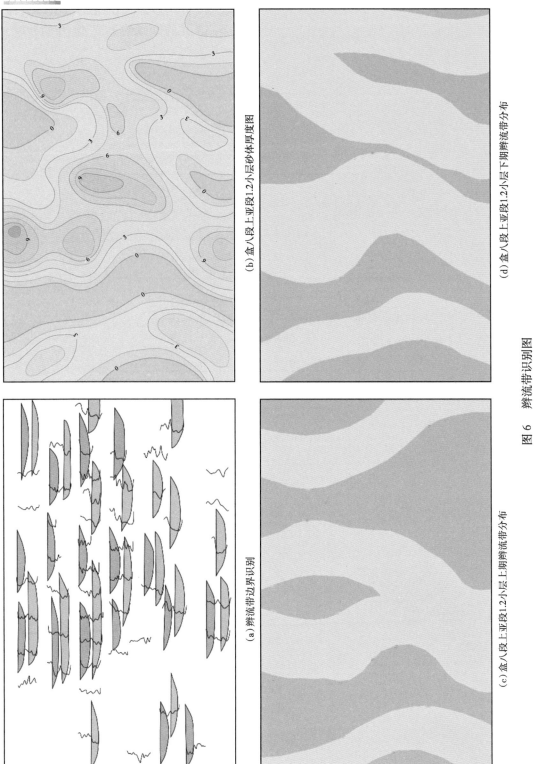

图 6 辫流带识别图

砂体,因此,可判断钻遇心滩的井,是位于心滩的边部还是中间部位,如图7所示,A1井与A3井同钻遇心滩砂体,A2井与A4井钻遇河道充填砂体。A1井的砂厚大于A2井的砂厚,即心滩顶部高于河道充填砂体,则A1井的位置为心滩中部;A3井的砂厚约等于A4井的砂厚,则A3井的位置为心滩边部。

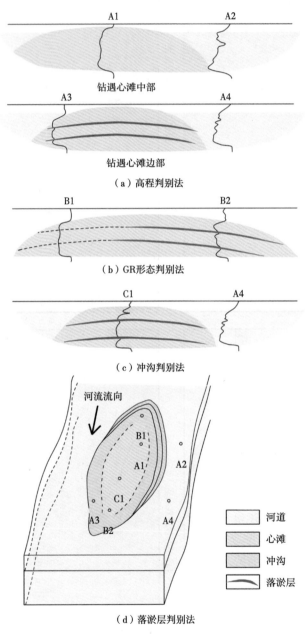

图7 心滩位置识别方法图

3.2.2 GR 形态判别法

此外,因心滩迎水流方向水流作用强,沉积粒度较粗,背水流面常形成静水面,沉积细粒沉积物,而在心滩的发育过程中,顺着水流方向逐渐推进叠置,沉积增生体,导致心滩尾部出现反旋回沉积[16],GR 曲线呈现轻微的漏斗形,而心滩头部垂向粒度变化不大,GR 曲线常呈现箱形。如图 7 所示,B1 井与 B2 井同钻遇心滩砂体,B1 井的 GR 曲线形态为箱形,B2 井的 GR 曲线形态呈现轻微的漏斗形,则 B1 井位于心滩前部,B2 井位于心滩尾部。

3.2.3 冲沟判别法

心滩顶部因露出水面遭受水流的冲刷而发育冲沟,冲沟的延伸方向多为顺水流方向,因而,钻遇冲沟的井,位置应在心滩中部靠近中线附近。冲沟在测井曲线上较容易识别,表现为心滩砂体顶部发育冲刷面,有一个小规模的正旋回沉积,向上逐渐过渡为泥岩。如图 7 所示,C1 井钻遇冲沟,则其位置应接近于心滩沿长轴方向的中线。

3.2.4 落淤层判别法

因滩头水动力较强,沉积的细粒泥质不易保存,因而心滩头部(迎水面)多不可见落淤层,而尾部(背水面)及两翼部位可见落淤层展布[7,17]。图 7 中,A1 井与 B1 井 GR 曲线皆未见回返,表示无落淤层发育,因而 A1 井与 B1 井位于心滩头部;A3 井、B2 井与 C1 井 GR 曲线可见回返特征,表示有落淤层发育,因而 A3 井、B2 井与 C1 井位于心滩尾部。

根据上述原则,可较好地确定心滩的位置与展布(图 8、图 9),同时,利用单期河道划分成果,按照盒八段上亚段 1.2 小层的上下两期河道展布,分别识别其内部心滩发育位置,得出上下两期沉积微相展布图,如图 8 所示。

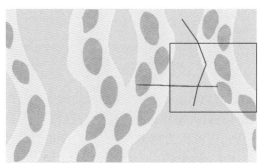 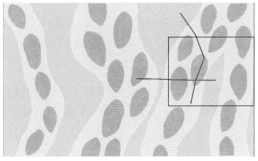

(a)盒八段上亚段1.2小层上期沉积微相展布　　　　(b)盒八段上亚段1.2小层下期沉积微相展布

图 8　盒八段上亚段 1.2 小层沉积相平面分布图

3.3 河道充填识别

河道在 GR 曲线上多为钟形、齿状钟形(图 4),发育板状交错层理、小型槽状交错层理,顶部可见波纹层理等。河道充填沉积的粒度小于心滩沉积,多为中砂岩、中细砂岩,底部发育冲刷面及滞留沉积,由下及上为典型的正韵律,顶部发育泥岩。

在构型连井剖面上,还可以识别废弃河道。河道在废弃过程中继续有粗的沉积物供应,但曲线显示出幅度相对下部较小,表明沉积物中细粒成分增加,物性变差[14]。同时,根据等高程

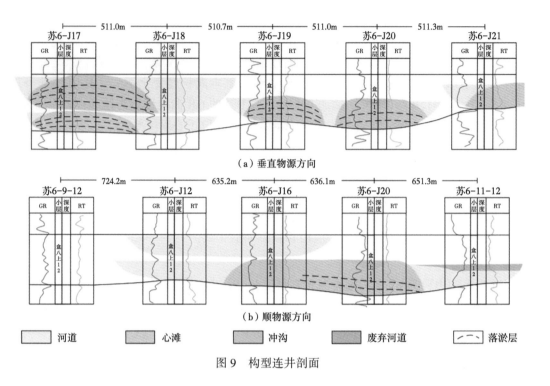

图 9 构型连井剖面

原理,在顶部拉平的剖面上,同期河道充填应具有相同的高程,据此可判断出废弃河道的发育位置。苏 6-11-12 井钻遇了一套河道充填沉积,在 GR 曲线上表现为钟形,但河道充填的顶部较同期河道充填沉积,有一段高程差异,表现为该河道充填沉积上部发育了一套细粒沉积物,为河道因改道而被废弃后,泥质沉积而形成的废弃河道。

4 构型规模表征及验证

在明确构型展布特征的基础上,可开展构型规模的研究。构型规模的表征的方法有野外露头、井间对比、经验公式等,由于辫状河发育规模具有较大差异,直接套用经验公式推算研究区构型展布参数,结果的准确性难以保证,但经验公式仍可作为参考,约束研究区构型参数展布。因此本文采用经验公式约束、井间精细对比统计法,来确定研究区构型参数范围。

4.1 构型参数表征

Bridge 等[3]根据大量实测数据建立了单一辫流带最小宽度及最大宽度的预测公式:

$$w_{\text{cbmin}} = 59.9 \times h_a^{1.8} \tag{1}$$

$$w_{\text{cbmac}} = 192 \times h_a^{1.37} \tag{2}$$

$$h_a = 0.55 \times h_d \tag{3}$$

式中 w_{cbmin} ——单一辫流带最小宽度,m;

w_{cbmax} ——单一辫流带最大宽度,m;

h_a——平均单河道满岸深度,m;

h_d——单河道满岸深度,m。

前人研究结果表明,单河道满岸深度与心滩的厚度相当[18],则可用单井统计出的心滩厚度代替单河道满岸深度(图9)。统计结果表明,研究区心滩厚度主要分布于4~12m,利用式(3)可得,平均单河道满岸深度为2.2~6.6m,平均值为4.4m,由式(1)、式(2)可得,单一辫流带最小宽度为862m,单一辫流带最大宽度为1462m。

Kelly 等[2]建立起辫状河心滩宽度与单河道满岸深度、心滩长度与心滩宽度之间的经验公式:

$$w_b = 11.413 \times h_d^{0.14182} \quad (4)$$

$$L_b = 4.9517 \times w_b^{0.9676} \quad (5)$$

式中 w_b——心滩宽度,m;

L_b——心滩长度,m。

根据统计得到的单河道满岸深度,利用式(4)、式(5)可得,心滩宽度分布于81~387m,心滩长度分布于350~1580m。

据井间精细对比绘制出的单一辫流带展布图(图6),以及根据上述心滩识别方法,绘制出的小层沉积相平面图(图8),测量辫流带宽度、心滩宽度及长度等相关参数,统计统计出研究区盒八段辫流带宽度、心滩长度、心滩宽度柱状图如下(图10)。

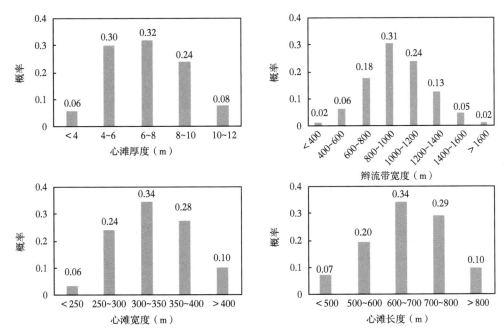

图10 苏6区块密井网区辫状河构型单元参数

由数据统计可知,辫流带宽度主要分布于600~1400m,心滩长度主要分布于500~800m,心滩宽度主要分布于250~400m。该统计数据与经验公式法计算得出的参数分布具有较好的吻合性,因经验公式法得出的数据区间较大,精确度较差,可将井间精细对比统计出的参数作为研究区构型参数分布范围。

4.2 构型展布准确性验证

为检验上述构型表征方法是否准确,选取干扰试井的方法,以盒八段上亚段1.2小层为例,验证其构型展布的准确性。

干扰试井试验是验证两口井是否连通的有效方法,其原理是,改变一口井的工作状态,使地下压力产生一个波动,在临井中监测是否发生压力变化,即是否发生干扰,若产生了干扰,则代表两井连通,反之亦然。

由于两口井的射孔层段可能不同,在判断两井在目的层是否连通时,需要先确定两口井的射孔层位。若两口井具有多个相同的射孔层段,如果干扰试井显示干扰,则不能确定两口井在目的层段是连通的,如果干扰试井显示未见干扰,则两口井在目的层段不是连通的;若两口井只有一个相同的射孔层段,如果干扰试井显示干扰,则两口井在目的层段连通,如果干扰试井显示未见干扰,则两口井在目的层段是不连通的;若两口井没有共同的射孔层段,则两口井之间大概率为不连通。

在准确判断目的层是否连通后,可进一步判断沉积相带的划分是否准确,在通常情况下,其判别标准为:处于不同河道带的井,由于泛滥泥岩的阻隔,两口井不连通;位于同一心滩内的井,或位于同一河道充填且距离较近的井,由于受同一沉积微相控制,岩性相近且砂体展布具有连续性,因而大概率是连通的;分属于心滩与河道充填的两口井,虽同为砂岩储层且连续分布,但因受不同沉积微相的控制,岩性略有差异,所以井间连通性无法确定[11]。

为验证上述河道带及河道充填、心滩位置划分的准确性,以盒八段上亚段1.2小层为例,选取若干口井做干扰试验(图11)。苏38-16-5与苏6-J21井,目的层为其唯一的共同射孔段,且干扰试验结果显示为未见干扰,则两口井在目的层不连通,根据相图可知(图8),两口井在目的层段钻遇了不同的心滩,符合判别标准。苏6-J20井与苏6-J21井具有包括目的层在内的多个射孔段,但井间干扰试验表明,井间未发生干扰,即目的层段不连通,根据相图可知,两口井分属于不同的河道,符合判断标准。苏38-16-5井与苏6-J16井,目的层为其唯一的

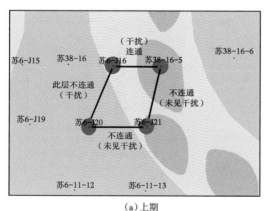
(a)上期

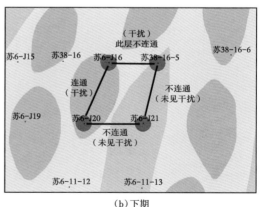
(b)下期

图11 干扰试验测试

共同射孔段,干扰试验结果显示为干扰,在上期沉积相图内,两口井分布于同一单河道带内,且一口井钻遇心滩,一口井钻遇河道充填沉积,符合判断标准。苏 6-J16 井与苏 6-J20 井,目的层为其唯一的共同射孔段,干扰试验结果显示为干扰,在下期沉积相图中,两口井钻遇了同一心滩,符合判别标准。

5 结论

(1)根据野外露头解剖,结合 Miall 河流相构型划分,将研究区辫状河储层划分为五级构型:①复合河道带;②单一辫流带;③心滩与河道充填;④心滩内增生体及冲沟;⑤层系组。

(2)由岩心、测井资料划分单井构型级次,由密井网数据划分出不同期次的河道分布,根据辫状河心滩的发育特征,提出了"高程定中边,形态定前后,冲沟近中线,落淤层辅助"的心滩位置判别方法,干扰试验表明,这种划分方法具有较强的可靠性。

(3)由密井网构型参数表征可知,研究区心滩厚度主要分布于 4~12m,单一辫流带宽度主要分布于 600~1400m,心滩长度主要分布于 500~800m,心滩宽度主要分布于 250~400m。

参 考 文 献

[1] 于兴河,马兴祥,穆龙新. 辫状河储层地质模式及层次界面分析[M]. 北京:石油工业出版社,2004.

[2] Kelly S. Scaling and hierarchy in braided rivers and their deposits:Examples and implications for reservoir modeling[M]// Sambrook Smith G H, Best J L, Bristow C S, et al. Braided rivers:Process, deposits, ecology and management. Oxford, UK:Blackwell Publishing, 2006:75-106.

[3] Bridge J S. Fluvial facies models:recent developments[J]. Society for Sedimentary Geology, 2006, 84(1):83-168.

[4] 徐东齐,孙致学,任宇飞,阳成. 基于地质知识库的辫状河致密储层地质建模[J]. 断块油气田,2018,25(1):57-61.

[5] 孙天建,穆龙新,吴向红,等. 砂质辫状河储层构型表征方法——以苏丹穆格莱特盆地 Hegli 油田为例[J]. 石油学报,2014,35(4):715-724.

[6] 邢宝荣. 辫状河储层地质知识库构建方法——以大庆长垣油田喇萨区块葡一组储层为例[J]. 东北石油大学学报,2014,38(6):46-53+108.

[7] 牛博,高兴军,赵应成,等. 古辫状河心滩坝内部构型表征与建模——以大庆油田萨中密井网区为例[J]. 石油学报,2015,36(01):89-100.

[8] 孙天建,穆龙新,赵国良. 砂质辫状河储集层隔夹层类型及其表征方法——以苏丹穆格莱特盆地 Hegli 油田为例[J]. 石油勘探与开发,2014,41(1):112-120.

[9] 毕明威,陈世悦,周兆华,等. 鄂尔多斯盆地苏里格气田苏 6 区块盒 8 段致密砂岩储层微观孔隙结构特征及其意义[J]. 天然气地球科学,2015,26(10):1851-1861.

[10] 崔连可,单敬福,李浮萍,等. 基于稀疏井网条件下的古辫状河道心滩砂体估算——以苏里格气田苏 X 区块为例[J]. 岩性油气藏,2018,30(1):155-164.

[11] 李易隆,贾爱林,冀光,等. 鄂尔多斯盆地中—东部下石盒子组八段辫状河储层构型[J]. 石油学报,2018,39(9):1037-1050.

[12] 乔雨朋,邱隆伟,邵先杰,等. 辫状河储层构型表征研究进展[J]. 油气地质与采收率,2017,24(6):34-42.

[13] 温立峰,吴胜和,岳大力. 粗粒辫状河心滩内部泥质夹层分布新模式——以吴官屯野外露头为例[J].

石油地质与工程, 2016, 30(4): 5-7+145.

[14] 张昌民, 尹太举, 赵磊, 等. 辫状河储层内部建筑结构分析[J]. 地质科技情报, 2013, 32(4): 7-13.

[15] 张可, 吴胜和, 冯文杰, 等. 砂质辫状河心滩坝的发育演化过程探讨——沉积数值模拟与现代沉积分析启示[J]. 沉积学报, 2018, 36(1):81-91.

[16] 马志欣, 张吉, 薛雯, 等. 一种辫状河心滩砂体构型解剖新方法[J]. 天然气工业, 2018, 38(7): 16-24.

[17] 宋子怡, 陈德坡, 邱隆伟, 等. 孤东油田六区馆上段远源砂质辫状河心滩构型分析[J]. 油气地质与采收率, 2019, 26(2):68-75.

[18] Ashworth P J, Smith G H, Best J L, et al. Evolution and sedimentology of a channel fill in the sandy braided South Saskatchewan River and its comparison to the deposits of an adjacent compound bar[J]. Sedimentology, 2011, 58(7):1860-1883.

原文见于2019年第31届全国天然气学术年会报告.

低渗透—致密砂岩气藏开发中—后期精细调整技术

付宁海　唐海发　刘群明

(中国石油勘探开发研究院,北京 100083)

摘要:低渗透—致密砂岩气藏进入开发中后期,面临如何进一步提高储量动用程度与采收率问题,开展开发中后期精细调整是改善其开发效果的重要技术手段。针对该类气藏进入开发中后期地质、生产动态以及开发方式上相对于开发早期的变化,梳理了开发中后期在剩余储量描述、提高储量动用程度和井网适应性评价中面临的关键技术问题,围绕气田挖潜与提高采收率,有针对性地提出了技术对策,提出了以开发中后期精细气藏描述、储量动用程度评价和井网井距优化为核心的气藏开发中后期精细调整的技术思路及流程。以 C 气田为例,采用该技术方法进行了气田中后期的开发调整,解决了剩余储量的有效动用问题,整体提升了气田开发效果和经济效益。

关键词:低渗透;致密气;精细调整;储量动用程度;开发中后期

近年来,随着中国天然气业务的快速发展,国内低渗透—致密砂岩气藏储量和产量快速增长,尤其是产量所占比重越来越大,低渗透砂岩气藏是中国天然气未来增储上产最重要的领域之一[1]。据统计,2011 年,致密气产量约占国内天然气产量的 1/4[2],2014 年,约占全国天然气总产量的 32%[3],目前,低渗透—致密砂岩气藏是中国天然气开发最具规模、储量和产量贡献最大的一类气藏,其有效开发成为天然气上产稳产的重要支撑。

低渗透—致密气藏具有储层非均质性强,产能差异大的特点,随着气田开发的深入,尤其是在气田开发中后期,储量动用不均衡进一步加剧,气田剩余储量难以有效动用的问题日益突出,因此,开展中后期精细调整研究,总结出相应的技术方法,对于提高储量动用程度,改善气田开发效果具有重要意义,也是提高气田最终采收率的关键。随着进入开发中后期气田比例的增加,客观上也需要加强中后期气藏精细调整技术研究,进一步细化对气藏的认识,明确储量动用状况与剩余储量分布,以延长稳产期和进行后期挖潜。

国内开发中后期精细调整研究多集中于油藏[4-7],关于气藏中后期精细调整主要以气藏描述为主,或针对某一具体气田面临的特定问题开展研究[8-13],缺少普遍的适用性与技术流程的梳理。在前人研究基础上,通过对低渗透—致密气藏开发中后期面临的关键技术问题的分析,总结提出了一套适用于低渗透—致密气藏开发中后期精细调整的技术思路及流程,可较为快速准确地确定气藏加密潜力、优选井位,进行剩余气挖潜,可为同类气藏开发提供借鉴。

1 低渗透—致密气藏开发中后期面临的关键技术问题

气藏进入开发中后期,开发重点由开发初期快速上产与规模开发向提高储量动用程度和提高采收率转变。开发中后期的气藏大多已进入稳产后期或递减阶段,由于储层的非均质性及井网的不完善性,导致剩余气分布相对分散。在开发早期阶段以砂层组或小层为单元所做的储层描述,已不能满足研究剩余气分布状况的需求。早期井网的不完善性需要井网的调整,同时需要根据气藏中后期生产动态特征的变化进行相应的开发调整。在地质、生产动态及开发方式等方面面临的主要问题不同于开发早期。

1.1 开发地质

气藏进入开发中后期,剩余储量的描述和预测是气藏面临的关键问题。根据开发初期井资料建立的地质模型精度较低,影响剩余储量预测的准确度。需要不断更新对气藏的认识,对剩余储量进行精细描述与评价。利用开发过程中逐渐丰富的资料,对地质储量进行复算,在此基础上,进行剩余储量评价,开展储层精细地质特征描述,有效砂体分布描述,以精细沉积微相、微构造和储渗单元为主对剩余储量进行描述和预测,研究剩余储量的分布规律,为气藏进一步的开发提供依据。

国内低渗透—致密储层多属陆相沉积,单层厚度薄,横向变化大,以河流相、三角洲相储层居多。对于三角洲相储层,三角洲类型的差异、沉积特征的差异以及相组合的不同,导致储层特征的差异以及平面非均质性的差异,形成不同的剩余储量分布特征。精细研究三角洲沉积微相及其物性,对后期改善气田开发效果具有重要的意义。河流相储层在剖面上一般呈透镜状分布,横向连续性差。单砂体较小且分散,砂体平面上呈不规则带状,多以顶平底凸、两侧不对称的透镜体为主[14]。开展有效储层三维精细刻画,对有效储层及其连通性进行准确预测和评价,有利于后期提高剩余储量的动用程度。

低渗透—致密气藏按储层产状,可划分为三种主要类型[1]:块状、层状和透镜状。透镜体多层叠置气藏,以鄂尔多斯盆地苏里格气田为代表;多层状气藏,以川中地区须家河组气藏、松辽盆地长岭气田登娄库组气藏为代表;块状气藏,以塔里木盆地库车坳陷迪西 1 井区为代表[15]。气藏类型不同,其面临的问题与中后期开发重点也不同(表1)。

表1 不同类型气藏开发中后期气藏描述重点

气藏类型	主要特点	描述重点	典型气田
透镜状	非均质性强,单砂体较小	储渗单元、有效砂体连续性、连通性	苏里格气田
层状	构造控制不明显、气层连续性差、层间差异大	储层连续性、连通性	川中须家河组气藏、长岭气田登娄库组气藏
块状	断块背斜为主,连通性较好,具有边底水	构造、裂缝、气水关系	迪西气田

透镜状与层状气藏,构造对气层分布控制不明显,气层连续性差,中后期对剩余储量的评价和预测,以储层连续性、连通性评价为重点。以苏里格气田为代表的透镜状气藏,辫状河发育,河道侧向迁移、改道和切割频繁,造成心滩和边滩砂体在纵向上相互叠置、交错排列[16]。储层小透镜体、多层发育,区域上富集不均,大面积复合连片分布,储层连续性、连通性差。气藏开发前期研究,苏里格大型复合砂体分级构型描述技术、富集区和井位优选技术,使富集区内Ⅰ类+Ⅱ类直井比例保持在75%以上[15,17],保障了建产区的优选,实现了气藏的规模开发。

中后期面临如何进一步提高储量动用程度与采收率,需要以储渗单元描述为重点,开展密井网区井间精细对比,对有效砂体分布进行定量描述,建立精细地质模型,以此为基础评价储量动用程度和剩余储量分布,提高剩余储量预测的准确度。

块状气藏一般储层整体连通性好,储量动用程度相对较高。进入中后期,应以构造、裂缝、气水关系的描述为重点。需要基于新的钻井、地震资料,动静结合,加强对构造的认识,精细刻画微构造、裂缝、断层等。构造精细刻画关系到后期高效井位部署的成功率。同时由于气水关系复杂,面临防、控、治水问题,需加强裂缝、隔夹层空间配置关系研究,深化气藏水侵模式、气水关系的认识,实现均衡开采,提高最终采收率。

1.2 生产动态

低渗透—致密气藏进入开发中后期,生产动态特征的变化主要表现在:(1)压力、产量下降速度变缓,单位压降产气量增加,气井泄气半径后期因外围低渗透区补给有一定扩大,在较低产量水平上可保持较长时期的稳定生产。(2)气藏产能分布不均,不同气井采气速度和采出程度差异较大,造成储量动用不均衡。(3)气井产能与初期变化较大,随着生产阶段的变化需要核实与调整,以充分发挥气井生产能力。(4)低产气井增多,少数高、中产气井对气藏整体产能贡献比例增加,气藏逐渐进入多井低产阶段。

生产动态特征的变化加上储层本身的非均质性,导致储量动用不均衡程度进一步加剧。因此,在开发中后期,面临对气藏储量动用程度进行评价问题,以此确定开发调整的重点。一方面需要根据生产动态特征的变化对气井的产能进行核实,确定气藏合理的产能规模,生产潜力。尤其是对于产能贡献大的中、高产气井,保证这部分中、高产井的合理生产,对整个气藏的稳产具有重要意义。另一方面需要确定气井的动态储量,结合生产动态特征,评价气藏开发效果,确定各小层储量平面和纵向上的动用状况,确定剩余储量的潜力区域和层位。

1.3 开发方式

低渗透—致密气藏往往采用衰竭式开发,通过加密钻井弥补递减保持稳产。对于多层和透镜状气藏,由于井控储量少,单井泄气面积小,井间加密是提高采收率的关键。国内外低渗透—致密气藏的开发经验表明,对于多层叠置透镜状气藏,通过加密井网,天然气的采收率可大幅度提高[17-20]。气藏初期井网对储量控制程度不够充分,现有井网存在未控制住的储量。表现在:(1)低渗透—致密储层展布规模小,连通性差,单井产能低,不宜采用稀井高产的模式开发,一般采用边评价、边开发的思路,井间逐次加密,局部区域存在初期井网未控制住的储量。(2)井网平面分布不均,局部较密,局部过稀,造成储量动用不均衡。(3)井网控制住的区域,在开发过程中由于储层非均质性及气井产能的差异,泄气半径的差异,造成平面上储量动用不均衡。(4)多层生产气井,垂向上具有补孔潜力的未射孔小层,其储量尚未动用。

因此,在开发中后期面临对井网进行加密与调整的问题,以提高井网对储量的控制程度。需要开展井型与井网的适应性评价,论证现有井网对储量的控制程度与开发效果,论证合理的布井方式和井网密度,优选井型,优化井网井距。结合地质研究和生产动态特征,进行井位论证,确定开发调整井位的部署。

2 精细调整思路及流程

在开发中后期的气藏调整中,需要以剩余气分布研究为核心,以精细的小层划分与对比以及储层的定量评价为基础,充分利用各种静态和动态资料,进行精细的气藏描述,建立精细地质模型,并通过开发过程中气藏动态变化研究、剩余气分布规律研究,最终确定剩余气的空间分布情况,为开发调整与挖潜提供可靠的依据。

整体思路是针对剩余储量描述、提高储量动用程度和井网适应性评价中面临的关键问题,分别采用精细气藏描述技术、储量动用程度评价技术和井网井距优化技术,解决储量的空间分布、剩余储量潜力区层位与平面分布,以及调整井井位论证问题。具体流程分为三部分:首先,开展精细气藏描述,精细刻画小层砂体、有效砂体展布,复算小层地质储量,通过对气藏有效砂体发育规律的认识,精细刻画储层特征,落实气藏地质储量,落实调整基础。同时,应用多种气藏工程方法,论证气藏合理产量和动态储量,评价单井采出程度和小层储量动用程度,确定开发调整方向和重点。最后,综合考量地质、动态、经济三方面因素,评价开发井网适应性和开发效果,论证气田合理井网井距,制定气田调整技术对策,落实调整井位的部署(图1)。

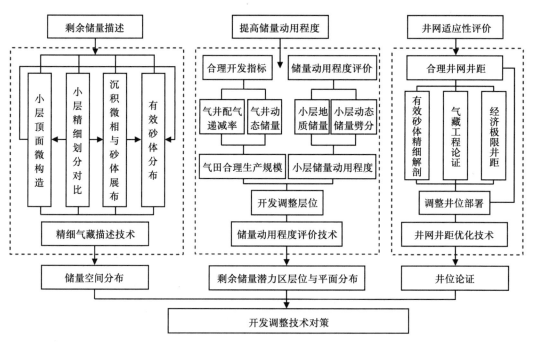

图1 气藏开发中后期精细调整技术流程图

3 开发中后期精细调整技术

精细调整前提是提高对气藏认识的深度。通过精细气藏描述技术、储量动用程度评价技术、井网井距优化技术,对气藏开发动态规律、剩余气特点以及分布规律进行深入研究,从而为下一步的调整挖潜提供依据,确定调整井位的部署,提高气藏最终采收率。

3.1 精细气藏描述技术

中后期的气藏描述,重点是提高描述精度。综合运用地质、地震、测井、测试等各方面资料,利用中后期增多的井资料,从区块到井间,通常以单砂体或流动单元为基本地层单元,与动态结合紧密,基本单元小,精细程度高。

3.1.1 精细地层结构描述

随着开发中后期资料的丰富,需要进一步细分小层,确定小层界限,落实微构造。主要包括小层精细划分与对比以及构造精细解释对比两方面内容。小层划分与对比是描述储层空间分布的前提,通过小层对比,对研究区内气井的原始地质分层数据进行复查,建立骨架对比剖面。在小层划分对比基础上,区域上结合地震资料、构造及断层解释成果,分小层编制顶面构造图。对比与前期认识上的变化,对前期认识的不足进行更新与调整。

3.1.2 沉积微相分析技术

综合区域沉积背景和单井相分析,确定沉积相类型,分析沉积特征,包括沉积环境、沉积相、沉积模式等。进行沉积微相分析,分析沉积微相类型、沉积微相垂向演化特征、沉积微相及砂体平面展布特征。

3.1.3 有效砂体描述技术

对于透镜状与层状气藏,开发中后期气藏描述重点是对有效砂体进行描述。开展密井网区精细地质解剖,刻画有效砂体的规模尺度、连通性及其分布规律。对密井网区单砂体进行解剖,通过垂直物源密井网连井剖面与顺物源密井网连井剖面,分析有效单砂体规模尺度。进而分析复合有效砂体的结构模式与规模,最后分析有效砂体平面、剖面在三维空间的分布特征。在此基础上,建立定量的精细地质模型,对剩余气分布进行预测。

以 C 气田为例,C 气田构造形态为一北北东向展布的背斜,以发育辫状河三角洲沉积相为主,东西向剖面单砂体相变快,连通差、延伸距离短,垂向上具有砂泥岩薄互层发育的特点。气藏有效砂体规模小、多呈孤立状分散分布,局部存在富集区。有效砂体钻遇率30%~60%,平均有效厚度4~9m。剖面上(图2),有效砂体分布密度小,90%以上有效砂体呈孤立状分散分布,横向范围局限,连通性差,垂向叠置模式以孤立状为主,少量垂向叠置型。平面上,小层有效砂体多呈孤立状分布。

密井网区单砂体解剖显示,单期河道砂体厚度一般在3~8m之间,三角洲平原相辫状分流河道宽400~700m,三角洲前缘相水下分流河道宽350~600m。有效单砂体厚度集中分布在2~4m之间,平均厚度为3.5m。垂直物源密井网连井剖面显示,有效单砂体宽度集中在300~400m之间。顺物源密井网直井连井剖面显示,有效单砂体长度主要分布在300~450m之间。

3.2 储量动用程度评价技术

在气藏地质认识与动态特征分析的基础上,进行储量动用程度评价,确定各砂层组和小层储量动用程度,落实已动用储量和剩余未动用储量,确定调整挖潜的潜力区。

3.2.1 气田开发指标计算

储量动用程度评价前,首先论证气井的合理开发指标。一方面,对气井目前的产能、递减等指标进行计算,评价气井的生产能力,确定开发调整时气田的合理生产规模。另一方面,重

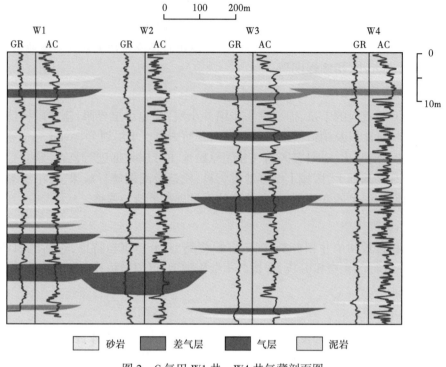

图2 C气田W1井—W4井气藏剖面图

点要计算与核实气井的动态储量,进而确定动用程度与开发潜力。不同的动态储量计算方法有其自身的适应性和局限性,针对不同生产动态特征的气井,应结合多种方法进行评价。对于有测压资料的气井,采用压降法计算动态储量较为准确;对于生产时间较长,采出程度较高,进入递减的气井,可通过产量累计法和油压递减法以及常规递减分析方法计算;气藏渗流达到或接近拟稳态,气井产量相对稳定的气井可采用流动物质平衡法[21];产量不稳定分析方法,对于计算低渗透气藏储量具有较大优势[22],其建立在常规的生产动态资料之上,对地层压力测试点的依赖程度较低,对产量和压力数据要求低,生产数据经过处理后,采用不稳定法进行图版拟合,得到气井动态储量,目前常用的有Blasingame、Agarwal-Gardner、NPI、Transient等方法。

3.2.2 储量动用程度评价

储量动用程度评价的目的是确定储量的动用状况和剩余储量的分布情况,从而确定挖潜的目标层位。在计算得到单井的动态控制储量,并对地质储量进行复核的基础上,将单井的动态储量和累计产量细化到小层,确定各个小层储量动用程度。目前小层产量劈分方法主要有地层系数法、产气剖面测试法、物理实验模拟法以及数值模拟方法等[23-25]。将各个单井的动态储量和累计产量劈分到各个小层,结合各个砂层组和小层的地质储量,就可以得到各砂层组及小层的储量动用程度和采出程度。计算气井在每个小层的泄流半径和动用面积。通过各小层储量动用面积与含气面积叠合图,确定储量在各小层平面上的动用情况。依此确定挖潜重点层位,明确挖潜的主力小层。

C气田自上而下,分为M2、N1、N2、N3、N4共五套砂层组。根据对C气田有效砂体的描述,N1、N2 砂层组有效砂体发育相对较好,局部存在富集区,N3、N4 砂层组有效砂体零星分布。根据砂层组储量动用程度和采出程度分析(图3),M2、N1 砂层组储量动用程度低(小于40%),采出程度低(小于25%),且剩余储量多;N2 砂层组储量动用程度高,但储量基数大,仍有较多的剩余储量;N3、N4 砂层组储量动用程度较高,且剩余储量少。因此,剩余储量潜力主要集中在 M2、N1、N2 砂层组,是下步气田挖潜的重点层位。从小层看(图4),M22、N13 小层地质储量动用程度低(小于35%),采出程度低(小于25%),剩余储量高,是挖潜的主力小层,其次是 N23、N22 小层。

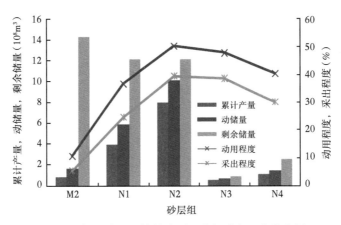

图 3　C 气田砂层组储量动用程度与采出程度分布图

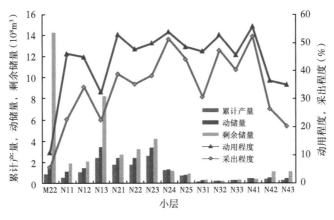

图 4　C 气田小层储量动用程度与采出程度分布图

3.3　井网井距优化与调整井位部署技术

从技术和经济方面确定合理井距,进行井网适应性论证,确定加密调整的空间和潜力,进而优选调整井位的有利目标区,确定井位及目的层。

3.3.1　井网井距优化技术

低渗透—致密气藏,不适合大井距开发,需采用密井网开发,以提高储量的动用程度和最终采收率。合理井距的论证主要有地质分析、气井泄气半径折算、井间干扰分析以及经济极限井距评价等方法。

(1) 地质分析。

选取密井网区,进行精细地质解剖,根据密井排有效砂体连井对比,分析有效砂体的规模尺度,研究砂体的连通程度,确定有效砂体规模大小。根据砂体的长度、宽度分布范围和频率,确定井排距范围。

(2) 气井泄气半径折算。

泄气半径计算方法主要可分为试井探测半径方法、不稳定产能分析图版方法、动静态储量结合反算方法等[26-28]。试井探测半径方法,多采用压降或压力恢复试井进行探边测试,以压力波传播的探测半径作为气井的泄气半径。不稳定产能分析利用气井的生产数据和地质参数,通过 Blasingame、Agarwal-Gardner、NPI、Transient 等图版拟合确定泄气半径。动静态储量结合反算方法,根据气井的动态储量,由容积法储量计算公式,反算气井泄气半径。

(3) 井间干扰分析。

同一气层上相邻两口气井同时生产时,某一口气井改变工作制度,对相邻气井的压力、产量产生影响,或是新井投产,在存在井间干扰情况下,邻近老井产量或压力发生改变。根据相邻气井压力产量的变化判断两口井间连通和干扰情况,以此来判断井距是否合理。

(4) 经济极限井距。

中后期调整涉及井网加密,低渗透—致密砂岩气藏一般属于边际效益气藏,经济的有效性是井网加密重要的考量因素,开发调整的井距应大于经济极限井距。根据经济极限井距计算公式[29],得到不同气价下的极限井距,以此来作为加密调整井距的下限。

综合以上几种方法,从技术、经济两方面确定合理的井距及井网密度。通过密井网区单砂体解剖、气井泄流半径分析以及经济极限井网密度计算,确定 C 气田合理开发井网井距:350m×400m,井控面积 0.14km²/口。从井网控制程度看,目前平均井控面积约 0.3km²/口,与合理井控面积 0.14km²/口相比,具有较大的加密空间。从单井的动用面积与储层的含气面积叠合情况以及小层储量动用状况的分析看,现有井网对储量控制不充分,气藏具有进一步加密调整的空间和潜力。

3.3.2 调整井位部署技术

在气藏挖潜主力层和加密潜力区域研究基础上,结合小层沉积相、砂体、有效砂体平面及剖面分布特征和邻井生产动态,优选加密井位。

加密井位部署时,依据"十图两表"(储量动用面积与含气面积叠合图、顶面构造图、沉积相平面图、砂体厚度图、有效砂体厚度图、孔隙度分布图、渗透率分布图、含气饱和度分布图、邻井砂体及有效砂体连井对比剖面图、邻井生产曲线图、储量动用程度与采出程度统计表、邻井生产动态统计表),重点分析部署位置的储层静态以及邻井生产动态特征,优选有利的位置和层位。具体分为四步:

(1)根据小层储量动用面积与含气面积叠合图,结合数值模拟剩余储量和压力分布确定加密调整井位部署的潜力区域。

(2)在确定的潜力区域基础上,进一步优选加密部署的有利目标区。根据区域地质特征,分析加密井及邻井的构造及储层分布情况,加密井的部署要满足三个基本条件:①处于微构造局部高点附近;②处于有利相带内,砂体发育厚度大,分布稳定,邻井可横向对比追踪;③储集物性好,有效砂体较发育。

(3)在此基础上,结合精细气藏描述中对小层砂体的精细刻画,从邻井砂体及有效砂体对比剖面,分析纵向上含气砂体发育状况以及有效砂体横向分布情况,结合小层储量动用程度、采出程度,确定加密井的目标开采层位。

(4)根据邻井生产动态及生产现状,分析加密井周围的储层生产情况,估算加密井所处井组的储量动用状况和剩余储量情况,进而预测加密井的生产能力及可采储量,判断加密井投产效果及对邻井可能产生的影响。

最终在动、静态特征综合分析的基础上,确定加密井位及开采目的层位。

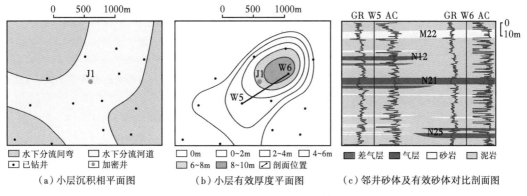

图 5 加密井及邻井沉积相与砂体分布图

以 C 气田加密井 J1 井为例,从储层物性分布图及构造图看,构造位置有利,物性较好,小层沉积相平面图及小层有效砂体厚度图显示(图 5a、b),该井处在水下分流河道有利相带内,砂体发育情况好,有效厚度 6~8m。邻井砂体及有效砂体对比剖面显示(图 5c),J1 井 N21 小层砂体、有效砂体横向发育好,分布稳定。从邻井动态来看,邻井以一类井居多,生产稳定,井距大大高于合理井距,具备加密的条件且加密位置较为有利。经过"十图两表"设计优选,该井投产后,产量达到 $2\times10^4 m^3/d$ 以上,生产稳定。

通过合理井网井距及储量动用情况的论证,确定了气田整体加密调整的技术思路。根据上述加密部署流程,对气田的加密有利目标区进行了优选,确定加密井位 49 口,目前已全部实施并投入开发。加密井投产以来,其产量占气田产量的 1/3 左右,加密效果良好,有效弥补了气田老井的递减,保持了气田稳产。预计加密井最终可使气田采收率提高 6%。

4 结语

(1)对低渗透—致密气藏开发中后期面临的关键问题进行了系统分析,以具体气田为例,提出以精细气藏描述为基础,综合储量动用程度评价及井网井距优化等关键技术手段,深入认识气藏,进行有效砂体精细刻画,落实储量动用程度,明确剩余储量分布与开发潜力并进行井位部署的具体做法,为气藏开发中后期的开发调整提供了现实可行的技术思路与流程。

(2)该技术方法可快速评价开发潜力区,实施调整井位的部署,有效解决了剩余储量潜力区优选以及调整井井位论证问题,取得了较好的现场应用效果,明确了气田的开发潜力,大幅提升了气田开发效果,可为同类气田开发调整提供方法借鉴。

参 考 文 献

[1] 雷群,李熙喆,万玉金,等. 中国低渗透砂岩气藏开发现状及发展方向[J]. 天然气工业,2009,29(6):1-3.

[2] 邱中建,赵文智,邓松涛. 我国致密砂岩气和页岩气的发展前景和战略意义[J]. 中国工程科学,2012,14(6):4-8.

[3] 胡俊坤,龚伟,任科. 中国致密气开发关键因素分析与对策思考[J]. 天然气技术与经济,2015,9(6):24-29.

[4] 李志鹏,林承焰,史全党,等. 高浅南区边水断块油藏类型及剩余油特征[J]. 西南石油大学学报(自然科学版),2012,34(1):115-120.

[5] 贾爱林,程立华. 数字化精细油藏描述程序方法[J]. 石油勘探与开发,2010,37(6):709-715.

[6] 刘敬强,邹存友,普明闯,等. 油田开发中后期加密潜力的计算方法[J]. 断块油气田,2011,18(4):498-501.

[7] 陈金凤,庞帅,吴辉,等. 唐家河油田馆陶油组剩余油研究及挖潜方法[J]. 西南石油大学学报(自然科学版),2011,33(5):79-83.

[8] 李东,张云鹏,张中伟. 复杂断块气藏精细描述研究[J]. 断块油气田,2000,7(4):11-15..

[9] 王勇飞,曾炎. 储层建模致密低渗气藏开发调整研究[J]. 断块油气田,2008,15(5):69-71.

[10] 王雯娟,成涛,欧阳铁兵,等. 崖城13-1气田中后期高效开发难点及对策[J]. 天然气工业,2011,31(8):22-24.

[11] 徐庆龙. 中浅层低渗透断块砂岩气田开发调整实践[J]. 大庆石油地质与开发,2013,32(4):67-70.

[12] 刘成川,卜淘,张文喜. 新场气田蓬二段气藏二次开发调整研究[J]. 油气地质与采收率,2004,11(4):46-48.

[13] 廖家汉,杜锦旗,谭国华,等. 户部寨复杂断块气藏剩余气分布及挖潜研究[J]. 吐哈油气,2005,10(2):127-132..

[14] 关富佳,李保振. 辫状河沉积气藏井网模式初探[J]. 天然气勘探与开发,2010,33(2):40-42.

[15] 马新华,贾爱林,谭健,等. 中国致密砂岩气开发工程技术与实践[J]. 石油勘探与开发,2012,39(5):572-579.

[16] 樊友宏,李跃刚. 河流相储层产能评价方法研究[J]. 天然气工业,2005,25(4):100-102.

[17] 何东博,贾爱林,冀光,等. 苏里格大型致密砂岩气田开发井型井网技术[J]. 石油勘探与开发,2013,40(1):79-89.

[18] Cipolla C L,Mayerhofer M. Infill drilling & reserve growth determination in Lenticular Tight Gas Sands[C]// SPE Armual Technical Conference and Exhibition,6-9 October 1996,Denver,Colorado,USA. DOI:https://doi.org/10.2118/36735-MS.

[19] McCain W D,Voneiff G W,Hunt E R,et al. A tight gas field study:Carthage(Cotton Valley)Field[C]// SPE Gas Technology Symposium,28-30 June 1993,Calgary,Alberta,Canada. DOI:https://doi.orga/10.2118/26141-MS.

[20] Cipolla C L,Wood M C. A Statistical approach to infill drilling studies:Case history of the Ozona Canyon Sands[J]. SPE Reservoir Engineering,1996,11(3):196-202.

[21] 王京舰,王一妃,李彦军,等. 鄂尔多斯盆地子洲低渗透气藏动储量评价方法优选[J]. 石油天然气学报,2012,34(11):114-117.

[22] 陈霖,熊钰,张雅玲,等. 低渗气藏动储量计算方法评价[J]. 重庆科技学院学报(自然科学版),2013,

15(5):31-35.

[23] 郭平,刘安琪,朱国金,等.多层合采凝析气藏小层产量分配规律[J].石油钻采工艺,2011,33(2):120-123.

[24] 史进,盛蔚,李久娣,等.多层合采气藏产量劈分数值模拟研究[J].海洋石油,2015,35(2):56-60.

[25] 李江涛,张绍辉,杨莉,等.涩北气田气层动用程度研究[J].油气井测试,2014,23(1):30-32.

[26] 何东博,王丽娟,冀光,等.苏里格致密砂岩气田开发井距优化[J].石油勘探与开发,2012,39(4):458-464.

[27] 庄惠农.气藏动态描述和试井[M].北京:石油工业出版社,2004.

[28] 刘海锋,王东旭,李跃刚,等.靖边气田加密井部署条件研究[J].低渗透油气田,2007(1):83-85.

[29] 汪周华,郭平,黄全华,等.大牛地低渗透气田试采井网井距研究[J].西南石油学院学报,2004,26(4):18-20.

原文刊于《西南石油大学学报:自然科学版》,2018,40(3):136-145.

鄂尔多斯盆地中—东部下石盒子组八段辫状河储层构型

李易隆[1]　贾爱林[1]　冀　光[1]　郭建林[1]　王国亭[1]　郭　智[1]　张　吉[2]

(1. 中国石油勘探开发研究院,北京 100083;
2. 中国石油长庆油田研究院,西安 710021)

摘要:基于国内外研究成果建立辫状河构型模式,并利用测井、野外露头及气井动态资料,对鄂尔多斯盆地中东部盒八段辫状河储层进行构型解剖。在野外露头中识别出河道带、心滩与河道充填,单元坝与心滩内水道,大型倾斜层,中型交错层、中型平行层和小型交错层五级构型单元。运用所测量交错层系厚度资料推算露头剖面中心滩长为 1163～1592m,宽为 282～390m,该推算值与实际测量值相符。对露头和苏里格气田密井网区盒八段储层构型解剖表明,河道带砂体、心滩砂体之间发育多成因的、连续的渗流屏障,连通性较差;而心滩内部渗流屏障规模较小,因此心滩可视为独立的储渗单元。运用成像测井资料推算出苏里格气田盒八段辫状河心滩长为 602～1243m,宽为 143～302m,大部分心滩规模小于目前气田的井距和排距;另外,气井动态资料也证明目前的井网密度下,有效储层井间连通性较差,因此苏里格气田具备加大井网密度的潜力。

关键词:辫状河;储层构型;鄂尔多斯盆地;苏里格气田;致密砂岩

鄂尔多斯盆地中东部下石盒子组盒八段致密砂岩为典型的砂质辫状河沉积[1],该套致密砂岩储层是多个气田的主力产层[2],其中苏里格气田作为国内最大气田,目前已进入稳产期,需要对井网进行加密调整[3]。因此,有必要通过精细的储层构型解剖了解地下储层展布和连通特征,为加密调整提供依据。近年来国内外对辫状河沉积构型进行了大量研究,尤其是运用遥感卫星和探地雷达(GPR)对现代辫状河的研究[4-5],使辫状河储层的精细构型解剖成为可能。本文基于国内外辫状河构型模式研究成果,以野外露头解剖为参考,对苏里格气田密井网区盒八段储层进行了精细构型解剖,提出了储层连通模式,并且对解剖结果和连通模式进行了实际验证。

1　区域地质概况

鄂尔多斯盆地是多旋回的叠合含油气盆地,经历了裂陷盆地、克拉通盆地、克拉通坳陷盆地、内陆坳陷盆地、平缓斜坡等多旋回叠合盆地发育阶段[6],并相应发育多套含油气系统。其中石炭统—二叠系煤系含气系统包含石炭系本溪组、下二叠统山西组与太原组、中二叠统下石盒子组等,该套地层发育于克拉通坳陷盆地阶段,沉积物由滨浅海碎屑岩过渡为陆相河流相碎屑岩[7](图1),并广泛分布于盆地中,其中下石盒子组盒八段是盆地低渗透—致密气藏主力产层。晚石炭世华北板块南高北低和早二叠世北高南低的"翘板式"构造运动使盆地填平补齐,盒八段沉积时期,盆地地形极为平缓,古坡度为 $0.1°～1.0°$ [8],在低缓的盆地斜坡上,发育多套由北向南推进的冲积扇—辫状河—辫状河三角洲沉积体系,其物源为北部的阿拉善和阴山古陆,盆地中东部主要受阴山古陆物源的影响[9-10]。苏里格气田位于盆地中部,其主力产层为辫状河沉积,而盒八段在盆地东部柳林地区也为辫状河沉积,且两个区域内盒八段沉积物源均为北部阴山古陆,因此对两个区域内盒八段辫状河沉积的构型解剖,具有可类比性。

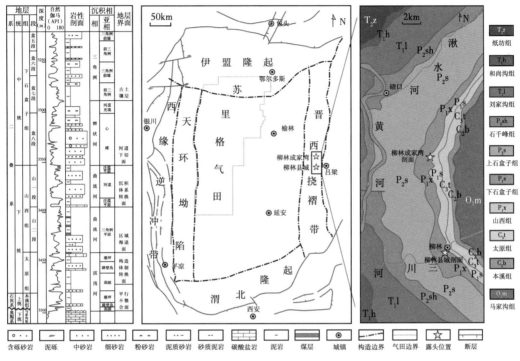

图1 鄂尔多斯盆地上古生界综合柱状图、研究区位置图与露头剖面地质图

2 辫状河沉积构型单元划分及特征

20世纪末,中外学者对辫状河沉积进行了大量研究,通过对现代沉积和露头剖面的观察与解剖,建立了辫状河沉积相模式[11-12],但这类模式多为二维的静态模型,未能反映辫状河的沉积过程和三维形态规模。21世纪以来,GPR更多地被应用于对现代辫状河沉积与辫状河相露头的研究中,在三维空间中描述了不同级次的辫状河沉积体细节[6],此外,历史遥感图像为研究辫状河演化特征提供了大量资料[4]。技术进步促使国外学者建立了具有三维尺度的辫状河分级构型模式[13-14],近年来,河流构型解剖也应用于油气储层表征中,取得了较大的进展[15-18],其中大庆油田萨中区块平均井距达50~60 m,是中国乃至世界上井网密度最大的地区之一,在该区域对辫状河沉积构型的解剖具有较大的参考价值[19]。

以国内外最新的辫状河构型模式为基础,结合鄂尔多斯盆地东部露头解剖以及苏里格气田密井网解剖经验,在辫状河沉积中,根据成因机理与规模,划分出五个级次的辫状河沉积构型单元,分别为:(1)河道带;(2)心滩/复合坝与河道充填;(3)单元坝与心滩内水道;(4)大型倾斜层;(5)中型交错层、中型平行层和小型交错层(图2)。河道带对应Miall的五级构型界面[20],由泛滥平原泥岩所分隔,底部发育冲刷侵蚀界面,并发育滞留沉积;河道带内心滩和河道充填顶部对应 Miall 的四级构型界面;单元坝是构成心滩和河道充填的基本单元,由季节性洪水所形成,呈朵页状;在心滩内低洼处可能发育心滩内水道(cross-bar channel),也称为串沟(chute)[21],心滩内水道可能继续扩大成为主水道,也可能废弃而被泥质充填[22];大型倾斜层界面对应 Miall 的三级构型界面,一般具有低缓的倾角;中型槽状交错层由沙坪的迁移而成,是心滩与河道充填底部最常见的构型单元,中型平行层一般在心滩顶部发育,小型交错层由流水波纹所形成,一般发育于心滩或河道充填的顶部,中型交错层、中型平行层与小型交错层均对应 Miall 的二级构型界面。

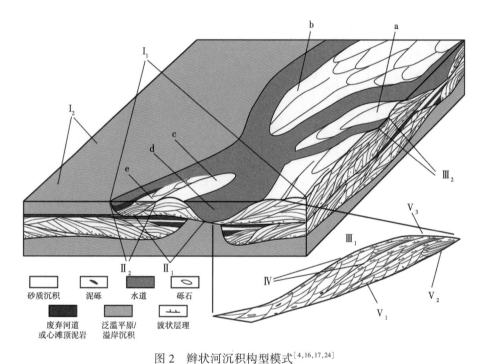

图 2　辫状河沉积构型模式[4,16,17,24]

I$_1$—河道带;I$_2$—泛滥平原/溢岸沉积;II$_1$—心滩;II$_2$—河道充填;III$_1$—单元坝;
III$_2$—心滩内水道;IV—大型倾斜层;V$_1$—中型交错层;V$_2$—中型平行层;V$_3$—小型交错层;
a—心滩头部朵体;b—卷形坝;c—汇流坝;d—转换河道;e—迁移河道(废弃阶段)

辫状河各类构型单元在井资料及野外剖面中表现出不同的垂向沉积序列和剖面展布特征。鄂尔多斯盆地盒八段河道带在测井垂向序列中厚度一般为 4~8m,在露头剖面中延伸范围,达数百米至千余米,底部发育明显的冲刷界面,界面之上发育滞留沉积,主要为含泥砾的砾石质砂岩(图3、图4、图5)。

心滩与河道充填在测井上表现出不同的垂向序列,同一河道带内心滩的粒度及砂岩厚度均大于河道充填(图3)。其中心滩上游部分垂向沉积序列整体上一般为块状粒序或向上变粗粒序,可能在沉积序列顶部出现粒度较粗的心滩头部朵体,自然伽马曲线一般为多个箱形曲线叠合,代表多个单元坝的垂向叠置,在上部可能出现较厚的箱形或漏斗形曲线,代表粒度较粗的心滩头部朵体(图3、图5),在朵体之间的低洼地带,可能会发育心滩内水道,具有向上变细

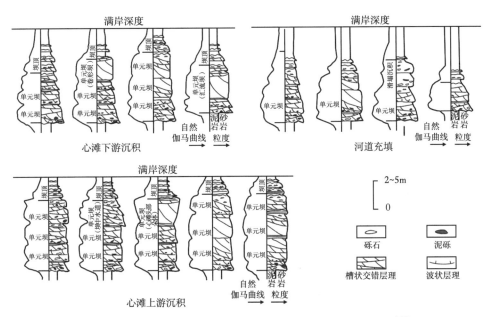

图3 心滩不同部位及河道充填的曲线特征与岩相垂向序列[17]

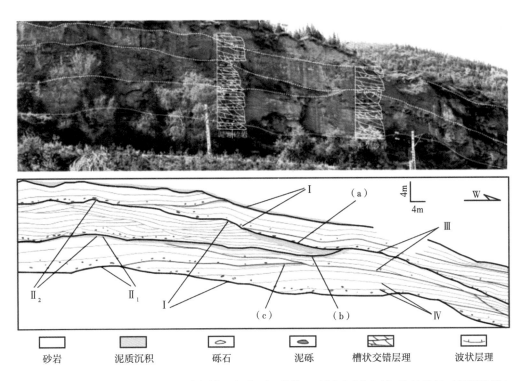

图4 鄂尔多斯盆地东缘盒八段露头剖面中叠置河道带、心滩与河道充填(柳林县城三川河北岸)
Ⅰ河道带;Ⅱ₁心滩;Ⅱ₂河道充填(迁移河道);Ⅲ单元坝(心滩头部朵体);Ⅳ大型倾斜层;
(a)废弃河道泥岩;(b)心滩坝顶斜列泥岩;(c)单元坝顶部泥岩(心滩内斜列泥岩)

的特征,底部发育小型冲刷面,自然伽马曲线为漏斗形(图3、图4);心滩下游部分一般具有粒度向上变细、单元坝厚度向上变小的沉积序列,自然伽马曲线为叠合的钟形和漏斗形,心滩下游可能发育侧向加积的卷形坝(图2b),或者河道汇合形成的汇流坝(图2c、图6 I_1),自然伽马曲线表现为单个较厚的箱形(图3),心滩下游一般发育心滩坝后部泥岩,形成与心滩后部的静水沉积中,剖面中心滩坝后泥岩一般较薄,厚度小于1m(图6c);心滩在演化的末期其上部多沉积泥质,在心滩顶部或平缓倾斜面上的泥质沉积易于保存,即为心滩坝顶斜列泥岩,在剖面中延伸范围较广,多数心滩顶部均覆盖厚度不等的斜列泥岩(图6b)。河道充填沉积发育明显的向上变细沉积旋回,自然伽马曲线多为单个钟形,幅度一般小于心滩(图3),在河道带中部,一般水动力较强,发育转换河道[23],在剖面中表现为深切对称底凸型(图6 II_2);在河道带边部,发育迁移河道[23],在垂向沉积序列顶部多见滑塌沉积(图3),迁移河道在剖面中表现为不对称的特点;辫状河中水道很可能因被阻塞而发生废弃,成为废弃河道,在剖面上废弃河道在顶部发育透镜状泥质沉积,其延伸范围局限于河道充填沉积内(图4a)。

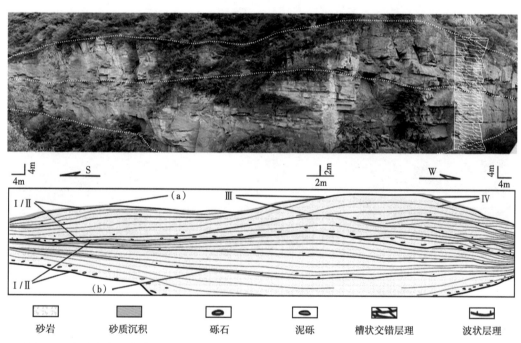

图5 鄂尔多斯盆地东缘盒八段露头剖面叠置河道带中心滩上游沉积(柳林县城三川河北岸)
Ⅰ河道带;Ⅱ心滩;Ⅲ单元坝(心滩头部朵体);Ⅳ大型倾斜层;
(a)心滩坝顶斜列泥岩;(b)单元坝顶部泥岩

露头剖面中可见心滩一般由数个单元坝叠瓦状构成,单元坝内部的大型倾斜层具有大体一致的倾向,不同单元坝间大型倾斜A层倾向不同(图6Ⅳ、图4Ⅳ、图5Ⅳ),心滩内斜列泥岩为单元坝顶部季节性洪水期内形成的泥质沉积,也称为"落淤层"[14-15],一般会被后期发育的单元坝所冲蚀,在剖面中观察到其厚度较小,一般为十余厘米,侧向延伸范围有限,小于单元坝规模(图6e、图4c、图5b)。

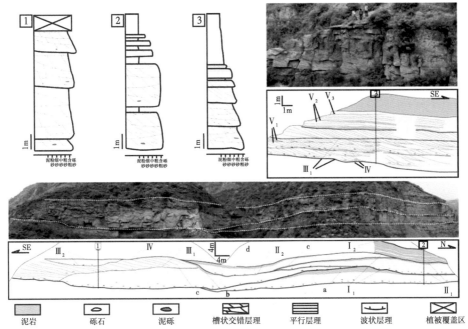

图6 鄂尔多斯盆地东缘柳林成家庄盒八段露头剖面中构型单元识别

1 心滩上游部分；2 叠置河道带；3 河道充填（顺古流向）；I₁河道；I₂泛滥平原/溢岸沉积；
Ⅱ₁心滩下游部分（汇流坝）；Ⅱ₂河道充填（转换河道）；Ⅲ₁单元坝；Ⅲ₂心滩内水道；
Ⅳ大型倾斜层；V₁中型交错层；V₂中型平行层；V₃小型交错层；
(a)泛滥平原泥岩；(b)心摊坝顶斜列泥岩；(c)心滩坝后泥岩；(d)废弃河道泥岩；(e)单元坝顶部泥岩

3 辫状河沉积露头剖面构型规模

前人对鄂尔多斯盆地东部多条露头剖面进行了观察测量[23-24]，但是对剖面的构型解剖较少。柳林地区露头剖面位于鄂尔多斯盆地东缘晋西挠褶带（图1），前人研究表明[10-12]，下石盒子组沉积时期，鄂尔多斯盆地中东部沉积体系的物源同为盆地北缘阴山古陆。苏里格地区和柳林地区盒八段均为辫状河沉积，且物源相同，因此，柳林地区盒八段露头剖面的构型解剖对苏里格气田盒八段的构型研究具有类比意义。

前人对全球大量现代辫状河的研究发现，交错层系厚度与辫状河的最大满岸深度具有较强的相关性，而后者决定了辫状河各构型单元的规模大小[25]，因此，通过在露头剖面中统计交错层系厚度，便可求取最大满岸深度，从而运用经验公式推算各构型单元规模尺度。

Bridge 等提出交错层系厚度与沙丘高度存在定量统计关系[13]：

$$b_m = \alpha\beta \tag{1}$$

$$\beta = s_m/1.8 \tag{2}$$

Lunt 等运用GPR对现代河流进行详细测量后，认为取 $\alpha = 4$ 比较合理[5]。在确定沙丘高度后，计算河流最大满岸深度为

$$d/b_m = 6\sim10 \tag{3}$$

一个完整的心滩沉积序列代表了河流的满岸深度，由于坝顶沉积通常是缺失的或较难判断，因此一般不用测量心滩厚度的方式来推测满岸深度，但是心滩上游部分坝顶沉积厚度一般较小，可用测量的露头剖面中心滩上游部分坝底沉积的平均厚度，来近似代替平均满岸河深$d_m(\approx d/2)$，从而反推d/b_m的可能取值范围。在柳林露头剖面测量了54个交错层系厚度，s_m为0.68m，$s_m/s_{sd}=0.64$，将s_m代入式(1)—式(3)，得到d值9.06~15.11m，d_m值为4.53~7.56m。在柳林露头测量了10余个心滩上游部分的坝底沉积厚度，平均值为6.21m，反推d/b_m值约为8，而心滩上游部分坝底厚度应是略小于满岸深度的，因此d/b_m应大于8，在该地区d/b_m可取值范围为8~10。

确定河流满岸深度后，可根据经验公式预测河道带宽度[26]：

$$W_{cb} = 59.9 d_m^{1.8} \quad (4)$$

通过最大满岸深度可以推测心滩宽度[27]：

$$W_b = 7.3862 d^{1.4614} \quad (5)$$

心滩长度与宽度存在定量关系[29]：

$$L_b = 4.9517 W_b^{0.9676} \quad (6)$$

在山西柳林露头剖面测量了54个交错层系厚度，s_m为0.68m。根据式(1)和式(2)，得出b_m值分别为1.51m和1.29m，代入式(3)，d/b_m值取8~10，得出辫状河最大满岸深度d为12.09~15.11m。取d_m值为$d/2$，代入式(4)，得到河道带宽度W_{cb}为1527~2282m。将d值代入式(5)，得到心滩宽度W_b为282~390m。运用式(6)，得出柳林剖面心滩长度L_b为1163~1592m。

由于野外剖面的延续性较差，未测得河道带尺度资料，但测量了10余个心滩宽度数据，分布在260~430m，从而验证了通过测量交错层系厚度推算河道带和心滩尺度这一方法的有效性。

4 密井网区砂体分级构型解剖

鄂尔多斯盆地中，苏里格气田具有多个密井网实验区，为辫状河储层分级构型解剖提供了有利条件。密井网区井距(东西向)为350~500m，排距(南北向)为600~700m，在该井距、排距条件下，可依次对河道带、心滩与河道充填进行构型解剖。

4.1 构型规模推测

盆地东缘露头剖面距离苏里格气田较远，所测量的交错层系厚度不能直接用于对苏里格气田盒八段辫状河沉积构型规模的预测，且在岩心上较难识别交错层系，而成像测井资料解决了这一难题，在成像测井灰度图中，可以通过灰度的微细变化的线状界面显示出来。各种类型层理具有不同的特征，平行层理纹层面与层面平行，且可用正弦曲线拟合，倾角较小，倾向一致；交错层理纹层面与层面斜交，板状交错层理中，在同一层系内部，纹层面倾向相同，倾角一致或向下略变小；槽状交错层理多表现为倾向倾角的杂乱不稳定模式[28]。在成像测井资料中识别出不同类型层理后，可以通过纹层倾角和倾向划分交错层系，从而测量交错层系厚度。利用苏里格气田中部20余口井盒八段地层成像测井资料，测量了174个交错层系厚度，s_m为0.57m，且$s_m/s_{sd}=0.61$，满足$s_m/s_{sd}=0.88(\pm 0.3)$的统计条件，根据前文的计算方法，推算出河道带宽度为662~1663m,心滩宽度为143~302m,心滩长度为602~1243m(图7)。

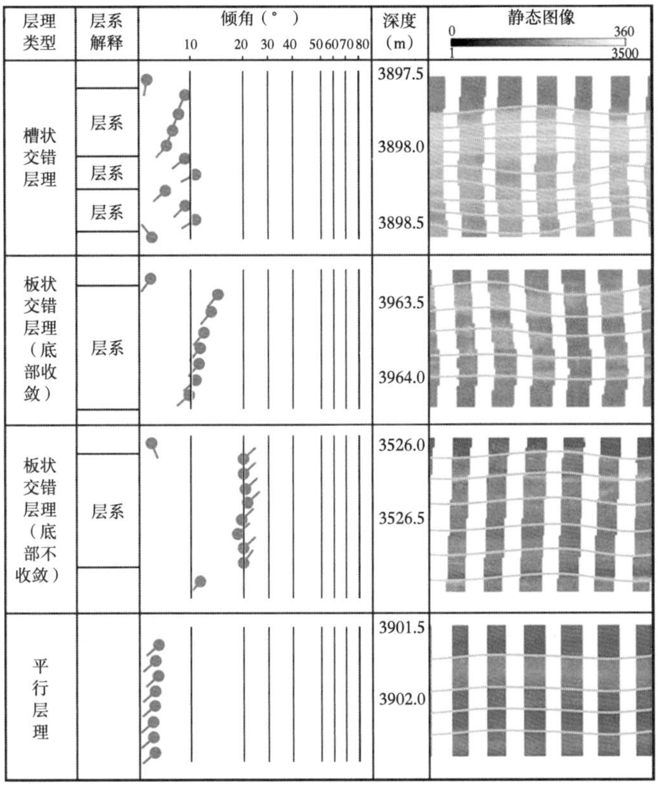

图7 苏里格气田盒八段地层成像测井层理解释与层系厚度测量

4.2 河道带的划分对比

4.2.1 单井划分

单井中粒序旋回的叠置样式可归纳为四类[16]:(1)两个向上变细序列之间为较厚的泥岩段分隔,代表两期河道带的垂向叠置(图8a);(2)两个完整且厚度相当的向上变细旋回上下叠置,无厚泥岩段分隔,极有可能为两期河道带的垂向叠置(图8b);(3)较薄的向上变细旋回叠置于较厚的向上变细旋回之上,这种情况可能代表同一河道带中坝中水道对心滩的切割,也有可能代表不同河道带的切割叠置(图8c);(4)较厚的向上变细旋回叠置于较薄的向上变细旋回之上,这种情况较难判断,只能通过井间对比进行划分(图8d)。可以看出,a、b两类叠置样式较易划分,因此,在进行井间对比前,可先在单井上识别a、b两类叠置类型。

4.2.2 井间对比

通过井间对比确定河道带的平面展布范围,关键点在于河道带边界的确定。河道带边界的确定主要方法为:

(1)通过泛滥平原泥岩确定河道带边界。较厚的泥岩段一般代表泛滥平原沉积,是确定河道带边界的重要标志,在进行河道带的井间对比时,如一个河道带砂体对应邻井中等高程段为厚层泥岩段,即可判断两井间为该河道带砂体边界(图9a)。

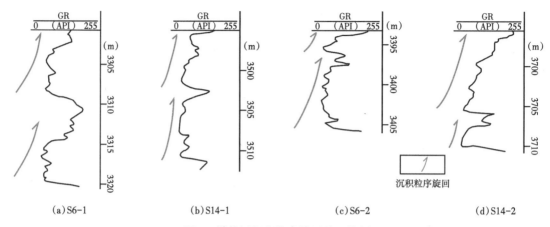

图 8 辫状河沉积粒序旋回叠置特征

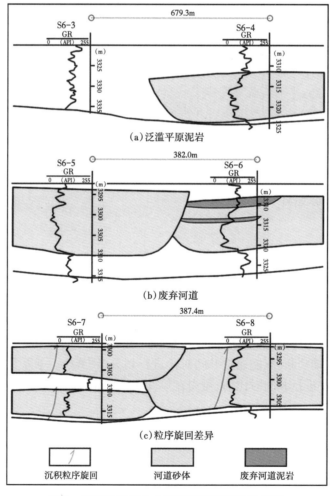

图 9 苏里格气田盒八段辫状河道带边界类型

(2)通过废弃河道沉积确定河道带边界。在辫状河沉积中,河道带边部常保留有废弃河道沉积,表现为砂岩段之上的厚层泥岩和粉砂质泥岩,与邻井相比,砂岩段明显较薄(图9b)。

(3)通过粒序旋回差异确定河道带边界。两井间存在明显粒序差异,如一井为两个等厚的向上变细序列,而邻井为单个向上变细序列,即可判断两井间为河道带边界(图9c)。

苏里格盒八段辫状河沉积体系的主要物源为北部变质岩—岩浆岩发育区,故该沉积体系的主要展布方向为近南北向[29]。因此,可以在栅状连井图中,将部分井中确定的垂向和横向河道带划分方案向南北邻井推演。

具体步骤为:(1)在单井上划分较易识别的叠置样式,在井间识别较明显的河道带边界。如前文所述,在单井中识别出 a 和 b 两种叠置样式,并通过部分井间特征在平面上判断河道边界(图10a)。(2)在近东西向上划分河道带砂体,密井网区井距为300~500m,而经推算河道带宽度为662~1663m,且河道带古流向为近南北向,故划分河道带时,应考虑单个河道带横向跨越2~5个井位。随后,可将所连接的河道带砂体向南北邻井推演,在推演中,应遵循的原则为:同一河道带砂体在南北方向上应具有近似的宽度(图10b)。(3)在以上划分结果的控制下,在近东西向上继续连接砂体,并向南北向推演,同样遵循以上原则(图10c、d)。

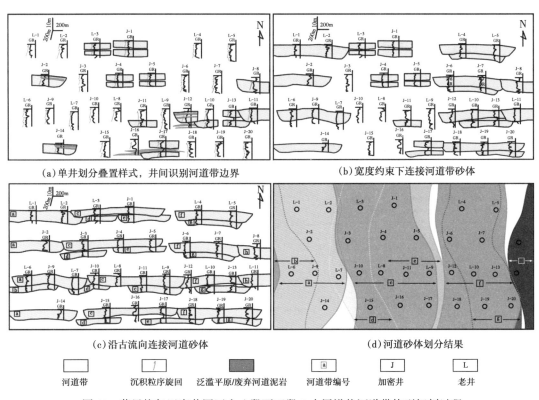

图10 苏里格气田密井网区盒八段下亚段1小层辫状河道带构型解剖过程

4.3 心滩与河道充填

辫状河沉积中,心滩沉积所占比重明显大于河道充填沉积[4]。如前文所述,心滩上游部分、心滩下游部分与河道充填沉积具有明显不同的垂向序列和测井曲线特征(图3),利用这些特征,在之前所建立的河道带划分框架中,对心滩和河道充填沉积进行井间对比。

4.3.1 心滩的井间对比

通过交错层系厚度推算心滩宽度为143~302m,心滩长度为623~1243m,而密井网区井距为300~500m,排距为600~700m,且心滩的展布方向为近南北向,因此在进行井间心滩对比时,在东西向剖面中,心滩较少跨越两个井位;在南北向剖面中心滩跨度应为1~2个井位。心滩的井间对比应利用心滩上游部分和下游部分不同的电性和岩性特征并结合剖面走向进行讨论。

在东西向连井剖面中,若通过测井曲线特征判别两井分别为心滩上游部分与心滩下游部分,则可判断两井中为不同的心滩,例如图11剖面3的河道带f中,J-12井自然伽马曲线为叠

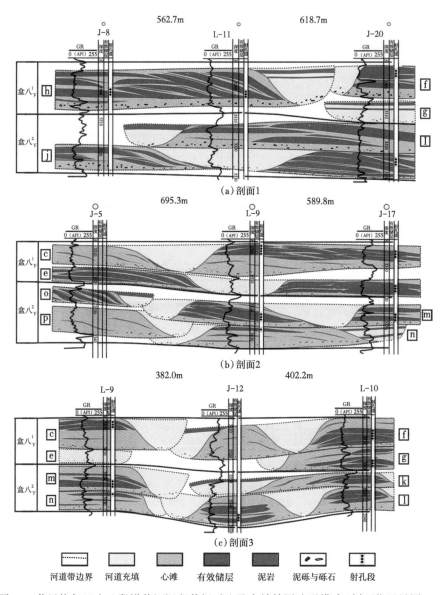

图11 苏里格气田盒八段辫状河沉积井间对比及有效储层连通模式(剖面位置见图12)

合的箱形与钟形,单元坝厚度具有向上变小的趋势,可判断为心滩下游部分,L-10井自然伽马曲线为叠合的箱形,幅度较大,可判断为心滩上游部分,则可判断两井中为不同的心滩;若判别两井均为心滩上游部分或均为心滩下游部分,在井距较小时(300m),可能为同一心滩,在井距较大时,为不同的心滩。

南北向上连进剖面中,若一井判断为心滩下游部分,而其南部邻井为心滩上游部分,则可确定两井钻遇不同的心滩,例如图11剖面2的河道带m中,L-9井为心滩下游部分,而其南部临井J-17井中为心滩上游部分,则可判断两井钻遇不同的心滩;若一井中判断为心滩下游部分,其南部临井为心滩下游部分,在密井网井距条件下,两井极有可能为同一个心滩,如图11剖面2的河道带c中,L-9井为心滩上游,J-17井为心滩下游,两井距离小于600m,即小于前文推测的心滩最小长度,故两井钻遇同一个心滩;若南北两井皆为心滩下游部分或皆为心滩上游部分,则应根据井距和周边其他井的判别结果进行综合判断。

4.3.2 河道充填井间对比

河道充填沉积序列与心滩下游部分序列均为向上变细旋回,但河道充填沉积中砂岩厚度与心滩砂体中相比一般较薄,粒度较细,在测井曲线上幅度较小,并发育废弃河道泥岩(图3、图11),如图11剖面1的河道带j中,J-8井与J-20井中自然伽马曲线幅度和砂体厚度均大于L-11井,前两井钻遇心滩沉积,而L-11井钻遇河道充填沉积。

5 致密砂体有效储层连通模式

苏里格气田盒八段发育典型的致密砂岩[30],致密砂岩储层覆压基质渗透率普遍低于0.1mD[31],但是在物性较差的致密砂岩中,仍存在物性相对较好、且在现有工艺技术条件下能够采出具有工业价值产气量的储层,这类储层被称为有效储层[32-33]。根据测井解释结果和实际生产经验,苏里格气田盒八段有效储层物性下限为:孔隙度5%,常压渗透率为0.1mD,含气饱和度50%;有效储层自然伽马测井值一般在70API以下。

5.1 有效储层发育特征

苏里格气田盒八段砂体中,相对较粗的砂岩是形成有效储层的主要岩相类型[34]。在辫状河沉积中,单元坝是最基本的向上变细沉积序列(图3),单元坝顶部岩石粒度较细,甚至有可能发育泥质沉积,单元坝底部可能部分发育泥砾和含砾砂岩(图4、图5),因此,单元坝顶部和顶部物性一般较差,但单元坝主体多为中粗砂岩,物性较好,多形成有效储层,因此可以认为单元坝的主体部分是辫状河致密储层中基本的渗流单元。对苏里格密井网区的解剖表明,有效储层多发育于粒度较粗、厚度较大的单元坝中,心滩上游部分形成于水动力较强的环境中,单元坝整体粒度较粗,垂向沉积序列中大多数单元坝形成有效储层,尤其是心滩头部朵体,是物性最好的储层(图11,剖面1L-11井中Ⅰ河道带,剖面2L-9井中c河道带),心滩下游部分中,卷形坝和汇流坝形成于水动力较强的环境中,多形成有效储层(图11,剖面2L-9井中m河道带)。河道充填沉积由于多发生废弃,且砂岩粒度多小于心滩,故一般不发育有效储层。

5.2 有效储层井间连通模式

致密砂岩气藏储层物性普遍较差,但有效储层具有较好的渗流能力,因此在讨论井间连通性时,一般讨论的是井间有效储层的展布连续性。辫状河沉积中,单元坝的主体部分是最基本的渗流单元,同一心滩内多个基本渗流单元叠置,不同渗流单元之间的渗流屏障为心滩内部斜列泥岩,这类泥岩层厚度较小,一般为数十厘米,且横向延伸范围有限,单个泥岩层仅局限于单元坝顶部,横向跨度多为数十米(图4c、图5b),因此,心滩内斜列泥岩并不能完全分隔各个渗流单元,因此不是心滩内有效的渗流屏障,在致密气藏中,若两井钻遇同一心滩,则井间有效储层渗流单元大多叠置连通(图11,河道带h,J-8井与L-11井;剖面2,河道带c,L-9井与J-17井)。但不同心滩之间可能发育多种渗流屏障,如心滩坝后泥岩沉积(图4河道c),另外河道充填沉积由于细粒沉积物较多,也可能成为不同心滩间的渗流屏障(图11,剖面1j河道带,J-8井与J-20井;剖面3l河道带,J-12井与L-10井)。不同河道带之间渗流屏障更多,包括泛滥平原沉积、河道带底部滞留沉积以及废弃河道泥岩等,因此,如两井钻遇不同河道带,则井间有效储层多不连通(图11,L-11井钻遇h河道带,J-20井钻遇f河道带,河道带分布见图10)。综上所述,可以得出井间有效储层连通的一般模式:在同一心滩中,发育于不同单元坝中的有效储层大多互相叠置连通;在不同的心滩间,有效储层一般不连通;不同河道带内的有效储层,大多不连通。

6 密井网构型解剖的验证

6.1 验证方法

苏里格气田密井网区进行了多次干扰试井试验,试验结果可用于直接判断井间连通性[35];另外,密井网区内的加密井均部署实施于老井生产数年之后,加密井投产前地层压力也可用于判断加密井与相邻老井间的储层连通性。

6.1.1 干扰试井验证

油气田干扰试井试验是了解井与井之间储层连通性的重要方式[36]。干扰试井的基本单元为两口井组成的井对,其中一口井称为"激动井",在测试中改变工作制度,对地层压力造成"激动",另一口井称为"观测井",在测试中关井,并下入高精度、高分辨率的井下压力计,记录从激动井传播过来的干扰压力变化。通过分析观测井的压力变化特征,即可判断井对间的储层连通性。如在J-19井与J-20井这一井对中,前者为激动井,后者为观测井,对这一井对的干扰试井试验表明,两井间储层连通。

6.1.2 加密井投产前地层压力验证

在加密井部署实施时,密井网区内老井已生产数年,如果加密井中有效储层与相邻老井连通,则老井的生产必然使加密井处地层压力下降,表现为加密井投产前地层压力小于该区内原始地层压力。苏里格气田盒八段压力系数平均约为0.82[37],根据地层垂深可推测原始地层压力,如苏里格苏6加密区内盒八段原始地层压力约为30MPa,多口加密井投产前压力接近该值,证明加密井与相邻老井间储层不连通,如J-12井于2008年投产,投产前地层压力为30.52MPa,证明该井与相邻老井(L-9井与L-10井均投产于2003年)间储层不连通(图12);相反J-8井投产前地层压力已下降到24.14MPa,小于原始地层压力,说明该井与相邻老井(L-11井,投产于五年前)间储层连通。

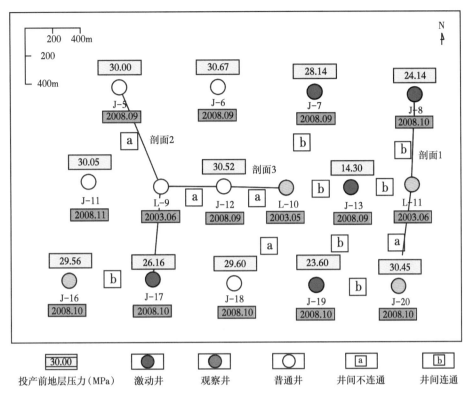

图 12 苏里格气田盒八段有效储层井间连通性验证方法示例

6.1.3 验证原则

在讨论井间储层连通性时,应按照射孔层位分三种情况讨论:(1) 射孔段不在同一小层的情况,这种情况下两井间大概率不连通,是由层间泥岩段的阻隔所造成,如剖面 3 中(图 11),J-12 井射孔层位为盒八段下亚段 2 小层,而 L-10 井射孔层位为盒八段下亚段 1 小层,动态验证表明 J-12 井与 L-10 井间不连通。(2) 两井有且仅有一个相同的射孔层位,这种情况下,动态验证的井间连通性即可认为是该小层中储层的连通性,如剖面 3 中(图 11),L-9 井与 J-12 井仅有盒八段下亚段 2 小层这一个相同的射孔层位,动态验证表明,两井间不连通,则说明两井的盒八段下亚段 2 小层中的有效储层在井间一定不连通,这一结论也验证静态解剖结果,因两井中射开的有效储层位于不同河道带中,如前文所讨论,不具备井间连通的地质基础;又如在剖面 1(图 11)中,J-8 井与 L-11 井仅有一个相同射孔小层(盒八段下亚段 1 小层),动态验证表明两井连通,则表明两井中盒八段下亚段 1 小层内的有效储层在井间连通,同样验证了静态解剖结果,即两井所钻遇该小层中的有效储层发育于同一心滩中。(3) 两井有多个相同的射孔层位,若动态验证两井间连通,则只能判断不同小层中的有效储层在井间为"可能连通",如在剖面 2 中,动态验证表明 L-9 井与 J-17 井连通,但两井在盒八段下亚段 2 小层与盒八段下亚段 1 小层两个小层中均有射孔,因此仅能判断两井中的多套有效储层在井间可能连通。

6.2 验证结果

共验证了苏里格密井网区内 57 个井对，排除了无相同射孔层位的情况后，主要结果为：(1)构型解剖结果为两井属同一心滩的情况，井间储层连通性较好，仅有 7.7% 井间不连通。(2)构型解剖为两井分属不同心滩，井间储层连通性较差，超过 85% 为井间不连通的情况。(3)在密井网区的井网条件下，井间储层总体连通性较差，超过 60% 的情况为确定的不连通，仅有 9.1% 为确定的井间连通情况。这一结果有效验证了前文井间有效储层的连通模式，即若两井钻遇相同的心滩，则井间有效储层连通的概率较大，若两井钻遇不同心滩或钻遇不同河道带，则有效储层井间连通的概率依次减小（表1）。

表1 苏里格气田盒八段有效储层井间连通性验证结果

构型解剖结果	连通数	可能连通数	不连通
同一心滩内	7	17	2
不同心滩（近南北向）	0	2	12
不同心滩（近东西向）	1	4	15
心滩与河道充填	0	1	4
不同河道带	0	0	23

以上验证结果证明了构型解剖方法的正确性，也证明有效储层连通模式的合理性。密井网区内井间储层连通性较差，也说明目前苏里格气田的整体井网密度可以进一步加大，可通过大井组立体开发技术提高气田采收率[38]。

7 结论

(1)辫状河沉积的构型解剖，应遵循河道带—心滩与河道充填—单元坝与坝中水道—大型倾斜层—中型交错层、平行层、小型交错层的五级级构型模式，目前苏里格密井网区中，单井中能解剖到单元坝级别，井间能解剖到心滩与河道充填级别，鄂尔多斯盆地东缘与苏里格气田盒八段辫状河沉积具有可类比性。

(2)构型解剖表明，河道带与河道带、心滩与心滩、河道带与心滩之间存在多种成因的连续渗流屏障，分隔有效储层；心滩内部渗流屏障规模较小，其内部单元坝可视为有效储层发育的基本渗流单元，多个有效储层渗流单元在同一心滩内叠置连通。

(3)通过交错层系厚度推算的心滩规模，小于苏里格气田的主流井距和排距，气井动态资料也证实了目前最大井网密度下有效储层连通性较差，因此，苏里格气田具有进一步加密井网的潜力。

符号说明

b_m——平均沙丘高度，m；

α——系数，$\alpha = 4\sim6$；

s_m——平均交错层系厚度；

d——河流最大满岸深度，m；

d_m——河流平均满岸深度,约为 $d/2$,m;
W_{cb}——河道带宽度,m;
W_b——心滩宽度,m;
L_b——心滩长度,m。

参 考 文 献

[1] 何顺利,兰朝利,门成全. 苏里格气田储层的新型辫状河沉积模式[J]. 石油学报,2005,26(6):25-29.

[2] 赵靖舟,付金华,姚泾利,等. 鄂尔多斯盆地准连续型致密砂岩大气田成藏模式[J]. 石油学报,2012,33(S1):37-52.

[3] 卢涛,刘艳侠,武力超,等. 鄂尔多斯盆地苏里格气田致密砂岩气藏稳产难点与对策[J]. 天然气工业,2015,35(6):43-52.

[4] Lane S N. Approaching the system-scale understanding of braided river behaviour[M]// Smith G H S, Best J L, Bristow C S, et al. Braided rivers: process, deposits, ecology and management. Oxford: Blackwell Publishing, 2006:107-135.

[5] Lunt I A, Smith G H S, Best J L, et al. Deposits of the sandy braided South Saskatchewan River: implications for the use of modern analogs in reconstructing channel dimensions in reservoir characterization[J]. AAPG Bulletin, 2013, 97(4):553-576.

[6] 杨俊杰. 鄂尔多斯盆地构造演化与油气分布规律[M]. 北京:石油工业出版社,2002.

[7] 李德生. 重新认识鄂尔多斯盆地油气地质学[J]. 石油勘探与开发,2004,31(6):1-7.

[8] 田景春,吴琦,王峰,等. 鄂尔多斯盆地下石盒子组盒八段储集砂体发育控制因素及沉积模式研究[J]. 岩石学报,2011,27(8):2403-2412.

[9] 王国茹. 鄂尔多斯盆地北部上古生界物源及层序岩相古地理研究[D]. 成都:成都理工大学,2011.

[10] 窦伟坦,侯明才,董桂玉. 鄂尔多斯盆地北部山西组—下石盒子组物源分析[J]. 天然气工业,2009,29(3):25-28.

[11] Robinson J W, McCabe P J. Sandstone-body and shale-body dimensions in a braided fluvial system: salt wash sandstone member (Morrison Formation), Garfield County, Utah[J]. AAPG Bulletin, 1997, 81(8):1267-1291.

[12] 葛云龙,逯径铁,廖保方,等. 辫状河相储集层地质模型——"泛连通体"[J]. 石油勘探与开发,1998,25(5):77-79.

[13] Bridge J S, Tye R S. Interpreting the dimensions of ancient fluvial channel bars, channels, and channel belts from wireline-logs and cores[J]. AAPG Bulletin, 2000, 84(8):1205-1228.

[14] Lunt I A, Bridge J S, Tye R S. A quantitative, three-dimensional depositional model of gravelly braided rivers[J]. Sedimentology, 2004, 51(3):377-414.

[15] Tye R S. Quantitatively modeling alluvial strata for reservoir development with examples from Krasnoleninskoye field, Russia[J]. Journal of Coastal Research, 2013(69):129-152.

[16] 孙天建,穆龙新,吴向红,等. 砂质辫状河储层构型表征方法——以苏丹穆格莱特盆地 Hegli 油田为例[J]. 石油学报,2014,35(4):715-724.

[17] 高兴军,宋新民,孟立新,等. 特高含水期构型控制隐蔽剩余油定量表征技术[J]. 石油学报,2016,37(S2):99-111.

[18] 蒲秀刚,赵贤正,李勇,等. 黄骅坳陷新近系古河道恢复及油气地质意义[J]. 石油学报,2018,39(2):163-171.

[19] 牛博,高兴军,赵应成,等. 古辫状河心滩坝内部构型表征与建模——以大庆油田萨中密井网区为例[J]. 石油学报,2014,36(1):89-100.

[20] 于兴河,马兴祥,穆龙新,等. 辫状河储层地质模式及层次界面分析[M]. 北京:石油工业出版社,2004.

[21] Li S L,Yu X H,Chen B T,et al. Quantitative characterization of architecture elements and their response to base-level change in a sandy braided fluvial system at a mountain front[J]. Journal of Sedimentary Research,2015,85(10):1258-1274.

[22] Lynds R,Hajek E. Conceptual model for predicting mudstone dimensions in sandy braided-river reservoirs[J]. AAPG Bulletin,2006,90(8):1273-1288.

[23] 郭英海,刘焕杰,权彪,等. 鄂尔多斯地区晚古生代沉积体系及古地理演化[J]. 沉积学报,1998,16(3):44-51.

[24] 金振奎,杨有星,尚建林,等. 辫状河砂体构型及定量参数研究——以阜康、柳林和延安地区辫状河露头为例[J]. 天然气地球科学,2014,25(3):311-317.

[25] Leclair S F,Bridge J S,Wang F. Preservation of crossstrata due to migration of subaqueous dunes over aggrading and non-aggrading beds:comparison of experimental data with theory[J]. Geoscience Canada,1997,24(1):55-66.

[26] Bridge J S,Mackey S D. A theoretical study of fluvial sandstone body dimensions[M]// Flint S S,Bryant I D. The geological modelling of hydrocarbon reservoirs and outcrop analogues. Oxford:International Association of Sedimentologists,1993:213-236.

[27] Kelly S. Scaling and hierarchy in braided rivers and their deposits:examples and implications for reservoir modelling[M]// Smith G H S,Best J L,Bristow C S,et al. Braided rivers:process,deposits,ecology and management. London:Blackwell Publishing,2006:107-135.

[28] 钟广法,马在田. 利用高分辨率成像测井技术识别沉积构造[J]. 同济大学学报,2001,29(5):576-580.

[29] 席胜利,王怀厂,秦伯平. 鄂尔多斯盆地北部山西组、下石盒子组物源分析[J]. 天然气工业,2002,22(2):21-24.

[30] 何东博,贾爱林,冀光,等. 苏里格大型致密砂岩气田开发井型井网技术[J]. 石油勘探与开发,2013,40(1):79-89.

[31] 中国石油勘探开发研究院. SY/T 6832—2011 致密砂岩气地质评价方法[S]. 北京:石油工业出版社,2011:1-10.

[32] 操应长,王艳忠,徐涛玉,等. 东营凹陷西部沙四上亚段滩坝砂体有效储层的物性下限及控制因素[J]. 沉积学报,2009,27(2):230-237.

[33] 何东博,贾爱林,田昌炳,等. 苏里格气田储集层成岩作用及有效储集层成因[J]. 石油勘探与开发,2004,31(3):69-71.

[34] 李易隆,贾爱林,何东博. 致密砂岩有效储层形成的控制因素[J]. 石油学报,2013,34(1):71-82.

[35] 郭智,贾爱林,冀光,等. 致密砂岩气田储量分类及井网加密调整方法——以苏里格气田为例[J]. 石油学报,2017,38(11):1299-1309.

[36] 廖红伟,王琛,左代荣. 应用不稳定试井判断井间连通性[J]. 石油勘探与开发,2002,29(4):87-89.

[37] 陈义才,张胜,魏新善,等. 苏里格气田下二叠统盒八段异常低压成因及其分布特征[J]. 天然气工业,2010,30(11):30-33.

[38] 朱亚军,李进步,陈龙,等. 苏里格气田大井组立体开发关键技术[J]. 石油学报,2018,39(2):208-215.

原文刊于《天油学报》,2018,39(9):1037-1050.

鄂尔多斯盆地东部盒八段致密砂岩储层特征
——以子洲气田清涧地区为例

郭 智[1] 冀 光[1] 王国亭[1] 彭艳霞[2]

(1. 中国石油勘探开发研究院 北京 100083；2. 中国地质大学 北京 100083)

摘要：清涧地区位于鄂尔多斯盆地东部，是子洲气田稳产的主力接替区块。研究区面积大，钻井数目少，地质认识程度低，尚处于开发评价阶段。以盒八段为研究对象，开展了沉积、储层等精细地质研究工作，并与苏里格致密砂岩气田进行综合对比，落实了有效砂体的厚度、规模、发育频率，总结了有效砂体在空间的分布规律，认识到区内有效砂体分布零星，连续性差，与心滩等优势相带对应关系较好，平面上主要集中在研究区的西砂带，垂向上在盒八段上亚段2小层、盒八段下亚段2小层相对发育。结合地质与试气资料，以"连续性有效厚度"为主要依据，将储层分成好、中、差、干层四种类型，优选了富集区，按照开发级次将研究区划分为三类区，建议在一类区、二类区优选直井开发，不建议部署水平井开发。本研究为气田开发方案编制提供了地质依据，同时也可对类似气田的地质工作起到借鉴作用。

关键词：子洲气田；清涧地区；盒八段；有效储层；连续性有效厚度

0 引言

致密砂岩气是全球非常规油气资源的重要组成部分，美国、加拿大等国家进行相关研究起步早，技术、经验较为成熟，这些国家比较重视储层岩石学和渗流机理等基础研究[1]，具有完善的致密砂岩分析实验室，能有效评价致密砂岩储层渗透率，在开发中充分利用地球物理技术，刻画储层的空间分布，为井位部署提供依据。中国致密气开发起步晚，近十年来发展迅速，2015年致密气产量已占到全国天然气总产量的1/3，建成了以鄂尔多斯盆地为代表的致密砂岩气大型产业基地。与国外同类气藏相比，中国致密砂岩气藏为陆相沉积，储层埋藏深度大、厚度薄、连通性差、分布分散，气藏条件更为复杂。

清涧地区位于鄂尔多斯盆地伊陕斜坡东部（图1），面积约3000 km^2，是子洲气田稳产最具潜力的接替区块之一。区块物性差，不经储层改造未有自然产能，属于致密砂岩气藏。截至2015年底，研究区共钻井79口，井网密度小（约40 km^2/口），井网井距差异大，地质认识程度低，尚处于开发评价阶段。山二段本为子洲气田主力产层，该层段储量占气田总探明储量的80%，但研究区山二段出水严重，无法按规划生产。因此，需要对次主力层段—盒八段开展储层综合地质研究，评估开发潜力。研究中面临的主要问题有：盒八段非研究区的主力层段，取心、分析化验少，沉积环境尚无定论；有效储层成因及分布规律尚不明确，研究区地表对地震波反射弱，地震资料分辨率低，信噪比小，不能作为储层评价的有效手段；富集区尚未圈定，探明储量尚未提交，试气井数少，产量低，且井间差异较大，产建难度大。

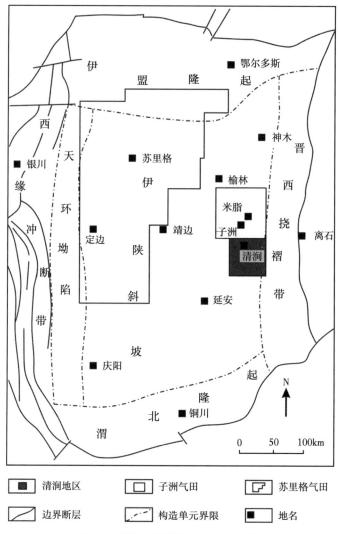

图 1　子洲气田清涧地区构造位置图

由于地质条件、开发技术及开发模式上的差异[2]，不可生搬硬套国外的致密气开发经验，需要探索适合国内条件的致密砂岩储层评价方法。前人研究成果表明清涧地区受控于米脂辫状河—三角洲沉积，在沉积背景调研的基础上，通过多种相标志分析，将清涧地区盒八段定为辫状河三角洲平原沉积，开展了相控下的有效砂体空间分布规律研究，结合地质静态和试气动态数据，提出"连续性有效储层"的概念，对储层进行了综合分类评价，优选了富集区，评估了开发潜力。考虑到研究区资料少，而苏里格气田是国内致密砂岩气田的典型代表，将研究区与苏里格气田沉积环境、岩性、物性、储层分布、储量丰度、试气产量等方面进行了综合对比，以期得出规律性认识。

1 沉积特征

晚古生界,鄂尔多斯盆地受海西构造运动的控制,石炭纪末期海水开始退出盆地,沉积环境由石炭纪的陆表海演变为二叠纪的陆相湖盆。北高南低的古地形控制着北河南湖的总体沉积格局[3-5],在宽缓的古构造背景下,形成了多个分布广泛的河流—三角洲沉积体系。清涧地区盒八段沉积受控于米脂河流—三角洲沉积,距离北部物源 210~270km。

研究区砂岩岩性以中、粗砂岩为主,测井曲线呈箱形、钟形,储层碎屑颗粒分选好—中等,磨圆呈次棱角、次圆状,结构成熟度中等,沉积构造常见反映强水动力条件的板状交错层理,平行层理,可见底冲刷构造,发育底砾岩。泥岩中可见植物叶片印模,指示温暖潮湿的水上沉积环境。研究区粒度 C—M 图主要发育 PQ、OP、NO 段,M 值 200~1000μm,C 值 400~2000μm,粒度较粗,反映悬浮搬运和滚动搬运相混合的沉积环境,指示强水动力下的牵引流沉积。

多种相标志标明,清涧地区盒八段沉积水动力强,符合河道沉积模式,将其沉积环境判断为辫状河三角洲平原沉积,划分为心滩、辫状河道、分流河道间三种微相。心滩位于辫状分流河道的中心,沉积物粒度粗,砂体厚度大,单层可达到 5m 以上。辫状河道是辫状河三角洲平原相的沉积主体,以中砂岩、细砂岩为主,自然伽马曲线以齿化钟形为主。分流河道间位于河道间,以泥质粉砂岩、粉砂质泥岩、泥岩等细粒沉积为主[6],沉积水动力弱,可见水平层理,波纹层理。自然伽马曲线幅度低,接近泥岩基线。

研究区盒八段由下至上发育两套正旋回沉积,分别对应盒八段上亚段、盒八段下亚段地层,在两套正韵律旋回的底部,多期分流河道砂体分流河道与心滩砂体之间互相切割叠置,形成了厚度大、横向连续性相对好、钻遇率高的砂体;在旋回的顶部,随着水动力条件的减弱,河道限制性迁移[7],多呈孤立状,延伸范围小,砂体规模小。

2 储层特征

鄂尔多斯盆地发育两大物源区。苏里格气田主要受盆地西北部元古宇物源控制,岩屑含量较低,而研究区受盆地东北部太古界物源控制,岩屑含量较高,石英含量相比于盆地中部储层较低。储层岩石类型以岩屑石英砂岩、岩屑砂岩为主。储层在沉积后经历了强烈的压实、胶结等成岩作用,石英次生加大普遍发育[8],可见长石高岭石化。孔隙类型以粒间溶孔、粒内溶孔等次生孔隙为主,两者大约占孔隙总体积的 65%。

根据 31 口井 877 块岩样的分析化验结果,研究区储层物性较差,制约了储集天然气的能力。孔隙度主要分布在 2~8% 区间内,平均 5.51%,中位数 4.85%,渗透率主要分布在 0.01~0.5mD,平均 0.258mD,中位数 0.113mD。将清涧地区与苏里格气田储层物性进行对比,表明清涧地区盒八段孔、渗条件劣于苏里格气田的中区和东区,甚至劣于苏南。这主要受物源控制,研究区储层沉积时搬运距离远,水动力相比与苏里格沉积时弱,原生孔隙度小;岩屑含量高,抗压能力弱,在压实作用中孔隙度、渗透率进一步急剧减小。

3 有效砂体特征

3.1 有效砂体发育规模及频率

二叠纪以来,鄂尔多斯盆地沉降中心由东向西迁移,清涧地区处在盆地东部,盒八段沉积时储层规模小,发育频率低,连续性差。经统计,研究区 79 口井共钻遇 460 个单砂体,其中以干层为主,有效砂体(气层及含气层)仅为 128 个,占砂体总数的比例为 27.8%。各井盒八段发育有效单砂体 1~5 个,主要发育 1~2 个,井均发育 1.62 个。

研究区有效单砂体厚度较薄,规模较小。经统计,有效单砂体厚度分布范围 1~5m,在 1~3m 较为集中。按照 40~120 的宽厚比数据,折算有效单砂体宽度 80~360m;按照 1.5~3 的长宽比数据,折算有效单砂体长度 200~700m。作为对比,苏里格气田有效单砂体厚度主要分布在 2~5m,有效单砂体宽度 100~500m,有效单砂体长度 300~800m。

从各小层来看(表1),砂体厚度主要分布在 4.5~6.5m,平均 5.49m,砂地比分布在 0.28~0.44,平均 0.35,即砂体厚度约为地层厚度的 1/3。有效砂体厚度分布在 1~2m,平均 1.42m,净毛比分布在 0.17~0.32,平均 0.26,即有效砂体厚度仅为砂体厚度的 1/4。盒八段上亚段 2 小层与盒八段下亚段 2 小层有效砂体相对发育。

表 1　清涧地区盒八段储层厚度统计表

层段	地层厚度(m)	砂体厚度(m)	有效砂厚(m)	砂体钻遇率(%)	有效砂体钻遇率(%)	砂地比	净毛比
盒八上^1	16.05	4.83	0.83	77.22	22.78	0.29	0.17
盒八上^2	14.97	6.49	2.08	86.08	46.84	0.44	0.32
盒八下^1	16.29	4.60	1.28	68.35	30.38	0.28	0.28
盒八下^2	15.02	6.02	1.48	79.75	36.71	0.39	0.25
平均	15.58	5.49	1.42	77.85	34.18	0.35	0.26

从整个盒八段来看,砂体厚度 21.9m,约为苏里格气田平均砂体厚度的 80%,有效厚度 5.7m,约为苏里格平均有效砂体厚度的 75%。盒八段上亚段、盒八段下亚段两亚段有效砂体分布局限,钻遇率仅为 56.9%,在平面上叠合形成一定的富集区,盒八段有效砂体钻遇率 83.7%。各井盒八段有效厚度差别较大且一般小于 5m,有效厚度为 0、0~5m、5~10m、大于 10m 的井数分别为 13 口、30 口、26 口及 10 口,井数占比分别为 16%、38%、33%、13%。

3.2　有效砂体分布规律

(1)有效砂体分布零星,在空间分布可分为四种模式,以单期孤立薄层型为主。

砂体及有效砂体呈"砂包砂"二元结构,砂体具有一定的连续性,垂向上多期叠置,平面上叠合连片,有效砂体在空间分布零星,连续性差。通过精细地质解剖,总结了研究区盒八段有效储层在空间分布的四种模式:具物性夹层的多期叠置型、具泥质隔层的多期叠置型、单期块状厚层型和单期孤立薄层型(图2)。根据大量数据的统计分析,四类储层模式分布频率分别为:20%、15%、13%、52%,即研究区有一半以上的有效砂体为单期孤立薄层型,单层厚度小于 3m。

(2)有效砂体是低渗背景下相对高渗的"甜点"。

研究区有效砂体不等同于砂体,是普遍低渗背景下相对高渗的"甜点"[9-10],基本对应心滩中下部及河道底部等粗砂岩相。这是因为:心滩中下部、河道底部等粗砂岩相,石英含量高,

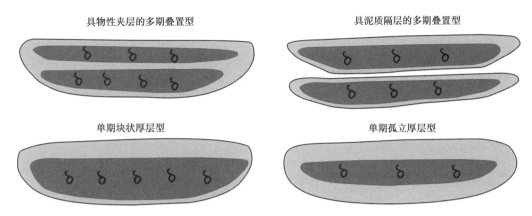

图 2　清涧地区盒八段有效储层分布模式

物性好,储层连续性强,原始孔隙在压实过程中得以最大程度地保存[11];较好的储层原始物性、较强的抗压性,为后期溶蚀作用提供了流体运移的有利通道,进一步了改善物性。

(3)在沉积、成岩双重控制下,有效砂体主要集中在西砂带。

沉积对储层具有很强的控制作用,河道的展布决定了储层分布的格局,受沉积环境的影响,储层分布集中在区内西、东两条主砂带内(图3a)。两条主砂体带,砂岩百分含量相对较高,整体大于40%。在沉积作用的基础上,后期压实、胶结、溶蚀等成岩作用深刻改变了储层面貌[12],塑造了有效储层的形态(图3b)。有效砂体与砂体平面图趋势上一致,又存在较大的差异。

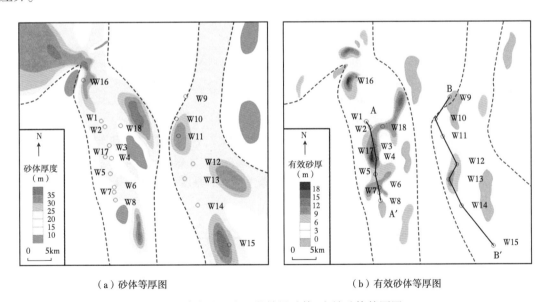

(a)砂体等厚图　　　　　　　　　　　　(b)有效砂体等厚图

图 3　清涧地区盒八段储层砂体、有效砂体等厚图

受沉积、成岩控制,盒八段有效砂体在平面分布具有很强的不均一性,在西砂带连续性强、厚度大(局部可达15m以上),东砂带次之,砂带间最差。西砂带有效砂体发育程度高,块状厚

层型、具物性夹层的垂向叠置型比例相对高(图 4a,对应图 3b 的 AA' 剖面),研究区试气的五口相对高产井(初期日产气 $>1\times10^4\mathrm{m}^3$)全部分布在西砂带;东砂带有效砂体是以孤立薄层型为主,块状厚层型发育较少(图 4b,对应图 3b 的 BB' 剖面);砂带间砂体发育零星,有效砂体基本不发育。

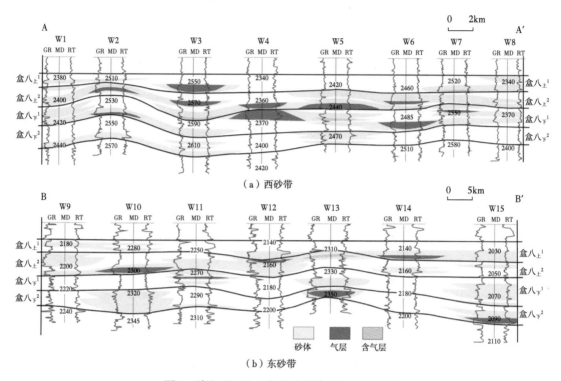

图 4　清涧地区盒八段有效砂体剖面发育特征

西砂带—东砂带—砂带间,砂体厚度、有效砂厚依次变薄,砂地比、净毛比、有效砂体钻遇率逐渐减小,开发潜力依次降低。

(4)有效砂体在盒八段上亚段 2 小层、盒八段下亚段 2 小层两小层相对发育。

总的来说,清涧地区盒八段有效砂体厚度薄,规模小,发育频率低。4 个小层中,盒八段上亚段 2 小层、盒八段下亚段 2 小层有效砂体相对发育,局部厚度达到 10m 以上,在平面上分布面积较广,连续性较强(表 1)。

盒八段上亚段 1 小层有效砂体分布零星,钻遇率仅为 22.7%,有效砂体主要分布在西砂带,厚度范围 1~3m,(图 5a);盒八段上亚段 2 小层有效砂体钻遇率 46.8%,在平面上分布范围较广,有效砂体厚度主要分布在 2~5m 之间(图 5b),在 W16、W17 等局部井区厚度可达 6m 以上;盒八段下亚段 1 小层有效砂体钻遇率 30.4%,有效砂体分布相对局限,厚度为 1~4m(图 5c),局部井区厚度可达 6m 以上;盒八段下亚段 2 小层有效砂体钻遇率 36.7%,在西、东砂带均有分布,有效砂体厚度主要分布在 1~5m 之间,在 W16、W18 等井区相对富集(图 5d)。

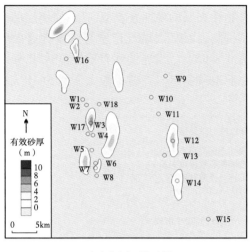
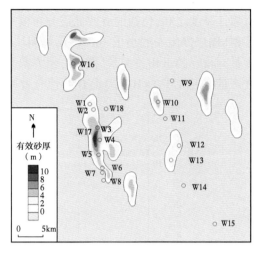

（a）盒八上亚段1小层　　　　　　　　　　　　（b）盒八上亚段2小层

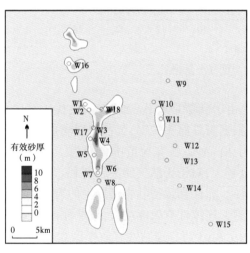
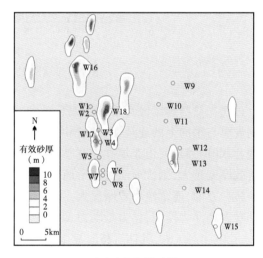

（c）盒八下亚段1小层　　　　　　　　　　　　（d）盒八下亚段2小层

图5　清涧地区盒八段各小层有效砂体等厚图

4　储层综合评价

结合地质资料和动态资料综合分析储层,是准确认识地下地质体、进行合理开发的有效手段。研究区投产井数目少,投产时间短,整体还处于开发评价阶段。在研究区共收集了22口井的试气数据,结合地质数据等静态资料,建立静态、动态数据的关联,进行储层的综合评价,使地质研究真正地能为生产服务。研究区试气产量低,仅五口井试气产量大于$1×10^4 m^3/d$。22口井最高产量$11.67×10^4 m^3/d$,最低$0.08×10^4 m^3/d$,平均$1.34×10^4 m^3/d$。

统计表明,试气产量与储层的孔隙度、渗透率、含气饱和度具有一定的正相关性,但相关性不高,这是由于储层普遍致密,储层参数值较小且差别不大等原因造成的。相比之下,试气产量与有效储层厚度关系较好,相关系数为0.5,但仍不够理想。

为此,提出"连续性有效厚度"的概念,系在现有的压裂工艺水平下,某井可以沟通的有效储层累计厚度之和,包括的有效储层分布模式为具物性夹层的多期叠置型、具泥质隔层的多期叠置型(隔层厚度一般小于3m)以及单期块状厚层型。经统计,试气产量与连续性有效储层厚度具有较好的正相关关系(图6),相关系数可达0.8以上。当连续性有效厚度大于5m时,试气产量一般大于$1×10^4 m^3/d$;连续性有效厚度大于3m时,试气产量一般大于$0.5×10^4 m^3/d$。

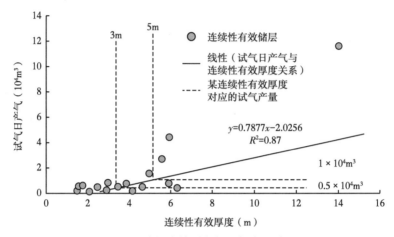

图6 连续性有效厚度与试气产量关系

试气产量与连续性有效厚度关系好,一方面是由于储层是地下三维地质体,优质储层在平面的规模与垂向上的连续性有效厚度密切相关。根据野外露头观察,沉积物理模拟,鄂尔多斯盆地盒八段心滩、河道充填砂体宽厚比范围一般为40~120,长宽比范围一般为1.5~3。连续性有效厚度越大,则优质储层在平面分布的规模也越大,对于天然气储集越有利。

另一方面,连续性有效厚度大,天然气优先充注。充注到致密砂岩储层中的天然气主要受毛细管力和浮力作用,浮力是动力,毛细管力是阻力。连续性有效厚度大,储层往往粒度粗、物性好,孔喉半径r大,气体运移时克服的毛细管力小[式(1)]。连续性有效厚度大,则H值大,气柱的浮力大[式(2)]。因此,连续性有效厚度大的储层流体势能低,是天然气优势运移通道和优先聚集场所。

毛细管力:

$$P_{cR}=\frac{2\sigma_{gw} \cdot \cos\theta_{gw}}{r} \tag{1}$$

浮力:

$$F_w=\rho g H \tag{2}$$

式中 P_{cR}——毛细管力;

σ_{gw}——气水两相界面张力;

θ_{gw}——润湿接触角;

r——孔喉半径;

F_w——浮力;

H——气藏高度;

ρ——水、气密度差;

g——重力加速度。

以连续性有效厚度为主要依据,结合物性、含气性、有效储层分布模式及试气动态动态参数,将盒八段储层分为四种类型(表2)。Ⅰ类储层基本对应气层,连续性厚度大于5m,有效储层分布模式主要为块状厚层型或具物性夹层的多期叠置型,孔隙度一般大于7%,渗透率一般大于0.3mD,含气饱和度大于50%,初期产气大于$1×10^4m^3/d$,无阻流量大于$2×10^4m^3/d$,为研究区最好的储层。Ⅱ类、Ⅲ类储层对应含气层及部分气层,连续性有效厚度1~5m,有效储层分布模式为具泥质隔层多期叠置型和孤立薄层型,孔隙度5%~7%,渗透率0.1~0.3mD,含气饱和度45%~50%,为中等—差储层。Ⅳ类储层达不到有效储层标准,为干层。

表2 清涧地区盒八段储层分类评价

储层类型	地质参数				动态参数			储层评价
	孔隙度(%)	渗透率(mD)	含气饱和度(%)	连续性有效厚度(m)	有效储层分布模式	产气量($10^4m^3/d$)	无阻流量($10^4m^3/d$)	
Ⅰ	>7	>0.3	>50	>5	块状厚层型/具物性夹层多期叠置型	>1	>2	好
Ⅱ	>6	>0.2	>48	3-5	具泥质隔层多期叠置型	0.5-1	1-2	中等
Ⅲ	>5	>0.1	>45	1-3	孤立薄层型	<0.5	<1	差
Ⅳ	<5	<0.1						干层

从Ⅳ类储层到Ⅰ类储层,随着储层品质的变好,分布频率依次降低(图7a)。研究区以Ⅳ类储层(干层)为主,分布频率为67.5%,Ⅲ类储层分布频率为17.4%,Ⅰ类、Ⅱ类储层仅为7.4%~7.7%。考察各类储层在平面的分布特征,Ⅳ类储层分布面积广,在平面上呈片状分布;Ⅲ类储层呈条带状;Ⅰ类、Ⅱ类储层分布较局限,呈串珠状、豆荚状。

5 富集区优选

5.1 富集区优选

清涧地区气井日产气与单井累计有效厚度存在正相关关系。有效厚度在5m以上时,产气量一般大于$0.5×10^4m^3/d$,无阻流量一般大于$2×10^4m^3/d$。苏里格气田一般将有效厚度大于6m定为富集区优选标准。依据试气分析,考虑到研究区储层品质差于苏里格气田,将有效厚度大于5m的区域定为富集区。

根据富集区优选标准,综合考虑有效砂体厚度、平面分布、叠置样式[13],优选富集区227.1km²(图7b)。总的来说,富集区发育比例低,仅占研究区面积的7.6%。

研究区气藏平均地层压力20.96MPa,平均气层温度365.15K,气体压缩因子0.927,富集区内储层平均有效厚度9.08m,平均孔隙度7.75%,平均含气饱和度57.68%。利用容积法,估算富集区地质储量为$167.34×10^8m^3$,储量丰度为$0.74×10^8m^3/km^2$。

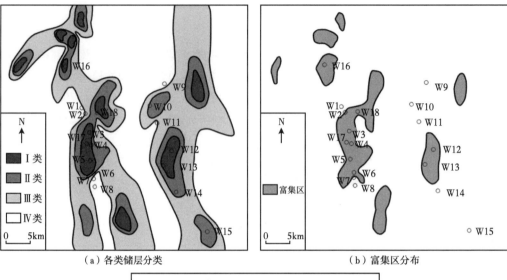

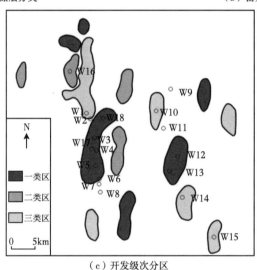

图 7 储层分类、富集区分布、开发级次分区对照图

分小层来看,富集区内四小层孔隙度、含气饱和度等参数差异不大,但平均有效厚度差异明显。盒八段上亚段 2 小层、盒八段下亚段 2 小层有效储层相对厚,储量相对集中,其中盒八段上亚段 2 小层储量 $60.8×10^8 m^3$,占整个盒八段储量的 36%,盒八段下亚段 2 小层储量 $46.6×10^8 m^3$,占盒八段总储量的 28%。

5.2 开发级次分区

结合富集区优选和储层分级评价成果,对研究区的预期开发级次进行了分区,分为一类区、二类区和三类区(图7c)。一类区、二类区主要位于研究区的西砂带,三类区在西、东两条砂带均有分布。

一类区为富集区与Ⅰ类储层的叠合区,储层物性好,有效厚度大,储层品质好,连续性强,是重点开发区。该区面积 164.7km², 地质储量 132×10⁸m³, 以地区面积的 5.5%,集中了研究区 80%的储量,储量丰度 0.80×10⁸m³/km²。

W4井在研究区西砂带的一类区内,盒八段共发育两套具物性夹层的多期叠置型有效储层,连续性有效砂厚 11.7m,储层品质相对好,试气日产量 11.67×10⁴m³,无阻流量 15.46×10⁴m³/d,为清涧地区盒八段开发潜力最好的井。该井于 2014 年 10 月 25 日投产,投产前套压 14.6MPa,目前套压 7.4MPa,平均日产气 0.96×10⁴m³。选取生产特征曲线符合衰竭递减规律、开井天数较为集中段进行生产特征分析,159 天内的单位压降采气量为 21.69×10⁴m³/MPa。

根据气藏工程"一点法"原理,在长庆气区大量生产井统计的基础上,制作了不同生产天数的生产井单井控制储量与单位压降采气量的关系图版(图8)。将 XX 井单位压降采气量代入图版中生产天然最接近的180天曲线,得出该井单井控制储量为 1054.68×10⁴m³,长庆气区最终累计采气量/单井控制储量平均为 0.85,计算最终累计采气量为 897×10⁴m³。按此方法计算,一类区内井平均单井累计采气量为 809×10⁴m³/d,对应苏里格气田的Ⅲ类井。

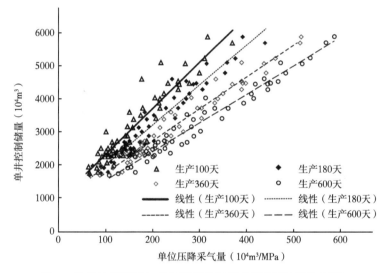

图8 单井控制储量与单位压降采气量关系图版(据吕志凯)

二类区为富集区与Ⅱ类储层的叠合区,含少量Ⅰ类储层,累计有效厚度虽然达到富集区标准,但与一类区相比:连续的有效储层垂向上厚度薄,平面上规模小,气层连通性差;以含气层为主,储量规模小;需多层段射孔,对经济开发效益有一定的影响。二类区面积 62.4km², 地质储量 35×10⁸m³, 储量丰度 0.56×10⁸m³/km², 是普通开发区,区内平均单井累计采气量 566×10⁴m³/d。

三类区为次富集区,累计有效厚度 3~5m,达不到富集区标准,以Ⅲ类储层为主,有效砂体薄,侧向连通性差,单独开发盒八段没有经济效益,可作为产能补充层段。

从整体来看,研究区与苏里格气田相比,距离物源远,岩屑含量高,孔隙度、渗透率低,有效砂体厚度薄,储量丰度小,仅为 0.74×10⁸m³/km², 试气无阻流量和单井产量低(表3)。在众多因素影响下,研究区单井平均累计采气量约为 748×10⁴m³/d,在目前条件下,经济有效开发难度大。

表 3　清涧地区盒八段与长庆气区其他区块参数对比

气田	区块	储层埋深(m)	静态参数					动态参数		
			孔隙度(%)	渗透率(mD)	砂厚(m)	有效砂厚(m)	储量丰度($10^8m^3/km^2$)	无阻流量($10^4m^3/d$)	单井产量($10^4m^3/d$)	单井累计产量(10^4m^3)
苏里格	中区	3300~3500	9.83	0.945	31.77	8.89	1.46	4.36	1.2	2200
	东区	3000~3200	8.62	0.667	30.43	7.98	1.35	3.44	1.0	1973
	南区	3750~3950	7.25	0.535	17.65	7.09	1.15	2.63	0.65	934
清涧地区		2300~2500	5.51	0.258	21.90	5.7	0.74	2.20	0.55	748

从局部来看,清涧地区存在一定的储量富集区和潜力区。建议重点开发一类区,谨慎开发二类区,进一步评价三类区。一类区单井累计采气量 $809×10^4m^3/d$,二类区单井累计采气量 $566×10^4m^3/d$,三类区达不到经济开发下限标准。一类区、二类区内井生产能力相当于苏里格气田Ⅲ类井,但埋深较浅在一定程度上节省了钻井成本,使得研究在单井采气量较低的情况下仍然可获得一定的经济效益。

根据实际生产情况计算,单独开发盒八段富集区,直井最终累计采气量小于 $900×10^4m^3$,具有一定的经济效益;水平井最终累计采气量小于 $2700×10^4m^3$,仅为苏里格水平井平均产量的38%,且盒八段有效砂体薄,横向连续性差,储量在垂向上较为分散,钻水平井风险大[14],另外水平井一般只能开发单一层段,而直井可对盒八段、山一段、山二段等多层段合采。在现有经济技术条件下,可在一类区、二类区优选直井开发,不建议布水平井开发。

6　结论

(1)清涧地区盒八段沉积环境为平缓构造背景下的辫状河三角洲平原沉积,主要发育西、东两条主河道,心滩是有利沉积相带。受物源和搬运距离影响,储层岩屑含量高,岩性以中、粗砂岩为主。储层沉积后遭受了强烈的压实、胶结等成岩作用,物性差,平均孔隙度5.51%,渗透率0.258mD。孔隙空间以粒间溶孔、粒内溶孔等次生孔隙为主。

(2)受沉积、成岩双重控制,研究区盒八段储层规模小,连续性差。研究区有效单砂体长度200~700m,宽度80~360m,厚度为1~3m,平均砂厚厚度为苏里格的80%,有效砂体厚度为苏里格的75%。有效储层平面上主要集中在研究区的西砂带,垂向上在盒八段上亚段2小层、盒八段下亚段2小层相对发育,在空间分布可分为四种模式,其中单期孤立薄层型占50%以上。

(3)"连续性有效厚度"与试气产量相关性好,是沟通地质静态研究和试气动态分析的桥梁,以此参数为主要依据,将研究区储层分成了好、中、差、干层四种类型。结合储层分类和富集区优选成果,按开发级次将研究区分成三类区:一类区面积164.7km²,单井平均累计采气量 $809×10^4m^3$;二类区面积62.4km²,单井平均累计采气量 $566×10^4m^3$;三类区147km²,暂时无法有效动用。

(4)在现有气价条件下,综合考虑地质条件、产气量、钻井和储层改造工艺及成本,建议针对一类区、部分二类区部署直井开发,不建议水平井开发。本研究为气田开发方案编制提供了地质依据,同时对于类似气田开发具有借鉴意义。未来应针对上古生界气藏盒八段、山一段、山二段等地层,开展多层段综合评价研究,多学科协调,联合攻关"分散薄层型致密砂岩气藏"开发配套技术,发展排水采气工艺,进一步引进市场合作机制,全面提升该类气藏开发效益。

参 考 文 献

[1] Munn K, Smith D M. A NMR technique for the analysis of pore structure:Numerical inversion of relaxation measurements [J]. Journal of Colloid and Interface Science, 1987, 119(1):117-126.

[2] Spencer C W. Review of characteristics of low-permeability gas reservoirs in western United States [J]. AAPG Bulletin, 1989, 73(5):613-629.

[3] 郑和荣,胡宗全. 渤海湾盆地及鄂尔多斯盆地上古生界天然气成藏条件分析[J]. 石油学报, 2006, 27(3):1-5.

[4] 赵靖舟,王力,孙兵华,等. 鄂尔多斯盆地东部构造演化对上古生界大气田的控制作用[J]. 天然气地球科学, 2010, 21(6):875-881.

[5] 赵忠英,柳广弟,孙明亮,等. 鄂尔多斯盆地上古生界天然气藏类型辨析[J]. 现代地质, 2010, 24(4):703-708.

[6] 王丽娟,何东博,冀光,等. 阻流带对子洲气田低渗透砂岩气藏开发的影响[J]. 天然气工业, 2013, 33(5):56-60.

[7] 樊太亮,郭齐军,吴贤顺. 鄂尔多斯盆地北部上古生界层序地层特征与储层发育规律[J]. 现代地质, 1999, 13(1):32-36.

[8] 王若谷,李文厚,廖友运,等. 鄂尔多斯盆地子洲气田上古生界山西组、下石盒子组储集层特征对比研究[J]. 地质科学, 2015, 50(1):249-261.

[9] 赵文智,汪泽成,朱怡翔,等. 鄂尔多斯盆地苏里格气田低效气藏的形成机理[J]. 石油学报, 2005, 26(5):5-9.

[10] 郑浚茂,游浚,何东博. 渤海湾盆地与鄂尔多斯盆地陆相优质储层形成条件对比分析[J]. 现代地质, 2007, 21(2):376-386.

[11] 郭智,贾爱林,薄亚杰,等. 致密砂岩气藏有效砂体分布及主控因素——以苏里格气田南区为例[J]. 石油实验地质, 2014, 36(6):684-691.

[12] 杨华,付金华,刘新社,等. 鄂尔多斯盆地上古生界致密气成藏条件与勘探开发[J]. 石油勘探与开发, 2012, 39(3):295-303.

[13] 郝玉鸿,王永强,雷小兰,等. 子洲气田开发地质研究及有利区筛选[J]. 石油化工应用, 2011, 30(2):41-45.

[14] 余淑明,刘艳侠,武力超,等. 低渗透气藏水平井开发技术难点及攻关建议——以鄂尔多斯盆地为例[J]. 天然气工业, 2013, 33(1):65-69.

原文刊于《现代地质》,2016,30(4):880-889。

Water and gas distribution and its controlling factors of large scale tight sand gas fields: A case study of western Sulige gas field, Ordos Basin, NW China

Meng Dewei   Jia Ailin   Ji Guang   He Dongbo

PetroChina Research Institute of Petroleum Exploration & Development, Beijing 100083, China

Abstract: The chemical characteristics and occurrence state of formation water in western Sulige gas field were analyzed based on geological features of reservoirs, the gas and water distribution regularity and the main controlling factors were determined. Aquifer zones are widely distributed in the whole Western Sulige gas field, gas zones developed poorly and distributed limitedly. Vertically, gas and water zones are of poor continuity and distributed in an isolated and staggered pattern. Inside the gas reservoir, there is not a uniform gas and water interface because of the hard differentiation between gas and water. On the whole, the lower segment of H8 Member in Shihezi Formation and S1 Member in Shanxi Formation are both better than the upper segment of H8 Member. Gas and water distribution of Western Sulige gas field is mainly controlled by gas-generating intensity, the distances between reservoirs and hydrocarbon source rocks, configuration relationship of sand shale and physical property differences inside sand body complex. Among them, gas-generating intensity mainly controls the macro pattern of gas and water distribution, with the decrease of gas-generating intensity, good gas accumulation gradually changes to be water associated gas reservoir. The closer the distance between reservoirs and hydrocarbon source rocks, the more developed the gas zones, on the contrary, the gas-bearing water zone and coexisting gas and water zone are more developed. Sand shale configuration relationship and physical property differences inside sand body complex mainly control the local gas filling and accumulation. Five types of gas and water distribution patterns are summarized: pure gas, thick reservoir with mixed gas and water, water above the gas, gas above the water, and gas sandwiched between water.

Key words: Ordos Basin; western Sulige gas field; tight sand gas; formation water; gas and water distribution; controlling factors

1 Introduction

Sulige gas field, the biggest tight sand gas reservoir in China, is a gentle west-leaning monocline located in the northwest of Yishaan slope in Ordos Basin. From top to bottom, there developed Carboniferous Benxi Formation, Permian Taiyuan Formation, Shanxi Formation, lower Shihezi Formation, upper Shihezi Formation and Shiqianfeng Formation. The coal layer at the top of Benxi Formation, is the major source rock in Sulige area, wide spread and stably distributed, it is also a marker bed in this region. He8 Member of Shihezi Formation and Shan1 Member of Shanxi Formation in the Permian are the main pay zones of the gas field with a total thickness about 100m, in which

He8 Member is subdivided into Upper He8 and Lower He8 submembers from top to bottom[1] (Fig. 1). Sulige gas field is a typical low pressure, low permeability, low abundance field[2-4] with strong reservoir heterogeneity. During exploration and development, Sulige gas field is divided into four parts: eastern district, central district, western district and southern district. By the end of 2015, 161 wells had been long shut down due to problems such as liquid loading in wellbore, no production or low productivity, choke damage or lost, high content of H_2S, fracturing failure, waiting for workover and secondary operation. Among them, wells with liquid loading account for as high as 49.7%, and most of these wells are distributed in the western district. The western district is the newly developed and backup reserve area, where complex gas-water relationship has seriously hindered production well deployment and effective field development. Therefore, figuring out gas-water distribution pattern and controlling factors in this area is of great practical significance for high efficient development of the gas field. In this study, based on the reservoir geological characteristics, chemical characteristics and occurrence state of formation water are analyzed to find out water distribution pattern and its controlling factors in western Sulige gas field.

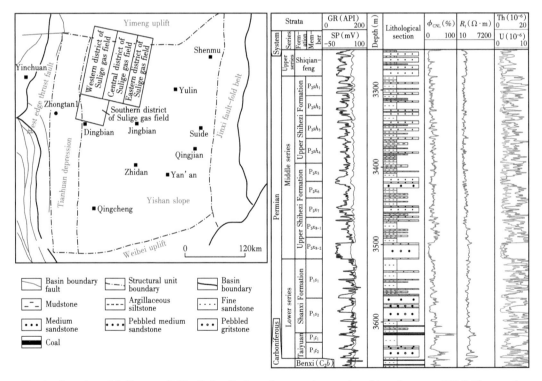

Fig. 1 Location of Sulige gas field, Ordos Basin and composite stratigraphic columnar of Well Zhongtan1.

2 Reservoir geological characteristics

2.1 Lithologic features

In western Sulige gas field, He8 Member and Shan1 Member are dominated by lithic quartz sandstone and quartz sandstone with a small amount of lithic sandstone. Statistical analysis of 1238 thin slices shows that, the reservoir contains primarily quartz and flint clast, with an average quartz content of 83.4% in Upper He8, 85.1% in Lower He8, and 82.3% in Shan1 Member; less metamorphic clast (accounting for 64.3% of the total clast), in which the main components are quartzite and strongly plastic phyllite, with a small amount of metamorphic sandstone and very rare feldspar; and high interstitial substance (11.7% on average). In the interstitial substance, the cement accounting for 5.2% is mostly siliceous cement and iron calcite cement; the matrix, accounting for 6.5%, is mostly water mica and kaolinite, with rare chlorite.

2.2 Physical property and pore throat characteristics

Generally western Sulige gas field has low porosity and low permeability. According to statistics on core data, the reservoirs have a porosity from 4% to 12%, 7.24% on average; and permeability from 0.01 to 10.00mD, 0.52mD on average. There is an obvious positive correlation between porosity and permeability, which means permeability is mainly affected by porosity[5]. According to analysis of 741 thin slices, He8 Member and Shan1 Member have experienced strong diagenesis, so there are primarily dissolution pores of feldspar or debris, and intercrystalline pores of kaolinite; clastic particles are directionally aligned, and under microscope, compact cementation, plastic rock debris alteration, and common quartz overgrowth, and extinction of primary intergranular pores are seen. The reservoirs with mainly small pore throats, feature low expulsion pressure (1.20 MPa on average), low median pressure (6.76 MPa on average), and small median radius (0.25 microns on average), and have a pore throat sorting coefficient of 0.05~5.10, 1.10 on average; pore throat variation coefficient of 0.06~2.11, 0.44 on average; maximum mercury injection saturation of 0.63%~99.36%, 78.26% on average; and mercury withdrawal efficiency of 9.52%~9.52%, 35.00% on average. Overall the western Sulige gas field features strong diagenesis, poor sorting and connectivity and uneven distribution of pores and throats.

3 Chemical features and occurrence state of formation water

3.1 Chemical feature of formation water

Analysis of formation water in western Sulige gas field shows that the formation water with no carbonate ions and hydroxyl ions, has features of salt water, and high salinity (38600~53300mg/L, 43700mg/L on average), with part of the samples being brine[6] (with a salinity of more than 50000mg/L). In the formation water, cations in conventional ion components in descending order of content are ($K^+ + Na^+$), Ca^{2+}, and Mg^{2+}, and anions in descending order of content are Cl^-, HCO_3^-, and SO_4^{2-}, among them Cl^- content is 21505~34638mg/L. Salinity and main content of cations and anions of the formation water are all higher than those of the current ocean water. Ca^{2+}, Na^+, K^+ and Cl^- are dominant in the ion components, so the water is $CaCl_2$ type (Table 1).

Desulfurization coefficient, $SO_4^{2-}/(Cl^- + SO_4^{2-})$, reflects the degree of oil and gas preservation, the smaller the value is, the higher the degree of oil and gas preservation will be. This coefficient of less than 2% means the reservoir is not damage at all[7]. This coefficient of western Sulige gas field ranges from 0.61% to 1.15%, 0.83% on average, which means that the formation water is sealed very well, free from influence or contamination of surface water, and indirectly reflects the gas field belongs to undamaged reservoir, with oil and gas preserved in good condition.

Coefficient of sodium/chloride (Na^+/Cl^-) reflects the concentration & metamorphism degree of formation water and geochemistry environment. Generally the smaller the coefficient is, the better the oilfield water sealing will be, and the more favorable the preservation conditions for oil and gas will be. A sodium chloride coefficient of greater than 0.85 indicates that water can flow, while the coefficient of less than 0.5 means the water is detained, and the reservoir is undisturbed[7-10]. Sodium chloride coefficient in western Sulige field is 0.29 to 0.58, 0.42 on average, which indicates that the formation water is in good sealing condition.

Metamorphic coefficient, $(Cl^- - Na^+)/Mg^{2+}$, reflects the metamorphic degree of groundwater, and the formation's sealing indirectly. Overall the greater the coefficient is, the better the formation is sealed and deeper the metamorphism is. Generally if metamorphic coefficient is greater than 4, the reservoir is considered as primary reservoir[7-10]. This coefficient in western Sulige gas field is 58.8 to 127.9, 71.3 on average, which shows the reservoir is not damaged. Comprehensive analysis indicates this gas reservoir in western district of Sulige gas field is well sealed, and the formation water is mainly ancient sedimentary and metamorphic water isolated from outside water.

3.2 Occurrence state of formation water

The occurrence of formation water is mainly affected by the pore and throat and surface adsorption of rock particles. Wells producing water in Western Sulige are scattered on the plane and gas-water relationship is complex in vertical direction, indicating different occurrence state of formation water[11]. According to microscopic pore structure and strong hydrophilicity of quartz sandstone, formation water is divided into three types: free water, retained water and irreducible water. Nuclear magnetic resonance (NMR) experiment and gas water displacement dynamic experiment show that movable fluid saturation and irreducible water saturation have good correlation with rock permeability (Fig. 2), and the mercury injection curve is used to find out the occurrence state of formation water (Fig. 3).

Free water: typical water in pore, stored in interconnected pores, it is the water not displaced by nature gas during the process of gas charging due to insufficient gas source. This kind of water, existing in pure water layer or in the lower part of gas layer, can flow freely under the action of gravity and differentiates from gas to some extent. In the study area, this kind of water mainly exists in the reservoir with porosity of more than 10%, and permeability of greater than 1mD, and movable fluid saturation of above 70%; and the production rate of this kind of water is generally more than $12m^3/d$.

Retained water: existing in fine throats, and isolated pores, it is the formation water left after gas displaced water in reservoir. This kind of water, commonly coexisting with natural gas, is hard to

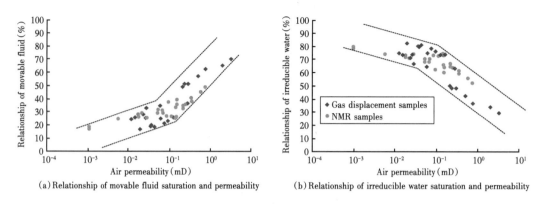

(a) Relationship of movable fluid saturation and permeability

(b) Relationship of irreducible water saturation and permeability

Fig. 2 Relationship of fluid saturation and permeability by NMR and gas displacement experiment.

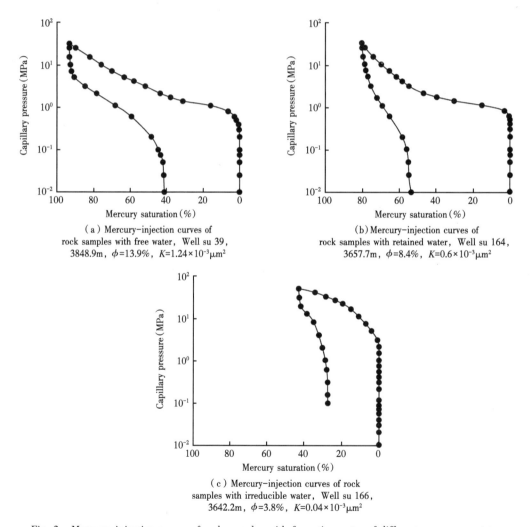

(a) Mercury-injection curves of rock samples with free water, Well su 39, 3848.9m, $\phi=13.9\%$, $K=1.24\times10^{-3}\mu m^2$

(b) Mercury-injection curves of rock samples with retained water, Well su 164, 3657.7m, $\phi=8.4\%$, $K=0.6\times10^{-3}\mu m^2$

(c) Mercury-injection curves of rock samples with irreducible water, Well su 166, 3642.2m, $\phi=3.8\%$, $K=0.04\times10^{-3}\mu m^2$

Fig. 3 Mercury-injection curves of rock samples with formation water of different occurrence state.

flow freely due to capillary force in the fine pore throat and can't differentiate with gas. In this gas field, retained water mainly exists in the reservoir with porosity between 5% and 10%, permeability of 0.3~1.0mD, and movable fluid saturation of 40%~70%. After reservoir fracturing, this kind of water can be produced with gas, at a production rate of 1 to 12m^3/d.

Table 1 Chemical features of the formation water in western Sulige gas field, Ordos Basin.

Well	(K$^+$+Na$^+$) (mg/L)	Ca^{2+} (mg/L)	Mg^{2+} (mg/L)	Cl$^-$ (mg/L)	HCO$_3^-$ (mg/L)	SO$_4^{2-}$ (mg/L)	Na$^+$/Cl$^-$	(Cl$^-$−Na$^+$)/Mg^{2+}	SO$_4^{2-}$/(Cl$^-$+SO$_4^{2-}$) (%)	Salinity (mg/L)	Water type
Su37	15 829	1 763	145	21 505	654	166	0.58	62.3	0.76	38 600	CaCl$_2$
Su38	8 623	7 134	267	26 440	463	188	0.29	70.3	0.71	45 600	CaCl$_2$
Su107	11 547	8 076	276	31 527	316	295	0.41	67.4	0.93	47 800	CaCl$_2$
Su154	9 271	6 782	198	29 583	364	204	0.52	71.7	0.68	44 900	CaCl$_2$
Su186	8 793	4 987	269	26 544	392	189	0.34	65.1	0.71	41 400	CaCl$_2$
Su309	8 835	5 031	287	27 235	534	178	0.38	58.8	0.65	39 500	CaCl$_2$
Su359	9 086	8 309	293	30 325	382	353	0.43	59.0	1.15	46 200	CaCl$_2$
Su360	8 962	4 538	286	26 655	697	164	0.30	65.2	0.61	39 000	CaCl$_2$
Su362	9 100	6 137	78	22 171	266	254	0.55	127.9	1.13	39 700	CaCl$_2$
Su392	10 395	9 156	302	34 638	179	334	0.43	65.4	0.95	53 300	CaCl$_2$

Irreducible water: mainly stored in tiny pores or adsorbed on surface of particles, it is difficult to flow freely, and does not differentiate from gas. In such reservoirs, porosity and permeability polarize seriously. Irreducible water is the main reason of low resistivity in both pure gas layers filled by natural gas and dry tight layer without gas charge. In western Sulige gas field, this kind of water mainly occurs in the reservoir with permeability of less than 0.1mD, and irreducible water saturation of more than 80%, and it is still difficult to be recovered after fracturing during field development.

Formation testing data of all 686 gas wells in Western Sulige gas field shows water production rate ranges from 0m^3/d to 46.5m^3/d, and the produced water is mostly retained water and free water.

4 Gas and water distribution pattern

Located in the lower position of the whole structure of the gas field, close to the gas reservoir boundary, neighboring to Tianhuan depression, Western district is in the gas water transition zone where most reservoirs as gas-water layers, pure gas layers are sparse, water layers with gas are common, and there is no natural gas enrichment region in the strict sense (Fig. 4). Vertically gas layers and water layers in lenticular shape are distributed alternately without uniform gas water interface; inside the same reservoir unit there is no apparent gas and water differentiation; gas layer development degree decreases gradually from bottom to top, while gas-water layers and water layer with gas increases (Fig. 5). Reservoirs in Upper He8 are generally water layers with no gas at all; in Lower He8 and Shan 1 Member, gas layers increase, but are still limited in distribution. From production data, all sections produce water. By the end of 2015, out of 577 gas wells tested in Wellblock Su 47, Su 48, Su 120, 240 wells produced water, with a water/gas ratio of $1.0 \sim 1.4 m^3/10^4 m^3$ on average, which is higher than that in the whole western Sulige district ($0.68 m^3/10^4 m^3$ on average). Due to water production, low yield, inefficient wells in the 3 wellblocks account for more than 60%, and even as high as 82.8% in Wellblock Su120 (Table 2).

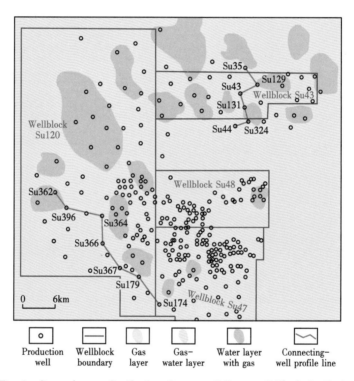

Fig. 4 Gas and water distribution of western Sulige gas field, Ordos Basin.

Table 2 Production situation of Wellblock Su47, Su48, Su120 in western Sulige gas field.

Well block	Producing well/				Instant shut-in	Long-term shut-in
	Total	I	II	III		
Su48	299	63	72	108	56	28
Su47	209	49	43	83	34	12
Su120	93	5	19	58	11	3

Note: type I wells: $q > 1.0 \times 10^4 m^3/d$; type II wells: $0.5 \times 10^4 m^3/d < q < 1.0 \times 10^4 m^3/d$; type III wells: $q < 0.5 \times 10^4 m^3/d$, q = gas rate, $10^4 m^3/d$.

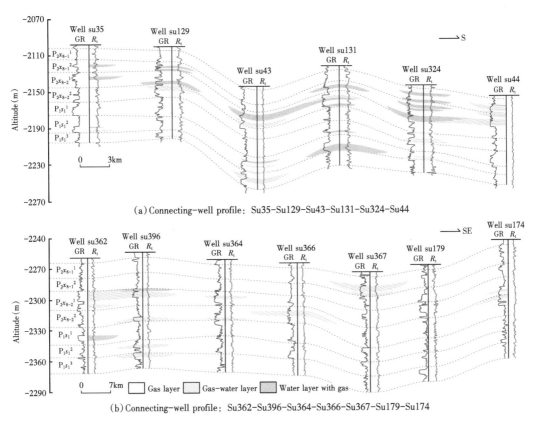

Fig. 5 Reservoir profile by connecting wells in western Sulige gas field, Ordos Basin.

5 Controlling factors of gas and water distribution

5.1 Hydrocarbon generating intensity

There are mainly two kinds of gas source rocks in the Carboniferous and Permian of the Paleozoic of Sulige gas field: humic coal and near humic marine carbonate. Distributed at the top of Benxi Formation, Taiyuan Formation and Shan2 Member, the source rocks feature universal hydrocarbon generation across the whole area, with a hydrocarbon generating intensity in most areas of $12\times10^8 \sim 28\times10^8 m^3/km^2$, while there are some parts with hydrocarbon generation intensity of less than $10\times 10^8 m^3/km^2$ in the central and north of western Sulige[11-14]. Previous studies showed that natural gas in Sulige gas field mainly migrated in short distance laterally and vertically to accumulate in nearby reservoirs, forming key productive layers of Shan1 Member and Lower He8 Submember[15-20]. Under this circumstance, areas with high hydrocarbon generation intensity are more likely to form gas enrichment zones, and areas with low hydrocarbon generation intensity are more likely to have gas reservoir associated with water. Bounded by the hydrocarbon generation intensity of $16\times10^8 m^3/km^2$, central district is rich in gas layer, while western and northern districts are rich in water-gas layer on the whole in Sulige gas field[21].

Though hydrocarbon generation intensity of western Sulige gas field is below 16×10^8 m^3/km^2, both static and dynamic data show it is still the major factor controlling gas and water distribution in this area. Taking Wellblock Su120 and Su43 as examples, comprehensive analysis of connected well profile and hydrocarbon generation intensity shows the south part of Wellblock Su120 with high hydrocarbon-generation intensity has much more gas layers than north-central part of the block with lower hydrocarbon generation intensity, where gas and water layers and water layers with gas gradually increase from south to north. In Wellblock Su43 with the lowest hydrocarbon generation intensity, there is hardly any pure gas layer, most reservoirs are gas-water layers or water layers with gas (Figs. 5 and 6). Vertically, Shan1 Member and Lower He8 Submember near source rock have more natural gas layers than Upper He8 Submember far away from source rock. In general, gas layers mainly occur in the lower formations, and gas-water layers and water layers with gas occur in the upper formations (Fig. 5), indicating the distance to hydrocarbon source rock affects the natural gas

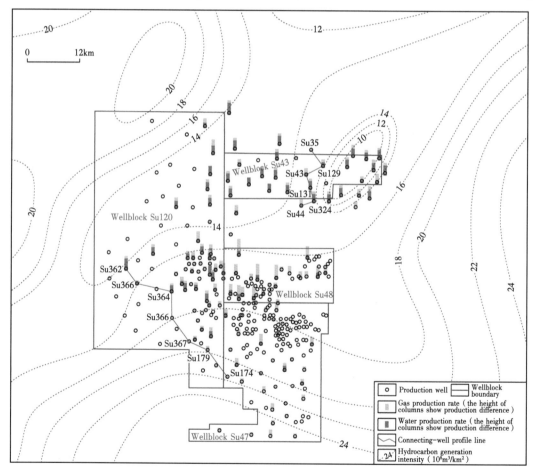

Fig. 6 Overlapping of hydrocarbon generation intensity and formation testing in western Sulige gas field, Ordos Basin.

accumulation to some extent. Formation testing results show with the hydrocarbon generation intensity gradually reducing from 24×10^8 m^3/km^2 to 14×10^8 m^3/km^2 from the southernmost to north, water produced from gas wells gradually increases, while gas produced gradually reduces. Wells in Wellblock Su43 with the lowest hydrocarbon generation intensity, have higher water production than wells in the surrounding wellblocks (Fig. 6). Inside Wellblock Su43, gas wells in the narrow strip with hydrocarbon generation intensity from 14×10^8 to 16×10^8 m^3/km^2, have much lower water production than wells in areas on both sides of the strip with a hydrocarbon generation intensity of less than 14×10^8 m^3/km^2. All the production dynamics indicate the hydrocarbon generation intensity controls the degree of water production in the gas wells.

5.2 Reservoir heterogeneity

5.2.1 Sand shale configuration

Since the whole hydrocarbon generation intensity is low in western Sulige gas field, sand bodies with high permeability in the sand-shale interbeds are preferable places for natural gas accumulation, but due to overall lower hydrocarbon generation intensity, natural gas is distributed in homogeneous thick sand layers at relatively low saturation instead of highly enriched in local sweet spots. According to real drilling statistics, the sand shale ratio of 35% ~ 50% is more conducive to gas layer formation. For example, gas in Well Su128 dispersing across the thick sand body, instead of gathering into gas layer (Fig. 7a), didn't produce gas in formation testing, but produced water at a high rate of 22 m^3/d. In contrast, in Well Su133, with moderate reservoir thickness, and good lithological and physical property, gas fully filling the reservoirs to form effective gas layers, so the well only produced gas without water in formation testing (Fig. 7b).

5.2.2 Physical property differences inside compound sand body

Sulige gas field is a large scale lithological reservoir with low porosity, low permeability, low abundance and low formation pressure, but strong heterogeneity, so natural gas accumulation has great uncertainty and difference there. Lab simulation experiment of gas enrichment in tight reservoir and related analysis show that during the process of natural gas accumulation, for good quality sandbodies with low shale content and good physical property, gas charges in at low initial pressure and meets low resistance, thus is likely to form gas layer; but for poor quality sandbodies with high shale content and poor physical property, gas charges at high initial pressure, meets high resistance, and is hard to drain away water and accumulate, so gas and water layers, water layers with gas, and dry layers are more likely to occur in these sandbodies[22]. Development and distribution of gas layers and water layers in western Sulige gas field are significantly affected by reservoir physical property difference. Provided the gas supply is relatively sufficient, sandbodies with good physical property have higher gas saturation and higher gas productivity in formation testing; while sandbodies with poor physical properties have higher water satuation and low gas productivity in formation testing. Taking Well Su190 as an example, the 1st tested interval with poor physical properties didn't produce gas and water, the 3rd interval with slightly better physical properties produced 0.2×10^4 m^3/d of gas and 1.2m³/d of water, and the 2nd interval with the best physical properties produced $4.2×10^4$m³/d of gas and no water in formation testing (Fig. 7c). Provided the hydrocarbon generation intensity is the same and low, reservoirs with high permeability and small displacement pressure, are filled by gas in priority, while reservoirs with poor physical property where formation water is hard to be displaced, are likely to form water layers with gas or gas-water layers. Furthermore, inside compound sandbodies with big physical property difference, gas layers and water-gas layers usually occur side by side, because natural gas would fill preferably sandbodies with with purer lithology and higher porosity and permeability, forming pure gas layers, in the rest sandbodies, gas and water would coexist (Fig. 7d-7f).

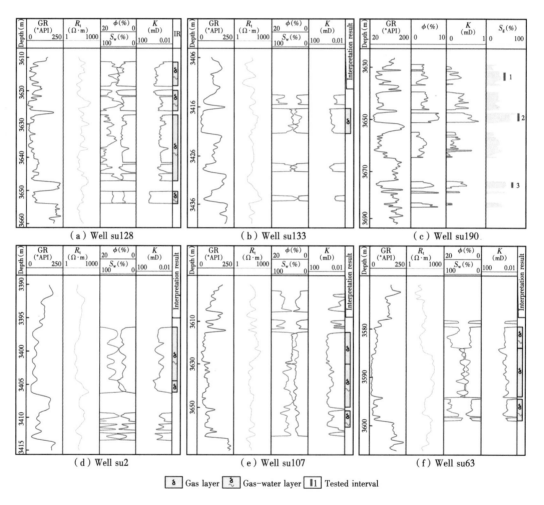

Fig. 7 Well logging curves and interpretation of some wells in western Sulige gas field, Ordos Basin.

According to the above analysis and logging interpretation results, five types of gas and water distribution patterns have been identified in western Sulige gas field: pure gas, gas and water coexisting in thick stratum, water above gas, water beneath gas, and gas sandwiched between upper and lower water. Among them, the pure gas and gas water co-existing types are controlled by sand-mudstone configuration, and the rest three types are controlled by physical property difference in compound sandbodies. The identification of the five types of gas-water distribution patterns can help the selection of layers to be perforated, to control water in the early stage of gas well production.

6 Conclusions

Formation water in western Sulige gas field is mainly residual water and metamorphic sedimentary water, and is divided into three kinds of occurrence state: free water, retained water and irreducible water. Among them retained water and free water take the majority of water produced in the study area.

Water and gas distribution are complicated in western Sulige gas field. Horizontally, water rich areas are widely distributed, especially in northern and eastern parts, pure gas areas in small scale only scatter in the south, generally gas and water co-exist with no strictly natural gas enrichment region in this district. Vertically gas layers and water layers are isolated and alternating with each other, and poor in continuity. Inside a reservoir, gas and water differentiation is not obvious, and there is no uniform gas water contact. Most gas wells produce water at higher rate, which seriously affects the normal production of gas and the overall development of the block.

The main factors controlling gas and water distribution in western Sulige gas field are hydrocarbon generation intensity, distance of reservoir to hydrocarbon source rock, sand shale ratio and physical property difference inside compound sand body. The higher the hydrocarbon generation intensity, the closer the distance between the reservoir and hydrocarbon source rock, the more developed gas layers will be. According to the sand shale configuration ratio and physical difference inside compound sand body, five different types of gas and water distribution patterns have been identified: pure gas, gas and water coexisting in thick stratum, water above gas, water below gas and gas sandwiched between water.

Nomenclature

G_R——natural GR, API;

K——permeability, mD;

S_g——gas saturation, %;

S_w——water saturation, %;

S_P——spontaneous potential, mV;

R_t——formation resistivity, $\Omega \cdot m$;

ϕ_{CNL}——neutron porosity, %;

ϕ——porosity, %.

References

[1] Xu Hao, Zhang Junfeng, Tang Dazhen, et al. Controlling factors of underpressure reservoir in the Sulige gas field, Ordos Basin[J]. Petroleum Exploration and Development, 2012, 39(1): 64-68.

[2] Ma Xinhua, Jia Ailin, Tan Jian, et al. Tight sand gas development technologies and practices in China[J]. Petroleum Exploration and Development, 2012, 39(5): 572-579.

[3] He Dongbo, Jia Ailin, Ji Guang, et al. Well type and pattern optimization technology for large scale tight sand gas, Sulige gas field[J]. Petroleum Exploration and Development, 2013, 40(1): 79-89.

[4] Fang Jianlong, Meng Dewei, He Dongbo, et al. Gas and water formation recognition and water producing well investigation in the western Sulige Gas Field, Ordos Basin[J]. Natural Gas Geoscience, 2015, 26(12): 2343-2351.

[5] Dou Weitan, Liu Xinshe, Wang Tao. The origin of formation water and the regularity of gas and water distribution for the Sulige gas field, Ordos Basin[J]. Acta Petrolei Sinica, 2010, 31(5): 767-773.

[6] Liu Jimin. The characteristics of underground water chemistry and its application in oilfield hydrology exploration [J]. Petroleum Exploration and Development, 1982, 9(6): 49-55.

[7] Liu Chongxi. Hydrogeochemical characteristics of non-marine oil-field basins in China[J]. Geochimica, 1982, 11(2): 190-196.

[8] Li Wei, Liu Jimin, Chen Xiaohong, et al. Characteristics of oil field water in TURUPAN depression and its petroleum geological significance[J]. Petroleum Exploration and Development, 1994, 21(5): 12-18.

[9] Wang Zeming, Lu Baoju, Duan Chuanli, et al. Gas-water distribution pattern in Block 20 of the Sulige Gas Field[J]. Natural Gas Industry, 2010, 30(12): 37-40.

[10] Fu Jinhua, Wei Xinshan, Ren Junfeng. Distribution and genesis of large-scale Upper Paleozoic gas reservoirs on Yi-Shaan Slope[J]. Petroleum Exploration and Development, 2008, 35(6): 664-667.

[11] Cao Feng, Zou Caineng, Fu Jinhua, et al. Evidence analysis of natural gas near-source migration-accumulation model in the Sulige large gas province, Ordos basin, China[J]. Acta Petrologica Sinica, 2011, 27(3): 857-866.

[12] Liu Xinshe, Xi Shengli, Fu Jinhua, et al. Natural gas generation in the Upper Paleozoic in E'erduosi Basin [J]. Natural Gas Industry, 2000, 20(6): 19-23.

原文刊于《Petroleum Exploration & Development》,2016,43(4):663-671.

砂质辫状河隔夹层成因及分布控制因素分析
——以苏里格气田盒八段为例

罗超[1]　郭建林[2]　李易隆[2]　冀光[2]　窦丽玮[3]　尹楠鑫[1]　陈岑[1]

(1. 重庆科技学院石油与天然气工程学院,重庆 400000;
2. 中国石油勘探开发研究院,北京 100083;3. 重庆师范大学,重庆 400000)

摘要:厘清砂质辫状河隔夹层成因、掌握其分布控制因素对该类储层的开发大有裨益。以苏里格气田盒八段砂质辫状河储层为例,综合现代沉积、露头、岩心及测井等动静态资料,分析隔夹层成因特征,建立各类隔夹层测井识别模板,采用密井网多井联动、平面剖面结合的分析思路,构建各类隔夹层的三维地质模型,明确了基准面旋回、构型界面及砂体叠置样式等对隔夹层分布的控制机理。结果表明:泛滥平原、坝间泥、废弃河道、落淤层以及冲沟是构成盒八段辫状河隔夹层的主要类型。泛滥平原规模较大,分布连续,宽度为100～1000m;单一坝间泥长宽比规律不强,宽度一般在300m以内;废弃河道样式由河道废弃方式决定,宽度为40～330m;落淤层分布受增生体大小以及落淤层发育位置决定,宽度为10～190m;冲沟分布分散,宽度一般小于100m。其中,基准面旋回升降控制隔层厚度,构型界面决定夹层分布位置及倾向倾角,砂质辫状河砂体叠置样式约束各类隔夹层比例及大小。

关键词:砂质辫状河;隔夹层成因;三维模型;控制因素;苏里格气田

现代沉积[1]和露头研究[2]表明,砂质辫状河隔夹层成因多样[3-4],几何特征[5]、大小[6]各异,组合关系不明确[7],而识别隔夹层不同成因[8-9]、建立隔夹层三维模型[10-11]并厘清其分布规律[12]是完善该类储层精细描述工作的重点。目前针对砂质辫状河隔夹层的精细研究还存在如下问题:(1)砂质辫状河沉积过程复杂,单一区域建立的隔夹层沉积模式并不具有普适性,在选择野外露头作为地下构型解剖的参考时,难以做到相同或相似沉积环境间的类比;(2)由于隔夹层并不是油气地质储量的组成部分,特别是对于气藏而言,气藏内流体的流动性明显强于油藏,对构型要素的研究往往集中于对有效储层大小[13]的分析,缺乏针对各类隔夹层定量规模的总结;(3)各类隔夹层分布控制因素认识不清,对后期开发调整、剩余储量挖潜提供的地质依据不充分。

苏里格气田是最为典型的砂质辫状河相致密气田,采用直井联合水平井开发方式,目前已完钻上千口水平井[14]。水平井实施过程中钻遇泥质隔夹层不可避免,复杂的隔夹层分布是影响水平井部署的重要因素。由于水平井钻井成本高,当钻遇较厚的泥质隔夹层时,井筒与储层的接触范围减小,降低了水平井段的利用率,制约了水平井整体开发效果。水平井高效开发应用,需要保证较高的砂岩钻遇率,提高砂岩钻遇率需减少水平井在隔夹层中的钻遇。明确隔夹层成因机理、分布控制因素对深入认识其储层非均质性和水平井高效开发尤为重要。因此,以苏里格气田盒八段砂质辫状河储层为例,综合岩心、露头、测井及开发动态等多种资料,通过野外露头分析,搞清了各类隔夹层沉积成因,结合地下构型解剖,明确地下地质条件下各类隔夹层规模特征及其分布控制因素,采用三维地质建模刻画了各类隔夹层在地下空间的分布特征,为相似气藏的隔夹层研究提供参考。

1 研究区地质概况

苏里格气田是国内已发现最大的致密砂岩气田,构造特征呈平缓的西倾单斜,属于典型的"低压、低渗、低丰度"气田[15]。下二叠统石盒子组盒八段沉积时期,由北向南发育的宽缓型砂质辫状河沉积体系构成了苏里格气田的主力产气层[16]。该时期河道频繁横向迁移,河道砂体侧向上多期叠置拼接,内部结构复杂,发育多层次隔夹层。

2 隔夹层成因

2.1 沉积特征认知

古水深一直是河流相储层沉积特征的一个重要方面,前人研究表明单期河道的沉积厚度与古水深近似,发育隔夹层的厚度、规模也与古水深关系紧密。Leclair[17-18]系统分析整理了多条现代砂质辫状河的水文数据,描述了砂质辫状河发育的交错层理系与古水深的定量关系[式(1)、式(2)]。取心井岩心资料表明,苏里格地区砂质辫状河交错层理系组平均厚度为13~32cm(图1),计算得到辫状河沉积古水深为5~7m。根据本次研究建立的精细

图1 苏里格地区河道内交错层理系厚度

(a)S36-J11井,3340.15~3340.28m,0.13m;(b)S36-J11井,3339.34~3339.66m,0.32m;
(c)S36-J11井,3358.88~3359.13m,0.25m;(d)S36-J11井,3339.97~3340.15m,0.18m;
(e)S6-J1井,3332.76~3332.90m,0.14m;(f)S36-J20井,3330.40~3330.53m,0.13m;
(g)S6-J1井,3333.44~3333.63m,0.19m;(h)S6-J6井,3318.25~3318.40m,0.15m;
(i)S171井,3634.2~3634.33m,0.13m

地层格架可知,盒八下亚段细分单层后,单层地层厚度约为4.6~6.8m,与经验公式计算的古水深结果吻合。综上所述,苏里格地区沉积古水深相对较深,有4~7m,发育的是中等规模的砂质辫状河。

$$H = (2.9 \pm 0.7)h^{[17]} \quad (1)$$

$$d_m = (H/0.086)^{0.84[18]} \quad (2)$$

式中　　h——交错层理系平均厚度,m;

H——沙丘高度,m;

d_m——辫状河古水深,m。

2.2 原型露头分析

砂质辫状河沉积过程复杂、储层结构规律不明显,在某一露头区建立的隔夹层沉积特征,特别是定量认识并不一定适用于其他区域,因此在确定隔夹层成因类型、分布特征时,需综合苏里格地区取心资料和露头分析,比较露头与研究区沉积水动力条件,特别是古水深参数,可以取得较好的效果。选取与研究目的层相同或相似沉积环境的野外露头进行类比,筛选了山西柳林、大同晋华宫地区的砂质辫状河沉积露头。其中山西柳林露头发育层位为下石盒子组盒八段,与苏里格气田主产气层段一致。而大同晋华宫地区砂质辫状河露头属于中侏罗统云冈石窟段,发育于现今盆地北边缘约10km,前人据区域地质资料分析推断[19],大同晋华宫地区侏罗系坳陷盆地边界范围与现今邻近,相距约30km,表明大同晋华宫地区露头剖面在盆地中所处的位置是比较靠近盆地边缘的部位,距物源区相对较近。与苏里格地区相似的物源条件[20],使得两区域发育的碎屑沉积物岩石学类型[21]、粒度特征[22-23]等较为相似。沉积水动力条件分析结果表明:露头区与地下沉积气候条件相近,同属于相对干旱的气候环境;山西柳林与大同晋华宫露头区单一期次砂体平均厚度为5.8m、6.7m,依据Miall[24-25]关于辫状河的分类标准均属于常年流水的深河型砂质辫状河,与苏里格地区盒八段发育的砂质辫状河沉积古水深相近。取心井观察显示,苏里格气田盒八段砂质辫状河内发育的隔夹层基本为泥质沉积物,这些泥质层应该沉积于流水速度较低的部位,对比野外露头发育的隔夹层类型(图2),主要包括泛滥平原、坝间泥、废弃河道、冲沟和落淤层五类。

(a)废弃河道、坝间泥岩及泛滥平原沉积　　(b)冲沟沉积、落淤层

图2　大同晋华宫辫状河露头剖面隔夹层类型

2.3 成因特征分类

结合研究区野外露头和取心资料,进一步确定了各类隔夹层的岩性、粒度等沉积特征,分析了野外露头发育的隔夹层沉积特征,描述了隔夹层在空间的分布特点,刻画了隔夹层与砂体的组合关系。

2.3.1 泛滥平原

该类隔层在洪水泛滥末期发育,分布于辫状河道顶部或边部,厚度较大,以块状灰色泥岩为主(图2a、图3a),内部可见典型的水平层理,偶有植物根茎及虫孔发育。洪泛期高能水流携带大量细粒沉积物质漫出水道,洪峰过后水体能量减弱,细粒物质在辫流带顶部、辫流带以外的洪泛区快速落淤堆积。和心滩的频繁迁移相比,整个辫状河道带的迁移频率要低得多,因此类似于大同晋华宫地区剖面,分布在辫流带边部的细粒沉积物保存条件更好,被后期水体破坏的概率更小,平面分布连续、沉积厚度更大,可作为单层划分与对比的主要标志。

2.3.2 坝间泥

现代沉积研究表明,活动心滩坝一般高于水面或与水面持平,在心滩坝的两翼形成有效的阻水区域,在坝后辫状河道中的小范围区域会形成相对静水区域,悬浮物质通常会在此聚集,形成坝间沉积。坝间沉积以粉砂质泥岩和泥质粉砂岩为主,也包括植物碎屑、黏土,沉积构造极少发育(图2a、图3d、图3f),厚度不稳定。然而,河床附近的低速水流有时会造就小型波状层理,心滩下游河道汇聚水流流速增加的区域,有时可见坝间泥岩与河道粗碎屑物质呈指状交互的情况。心滩的频繁迁移使得坝间泥的沉积时间往往较短,厚度一般小于1m,向着坝侧缘的方向厚度逐渐增大,平面上通常呈分散片状,与辫状河道沉积交织。

2.3.3 废弃河道

砂质辫状河道频繁废弃过程中,水动力减弱,河道内充填悬浮细粒颗粒形成灰色泥岩或泥质粉砂岩夹层。相比研究程度较高的高弯度曲流河,砂质辫状河废弃河道在废弃后往往可以重新复活,因此根据废弃河道废弃时间长短以及受后期水流改造程度的大小,可将该类沉积细分为砂质全充填、泥质半充填、泥质全充填三种亚类。砂质全充填、泥质半充填废弃时间较短,形态受废弃河道与复活河道共同控制,河道底部沉积较粗的中—细砂岩、细砂岩,受复活河道改造情况影响,辫状河道上半部被粉砂质或泥质充填。泥质全充填废弃时间长,改造程度小,形成夹层的厚度较大,呈现较规整的河道顶平底凹形态。泥质半充填亚类表现出上半部近基线或呈现小幅度锯齿状,下半部呈现低幅钟形特征,不同于砂质全充填亚类的中高幅钟形特征;泥质全充填亚类则表现出近基线的低幅度弱齿化特征,形成的辫状河道顶面与坝体顶面高程差明显。大同晋华宫地区剖面辫状河露头中废弃河道多为砂质全充填、泥质半充填亚类,一般为块状层理,泥质粉砂岩常发育小型流水沙纹(图2a、图3b)。

2.3.4 冲沟

冲沟发育于心滩坝中部、上部,为细粒悬浮物质沉积形成的泥岩夹层,剖面上呈透镜状,形成机制与废弃河道相似,侧向范围小于废弃河道。厚度相对较小,一般小于2m;平面上呈现细条带状,长度与心滩延伸距离近似。岩性以灰色粉砂质泥岩、泥岩为主,发育小型波状层理和水平层理(图2b、图3c)。当持续性水流不断冲刷心滩,冲沟不断拓宽加深,可以演化为辫状河道。

(a)泛滥泥,J11井,3370.13~3370.23m　　(b)废弃充填泥,J11井,　　(c)冲沟泥,J11井,3355.54~3355.81m
　　　　　　　　　　　　　　　　　　　　3354.65~3354.72m

(d)坝间泥,J11井,3357.83~3357.86m　　(e)落淤层泥,具成层性,　　(f)坝间泥,J11井,3362.78~3363m
　　　　　　　　　　　　　　　　　J11井,3344.89~3344.98m

图3　苏里格地区不同类型隔夹层岩心照片

2.3.5　心滩落淤层

心滩落淤层是洪峰过后憩水期在心滩核部、翼部、背水面尾部发育的近平行悬浮物[26],由于心滩落淤层为事件性沉积,接受沉积的时间较短,同时受后期辫状河道改造影响,落淤层厚度往往较薄。岩性主要由灰色泥岩、粉砂质泥岩构成,发育水平层理,有时可见波状层理。整体上,心滩落淤层在滩核、滩尾侧向上较为连续,类似于心滩的形状,呈现菱形或椭圆状近水平分布,倾角小于3°(图2b、图3e)。

从野外露头来看,各类型隔夹层具有不同的几何形态、岩性特征、规模大小,在空间上与心滩砂体有机组合,构成了完整的砂质辫状河沉积(图4)。除上述五类主要隔夹层外,辫状河内

图4　山西柳林辫状河露头隔夹层分布特征

还发育冲刷泥砾沉积,层厚为5~10cm,粒径在1cm左右,呈次圆状,具有定向排列的特点。由于冲刷泥砾横向变化快,连续性较差,因此不作为主要隔夹层类型进行单独研究。

3 地下隔夹层识别

3.1 测井识别标志

采用岩心标度测井[27]的方式建立了各类隔夹层识别模板,确定不同类别隔夹层岩性、厚度、典型电性特征以及与构型界面的关系。从识别的结果来看,泛滥平原泥与废弃河道细粒沉积的厚度在五种类型的隔夹层中厚度占优势,而岩性上二者又有明显差别。泛滥平原为较纯的泥质沉积,废弃河道沉积含有较多的泥质粉砂岩、粉砂质泥岩,二者在电测曲线上有较大差别,其中自然伽马曲线表现尤为典型,废弃河道自然伽马响应值一般在30.6~58.9API之间,而泛滥平原泥质沉积自然伽马值一般介于41.5~127.3API。坝间泥岩、冲沟以及落淤层三者均有厚度小的特点。从取心井统计结果来看,坝间泥岩厚度在0.2~1.1m之间,平均厚度为0.4m;冲沟厚度在0.5~2.1m之间,平均厚度为1.2m;落淤层厚度在0.2~1.3m之间,平均厚度为0.5m。在三者的识别过程中,需要借助单层(构型单元)的分层界限,由于坝间泥岩分布在心滩坝间,也就是坝间泥岩具有与四级构型界面伴生的特征,而四级构型要素和界面较易识别。因此,依靠构型界面这一特征,可以在测井曲线上将坝间泥岩与冲沟、落淤层区分开来。参考自然伽马、声波时差以及电阻率曲线,可以进一步将冲沟、落淤层区分开。

表1 取心井隔夹层测井识别特征

隔夹层类型	岩性	厚度(m)	自然伽马(API)	构型界面
泛滥平原	泥岩等	1.5~7.5/4.6	41.5~127.3	4、5、6
废弃河道	泥质粉砂岩、粉砂质泥岩、粉砂岩等	1.2~6.6/4.2	30.6~58.9	4
坝间泥岩	泥岩等	0.2~1.1/0.4	37.8~75.1	4
冲沟	粉砂质泥岩、泥质粉砂岩等	0.5~2.1/1.2	27.8~63.5	3、4
落淤层	泥岩等	0.2~1.3/0.5	30.3~125.5	3

综合各类隔夹层在岩心、构型界面上的多种特征,对单井进行了隔夹层的识别及分类。以S36-J11井为例,该井第二次取心层段为$H8x^{1-2}$、$H8x^{1-3}$、$H8x^{2-1}$,共识别了各类泥质层段7层,其中落淤层2层,冲沟1层,泛滥平原2层,废弃河道泥1层,坝间泥岩1层,坝间泥岩分布在$H8x^{1-2}$、$H8x^{1-3}$两个单层间(图5)。

3.2 隔夹层平面、剖面分布特征

各类成因类型的隔夹层在空间分布、边界条件和规模特征参数等信息上有不同显示,是野外露头分析、密井网解剖等手段得到的综合性地质资料集合,对同类构型单元的分析、预测有指导作用。采取平面及剖面相结合的思路,对S36密井网的储层构型开展分析,明确了各类隔夹层的空间分布,过S36-J1—S36-2-21—S36-J2井剖面可作为隔夹层分布特征的典型代表,$H8x^{2-1}$、$H8x^{2-2}$平面上呈现不同的分布特征。$H8x^{2-1}$上S36-J11井上发育坝间泥、S36-2-21井发育废弃河道;$H8x^{2-2}$在S36-J1井发育废弃河道,S36-2-21井发育冲沟,S36-J2井发育落淤层。

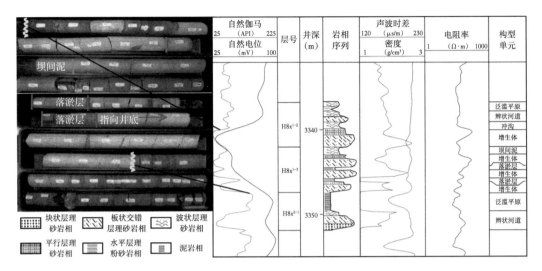

图 5 S36-J11 取心井隔夹层划分

3.2.1 泛滥平原

顶部的 $H8x^{1-1}$ 泛滥平原延伸范围较大,呈厚层块状(表2),自然伽马曲线表现为典型的靠近基线的线性或微齿线形,根据标志层高程差、废弃河道平面分布,从确定的泛滥泥岩边界来看,其分布范围较大、连续性好,一般延伸距离可超过两个井距(图6、图7),宽度最大可达到 1000m 左右。

表 2 盒八段各类隔夹层定量地质参数统计表

成因类型	野外露头实测			
	几何形态		规模尺度(m)	
	平面	剖面	宽度	长度
泛滥平原	不规则片状	厚层状	100~1000	800~2000
废弃河道	条带	顶平底凸透镜状	40~330	500~980
坝间泥岩	条带	透镜片状	10~300	265~480
冲沟	窄条带	小型透镜状	8~100	240~630
落淤层	椭圆	斜列薄板状	10~190	140~320

3.2.2 坝间泥岩

该类夹层在岩心、测井资料上特征不明显,可通过组合平面构型单元的相对位置加以识别。鄂尔多斯盆地由北至南的物源方向决定了盒八段心滩走向主要以南北向为主,苏里格气田水平井钻遇上下两期次心滩过程中,常钻遇这类夹层,由于其规模较其他夹层小,在水平井钻井过程中不需要做较大调整。该剖面上,S36-J1 井在 $H8x^{1-3}$ 心滩坝尾处所形成的坝间泥,在 $H8x^{2-1}$ 顶部发育,在 $H8x^{2-1}$ 小层对应为坝头位置。这类夹层仅发育在坝尾,呈椭圆或长条状(图6a),剖面上呈透镜状或楔形,且与心滩坝砂体呈指状互层,受后期河道沉积冲刷破坏所致,该类夹层长宽比规律性不强,长宽特征差异大,宽度一般在300m以内,长度有时可接近单

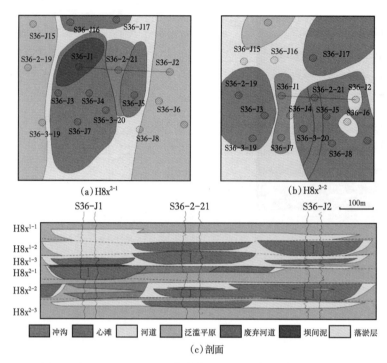

图 6 过 S36-J1—S36-J2 井多期次心滩叠合与隔夹层分布图

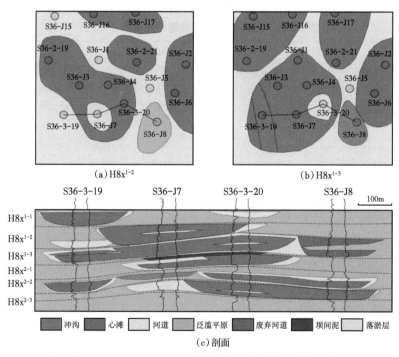

图 7 过 S36-3-19—S36-J8 井多期次心滩叠合与隔夹层分布图

3.2.3 废弃河道

过 S36-J1—S36-2-21—S36-J2 井剖面上,发育两处废弃河道。其中 S36-2-21 井 $H8x^{2-1}$ 小层发育砂质充填型,S36-J1 井 $H8x^{2-2}$ 小层发育泥质半充填型,侧向上延伸到南部的 S36-J7 井处。由于废弃河道为河道废弃后充填形成的,其规模、形态受辫状河道控制,多呈顶平底凸透镜状(表2)。考虑到现代辫状河沉积与研究区特征(特别是构型规模)的一致性,对西藏拉萨河、伊通河、雅鲁藏布江上游等15个中等规模的现代砂质辫状河道段水文地质条件进行调研,应用 Google Earth 软件对单河道平水期规模数据进行测量。通过单井确定的单一辫状河道水深,利用现代沉积、古露头建立的数学关系,结合苏里格地区丰富的钻井资料,建立苏里格地区砂质辫状河单河道宽度与辫状河水深的关系。计算结果表明,盒八段单层废弃河道的平均宽度为 40~330m,平均 155m。

3.2.4 冲沟

冲沟沉积发育于 S36-2-21 井 $H8x^{2-2}$ 层处,自然伽马曲线回返明显,幅度一般大于2/3,主要向着该井所处心滩长轴方向延展。由于冲沟是洪泛期辫状河道切割心滩坝顶部形成的小型水道,几何规模受心滩规模控制,与心滩的长宽有较好的相关性。利用柳林地区露头资料和卫星照片数据对22个典型心滩坝上冲沟与心滩长宽进行回归统计,建立了心滩宽度与冲沟泥岩宽度、心滩长度与冲沟泥岩长度之间的关系。结果显示冲沟泥岩宽度为 8~100m,平均宽度为 60m,沟道泥岩长度为 240~630m,平均长度为 425m;冲沟与所在心滩长轴方向基本一致,夹角小于 20°,多为 0°~10°。

3.2.5 心滩落淤层

心滩落淤层发育主要受心滩坝底形以及落淤层发育于心滩的部位(滩头、滩核、滩尾部)所决定。滩头部位受水流冲刷严重,落淤泥岩最不易保存;滩翼受对称环流作用影响,辫状河道的迁移冲刷使得低部位沉积的落淤层往往被破坏;滩核和滩尾分布位于心滩中央及缓部背水流位置,地势高且平坦,受水流冲刷作用影响很小,落淤泥岩发育、保存条件较好,一般呈水平状分布[28-29]。落淤层沉积过程受控于与之伴生发育的增生体,因此单一增生体的长度、宽度限制了落淤层的规模。根据盒八段各单层单一增生体厚度为 3~6m,根据经验公式确定增生体宽度一般在 100~400m,从而可得出心滩内部落淤层最大宽度范围为 100~400m,最大长度范围为 300~800m。在增生体规模约束下,进一步分析了密井网落淤层的解剖结果。S6、S36 密井网加密区内,平均1个心滩有近7口井钻遇,通过长、短轴多方位的连井对比,对心滩内落淤层进行井间匹配分析,可以识别空间分布规律,并统计落淤层厚、宽、长等几何参数信息。精细表征了12个心滩坝内部发育的35个落淤层,发现落淤层大部分位于心滩坝的翼部、后部,且多呈椭圆或菱形薄片状分布。例如 S36-J2 井、S36-J6 井的 $H8x^{2-2}$ 小层发育的落淤层,在自然伽马曲线上显示为小幅度回返,发育于该井所在心滩坝的核部及偏翼部。统计的35个落淤层宽度分布在 10~190m,平均为 110m;长度分布在 140~320m,平均为 238m,长度、宽度占增生体长宽的 15%~35%,以 20%~25% 最为集中。

3.3 隔夹层三维模型

针对苏里格气田,选取投入开发时间长、井排距小、动静态资料完善的 S6 加密区,该区储层钻遇率约为60%~90%,资料证实各类隔夹层发育情况良好,具有较好代表性,可作为精细刻画隔夹层的三维建模区。对于辫状河隔夹层这类离散变量的模拟,序贯指示方法作为离散变量模拟的最常用算法,其实用性已被广泛证实,由于序贯指示模拟过程过度依赖变差函数分析的结果,各类隔夹层空间几何形态及分布规律在模拟结果中难以体现,同时苏里格地区砂质辫状河隔夹层分布复杂,经过后期改造的隔夹层,难以用规则几何参数来定义。基于此,采用Petrel软件,以S6-J19井区为例,建立了该区域隔夹层分布模型,具体建模步骤如下:(1)通过S36-J11等取心井岩心刻度测井,在S6-J19密井网区构造模型的基础上,完成包含S6-J18、S38-16-4、S6-J15等11口单井上各类型隔夹层的识别;(2)进行三维网格划分,考虑隔夹层平面和垂向网格的划分要能够充分表现隔夹层复杂的三维形态,同时兼顾部分夹层(如S6-11-12井$H8x^{2-2}$发育的落淤层)厚度较薄的特征,要求平面网格要小于20m×20m,垂向网格不大于0.1m,最终整个密井网工区网格数为115×78×58=520260个;(3)以单井隔夹层识别结果作为建模硬数据,各类隔夹层的地质参数信息以及解剖结果为约束条件,采用序贯指示+人机互动的方式,最终建立了S6-J19典型井区的隔夹层三维分布模型,通过不断修正建立的隔夹层模型,使得模型中隔夹层的分布特征符合地质参数及分布规律。

综合纵横向对比以及综合地质认识结果,得到密井网区各井上隔夹层的厚度、长、宽、展布方向等信息,建立多个加密区隔夹层的分布模型。从模型可见,废弃河道、坝间泥、落淤层、泛滥平原、冲沟构成了砂质辫状河隔夹层的主体,冲沟分布在心滩坝顶端,废弃河道泥岩与心滩坝侧向叠置,坝间泥和落淤层充填其中,连片状泛滥平原沉积向辫状河体系间方向不断增厚,各类隔夹层与各类储层共同拼合了辫状河构型空间结构。从S6-J19井隔夹层三维建模的结果来看,除泛滥平原沉积外,各类夹层侧向规模均难以跨越1个井距(图8),这主要是受砂体的叠置样式所决定的,该井区为强水动力条件,砂体频繁迁移叠置,使得夹层的规模较小,难以有效保存。

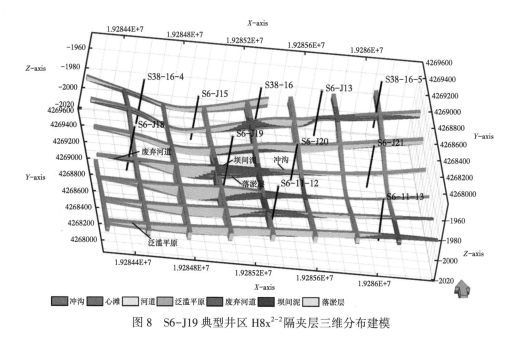

图8 S6-J19典型井区$H8x^{2-2}$隔夹层三维分布建模

4 隔夹层分布控制因素分析

通过隔夹层平、剖面分析结果,以及建立的隔夹层三维地质模型,分析了隔夹层分布控制因素。

4.1 沉积层序控隔层

隔层主要包括泛滥平原、废弃河道,二者具有厚度大、分布相对稳定的特点,特别是泛滥平原泥质沉积。其规模主要受基准面旋回控制(A/S 值低时,保存条件差,延伸井距短;A/S 值高,保存条件好,延伸井距长),A/S 除了控制砂体发育的规模、叠置关系以外,也同样控制隔夹层的保存条件。具体表现为 A/S 值低时,辫状河叠置带发育,沉积水动力最强,古河道持续发育,心滩厚层粗砂岩与河道充填砂岩切割叠置,泛滥平原泥岩较少发育。A/S 值高时,辫状河沉积过渡带、体系间较为发育,剖面上呈现砂泥岩互层交互特征,砂体规模小、连续性差,沉积的泛滥平原泥岩厚度较大。H8x 亚段作为 1 个完整的中期旋回,该中期旋回为上升半旋回优势发育的不对称旋回,内部细分的 $H8x^{2-3}$、$H8x^{2-2}$、$H8x^{2-1}$、$H8x^{1-3}$、$H8x^{1-2}$、$H8x^{1-1}$ 六个单层,其层间发育的隔层平均厚度分别为 2.0m、3.7m、4.0m、2.4m、5.9m、6.7m(图 9)。整体来看,从下部的 $H8x^{2-3}$ 到上部的 $H8x^{1-1}$,除去中间 $H8x^{1-3}$、$H8x^{1-2}$ 之间发育的隔层厚度为 2.4m 以外,隔层的厚度呈现不断增大的趋势,这完全符合 H8x 亚段的旋回结构。其中,$H8x^{1-2}$、$H8x^{1-1}$ 水动力

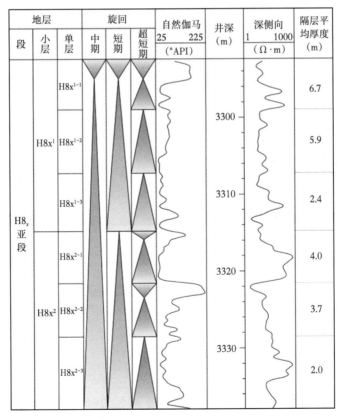

图 9 隔层厚度垂向分布特征

弱,叠置带分布范围相对较小,叠置带分布面积占研究区总面积的比例小于30%,过渡带和体系间展布面积优势明显,二者面积之和超过60%;$H8x^{2-3}$、$H8x^{2-2}$为该区砂体最为发育的层段,叠置带分布范围最广,过渡带和体系间展布范围萎缩,叠置带、过渡带和体系间占面积比例分别为58%、35%和7%。平面上,由辫状河叠置带向过渡带到体系间,沉积砂体厚度具有整体变小的侧向变化规律,反之泛滥平原泥岩具有相应增厚的趋势。盒八段各单层间泛滥平原在体系间呈连片状分布,过渡带多呈宽条带状,二者长宽延伸数千米乃至数十千米。而叠置带内的窄条带状泛滥平原泥岩规模最小,厚度一般小于3m,但宽度一般也在700~1500m,长度能达到1000~2000m之间。

4.2 构型界面控夹层

夹层主要包含落淤层、冲沟、坝间泥,对苏里格地区以及露头区辫状河夹层沉积与构型界面特征进行分析发现,四级乃至三级构型界面控制着夹层的发育位置,限制了各类夹层的发育规模。以大同晋华宫地区剖面露头东段为例,该露头位于辫状河沉积叠置带,砂体大范围叠置发育,剖面上夹层较少发育。该复合砂体为上下两期次河道叠置形成(图10),在两期砂体间,心滩坝沉积尾部发育坝间泥岩,四级构型界面控制了该坝间泥岩的展布范围,使得该坝间泥岩向西倾,倾角5°,剖面上分布范围较小。下部心滩内,仅仅能通过岩性、颜色上的变化,确定出三级构型界面的位置,而未见有夹层发育。而在上部心滩中,通过进一步完成三级构型界面的识别,确定了冲沟、落淤层发育的位置,冲沟、落淤层倾向向东,与坝间泥岩相反,倾角较小,只有1.5°。冲沟和落淤层保存情况较好,侧向上延伸范围大于坝间泥。可见由于四级、三级构型界面特征的差异,造成了不同夹层间倾向、倾角的差异。

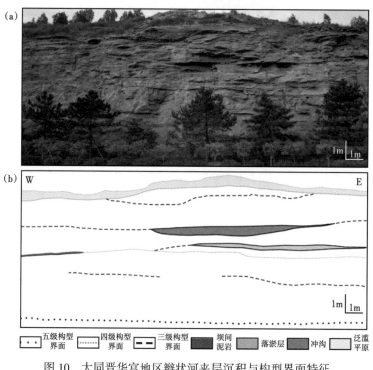

图10 大同晋华宫地区辫状河夹层沉积与构型界面特征

4.3 砂体叠置样式控比例

按照砂体叠置样式的不同,可划分出孤立型、侧向切叠型、垂向叠置型分布样式(图11),不同类型的砂体叠置样式,具有不同的隔夹层分布特征。统计了近360口水平井实钻隔夹层参数,平均1032m水平段一般钻遇隔夹层3~8段,对比了水平井在辫状河沉积孤立型、侧向切叠型、垂向叠置型内钻遇隔夹层所占的比例。其中孤立型中钻遇隔夹层长度341.2m,其中泛滥平原、废弃河道、坝间泥岩、冲沟、落淤层所占比例分别为63.4%、19.3%、0.3%、6.9%、10.1%;侧向切叠型钻遇隔夹层长度210.8m,各自所占比例分别为13.4%、23.1%、11.2%、18.7%、33.6%,垂向叠置型水平段钻遇隔夹层长度148.9m,各类隔夹层所占比例分别为8.5%、30.9%、3.4%、20.7%、36.5%。可见,孤立型砂体分布样式中隔层最为发育,而垂向叠置型夹层最发育,但夹层样式少于、规模也小于侧向切叠型。在苏里格地区水平井规模化应用中,筛选垂向叠置型砂体部署水平井,钻遇气层平均长度为829.3m,气层钻遇率为83.5%,日产$7.1×10^4 m^3$,可以取得较高的产量。

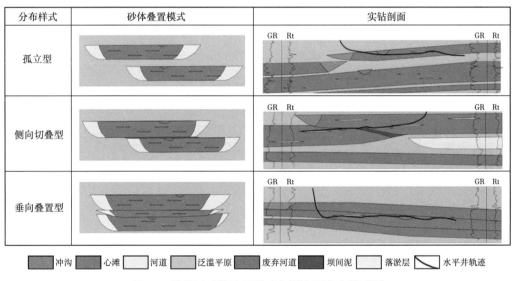

图11 辫状河砂体叠置样式与隔夹层分布模式图

5 结论

(1)通过苏里格气田S36-J11等井的取心资料以及山西柳林等地区相似古水深条件的野外露头解剖,确定了砂质辫状河隔夹层类型成因类型,划分出废弃河道、坝间泥岩、泛滥平原、冲沟以及落淤层五类。

(2)岩心标定测井,确定不同类别隔夹层岩性、厚度信息以及与构型界面的关系,结合S36-J1密井网区多井对比得到了各类隔夹层平剖面规模等特征,基于Petrel软件建立了能反映三维分布规律的隔夹层模型。

(3)分析了砂质辫状河内隔夹层发育控制因素,其中基准面旋回控制隔层厚度,盒八段不对称旋回结构使得隔夹层的厚度自下而上整体呈不断增大的趋势,最大可达6.7m;构型界面控制夹层分布、倾向及倾角,空间上距离相近的夹层往往存在倾向相反、倾角差异较大的情况;辫状河沉积砂体叠置样式控制隔夹层比例,孤立型砂体分布样式中隔层最为发育,而垂向叠置型夹层最发育,但夹层样式少于、规模也小于侧向切叠型。

参 考 文 献

[1] Lynds R, Hajek E. Conceptual model for predicting mudstone dimensions in sandy braided-river reservoirs[J]. AAPG Bulletin,2006,90(8):1273-1288.

[2] Best J, Woodward J, Ashworth P, et al. Bar-top hollows: A new element in the architecture of sandy braided rivers[J]. Sedimentary Geology,2006,22(5):241-255.

[3] Robinson J W, McCabe P J. Sandstone-body and shale-body dimensions in a braided fluvial system: Salt wash sandstone member (Morrison Formation), Garfield County, Utah[J]. AAPG Bulletin, 1997, 81(8):1267-1291.

[4] Ian A L, Gregory H S S, James L B, et al. Deposits of the sandy braided South Saskatchewan River: Implications for the use of modern analogs in reconstructing channel dimensions in reservoir characterization[J]. AAPG Bulletin,2013,97(4):553-576.

[5] Skelly R L, Bristow C S, Ethridge F G. Architecture of channel-belt deposits in an aggrading shallow sandbed braided river: the lower Niobrara River, northeast Nebraska[J]. Sedimentary Geology,2003,158(3):249-270.

[6] Hooke J. Coarse sediment connectivity in river channel systems: A conceptual framework and methodology[J]. Geomorphology,2003,56(1):79-94.

[7] Sambrook S G H, Ashworth P J, Best J L, et al. The sedimentology and alluvial architecture of the sandy braided South Saskatchewan River, Canada[J]. Sedimentology,2006,53(2):413-434.

[8] 马志欣,付斌,王文胜,等. 基于层次分析的辫状河储层水平井地质导向策略[J]. 天然气地球科学,2016,27(8):1380-1387.

[9] Best J L, Ashworth P J, Bristow C S, et al. Three-dimensional sedimentary architecture of a large, mid-channel sand braid bar, Jamuna river, Bangladesh[J]. Journal of Sedimentary Research, 2003,73(4):516-530.

[10] 牛博,高兴军,赵应成,等. 古辫状河心滩坝内部构型表征与建模——以大庆油田萨中密井网区为例[J].石油学报,2015,36(1):89-100.

[11] 袁新涛,吴向红,张新征,等. 苏丹Fula油田辫状河储层内夹层沉积成因及井间预测[J]. 中国石油大学学报(自然科学版),2013,37(1):8-12.

[12] 孙天建,穆龙新,赵国良. 砂质辫状河储集层隔夹层类型及其表征方法——以苏丹穆格莱特盆地Hegli油田为例[J]. 石油勘探与开发,2014,41(1):112-120.

[13] 罗超,罗水亮,贾爱林,等. 扶新隆起带东缘泉三段储层构型差异[J]. 中南大学学报(自然科学版),2016,47(5):1637-1648.

[14] 郭智,贾爱林,何东博,等. 鄂尔多斯盆地苏里格气田辫状河体系带特征[J]. 石油与天然气地质,2016, 37(2):197-204.

[15] 罗超,贾爱林,郭建林,等. 苏里格气田有效储层解析与水平井长度优化[J]. 天然气工业,2016, 36(3):41-48.

[16] 张吉,侯科锋,李浮萍,等. 基于储层地质知识库约束的致密砂岩气藏储量评价——以鄂尔多斯盆地苏里格气田苏14区块为例[J]. 天然气地球科学,2017,28(9):1322-1329.

[17] Leclair S F, Bridge J S. Interpreting the height of dunes and paleochannel depths from the thickness of medium scale sets of cross strata// AAPG Annual Meeting Expended Abstracts: AAPG,1999,80.

[18] Leclair S F, Bridge J S. Quantitative interpretation of sedimentary structures formed by river dunes[J]. Journal of Sedimentary Research, 2001, 71(5):713-716.

[19] 于兴河,马兴祥,穆龙新,等. 辫状河储层地质模式及层次界面分析[M]. 北京:石油工业出版社,2004.

[20] 蔺宏斌,侯明才,陈洪德,等. 鄂尔多斯盆地苏里格气田北部下二叠统山1段和盒8段物源分析及其地质意义[J]. 地质通报,2009,28(4):483-492.

[21] 肖建新,孙粉锦,何乃祥,等. 鄂尔多斯盆地二叠系山西组及下石盒子组盒8段南北物源沉积汇水区与古地理[J]. 古地理学报,2008,10(4):341-354.

[22] 田景春,吴琦,王峰,等. 鄂尔多斯盆地下石盒子组盒8段储集砂体发育控制因素及沉积模式研究[J]. 岩石学报,2011,27(8):2403-2412.

[23] 李易隆,贾爱林,何东博. 致密砂岩有效储层形成的控制因素[J]. 石油学报,2013,34(1):71-82.

[24] Miall A D. The geology of fluvial deposits: sedimentary facies, basin analysis and petroleum geology [M]. Springer-Verlag Inc., Berlin, 1996, 581-583.

[25] Miall A D. Reconstructing the architecture and sequence stratigraphy of the preserved fluvial record as a tool for reservoir development: A reality check [J]. AAPG Bulletin, 2006, 90(7): 989-1002.

[26] 余宽宏,金振奎,高白水,等. 赣江南昌段江心洲沉积特征[J]. 现代地质,2015,29(1):89-96.

[27] Bridge J S, Tye R S. Interpreting the dimensions of ancient fluvial channel bars, channels, and channel belts from wireline-logs and cores[J]. AAPG Bulletin,2000,84(8):1205-1228.

[28] 杨少春,赵晓东,钟思瑛,等. 辫状河心滩内部非均质性及对剩余油分布的影响[J]. 中南大学学报(自然科学版),2015,46(3):1066-1074.

[29] 印森林,吴胜和,许长福,等. 砂砾质辫状河沉积露头渗流地质差异分析——以准噶尔盆地西北缘三叠系克上组露头为例[J]. 中国矿业大学学报,2014,43(2):286-293.

原文刊于《天然气地球科学》,2019,30(9):1272-1285.

苏里格气田气井压力前缘到达渗流边界时间的判断

杨炳秀[1] 何东博[2] 王丽娟[2] 孟德伟[2]

(1. 中国石油勘探与生产公司,北京 100007;2. 中国石油勘探开发研究院,北京 100083)

摘要:致密砂岩气藏储层物性差,流体流动阻力大、渗流速度低,压力前缘经过较长时间才能到达气藏边界。在进行气井生产指标评价时,采用压力前缘到达边界前和到达边界后的生产动态资料所得到的结果存在较大差异。准确判断气井压力前缘传导到边界的时间对评价气井生产指标尤为重要。以中国苏里格致密砂岩气田为例,应用流线模拟技术、影响半径计算公式、产量不稳定分析法和气井动储量图版等四种方法进行了压力前缘到达边界时间的判断,并对判断结果进行了综合分析,得到该气田气井的压力前缘到达边界的时间为 200~300 天。该结果较好地解释了随着生产时间的延长,气井生产指标发生变化的原因,为科学合理评价气井生产指标提供了指导。

关键词:致密砂岩;气井;压力前缘;边界;时间

0 前言

致密砂岩气是目前我国开发规模最大的非常规天然气,在天然气产量中所占比例较大[1-2]。中国苏里格致密砂岩气田近年来得到了规模化有效开发,已达到年产 $230\times10^8 m^3$ 的生产规模,成为我国规模最大的天然气田[3-4]。

致密砂岩气藏储层物性差,储层非均质性强[5],气体流动阻力大,渗流速度低,经过较长时间,气井的压力前缘才能到达边界,储集体的整体特征才能通过生产动态体现出来[6]。压力前缘到达边界的时间判断,对科学合理地评价气井的生产指标较为重要。本文以中国苏里格致密砂岩气田为例,对气井压力前缘到达边界的时间进行判断。

1 气田概况

气田的构造位于属于鄂尔多斯盆地伊陕斜坡,主要产层为二叠系盒八段—S_1 段。气藏埋深为 3000~3600m,孔隙度主要分布在 3%~12%之间,平均 9%,平均覆压渗透率 0.07mD[7]。有效砂体规模小,单个砂体厚度为 2~5m,累计有效储层厚度约 10m,大小主要为 600m×800m,有效砂体呈孤立状分布在致密储层中[8]。

2002 年气田开始投产,2006 年开始规模建产,经过八年的持续开发,完成建产任务,年产能达到 $250\times10^8 m^3$,进入稳产阶段。由于储层物性差,非均质性强,气井控制储量和产能较低[9]。直井控制储量主要集中在 1000×10^4~$3500\times10^4 m^3$,单井配产 $1\times10^4 m^3/d$[10-11]。

2 压力前缘到达边界时间的判断

致密砂岩气藏的储层物性差,气体流动速度慢,同时储层非均质性强,内部发育细小的夹层,阻碍气体流动,导致气体流动前缘到达渗流边界的时间较长。在流动前缘到达渗流边界之前,随着生产时间的延长,气井控制储量会逐渐增加;当流动前缘到达渗流边界以后,气井控制储量保持不变。所以在流动前缘到达渗流边界前后,气井生产的评价指标会有较大差异。下面以苏里格致密砂岩气田为例,通过流线模型模拟、影响半径计算公式、产量不稳定分析法和气井动储量变化图板等四种方法计算压力前缘到达储层边界的时间。

2.1 流线模拟

流线模拟技术在早期流管方法基础上发展而来,它可以真实地反映出流体的实际运移路线,而在传统有限差分方法模拟中,流体沿着网格流动,网格尺度越大,流体流动轨迹与真实运移路线偏差越大。流线模拟技术将三维模型转化为一维流线模型,沿着压力梯度的方向,形成流体流动的流线。对于注入井向生产井追踪流线,通过流线分布可以清楚地显示油水驱替前缘的变化,同样对于存在渗透阻挡层和流动边界的流动单元,也可以清楚地看到渗透阻挡层和边界对气井生产动态的影响,并量化压力前缘到达渗透阻挡层和边界的时间。

本文选用比较成熟的数值模拟软件 Eclipse 的 Frontsim 流线模块设计了概念模型,依据苏里格致密砂岩气田的储层参数,设定了模型基本参数,具体如下:气水两相均质模型,网格大小 5m×5m×10m,网格数 240×240×1,孔隙度 0.09,渗透率 0.07mD,含气饱和度 0.7,束缚水饱和度 0.3。模型中间部署一口气井,气井采用定井底流压生产,模拟结果见图1和表1。

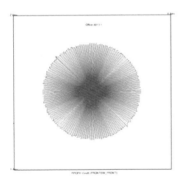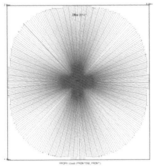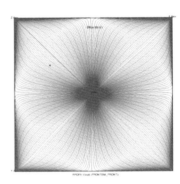

图1 压力前缘逐渐到达边界的流线模拟图

表1 影响半径与生产时间的关系表

开井时间(d)	影响半径(m)
2.6	50
10.3	100
35.8	200
56.3	300
198.7	400
298.3	500
395.8	600

从图 1 可以看出,流线直观地显示了压力前缘的变化过程,也将边界对流体流动的影响可视化。根据表 1 模拟统计结果,对渗透率为 0.07mD,半径为 300~400m 的均质储层,压力前缘完全到达边界的时间为 50~200 天。由于模型是理想的均质模型,模型计算结果要比实际储层的压力前缘到边界的时间要短。所以苏里格气田气井的压力前缘到达边界的时间要大于区间 50~200 天。

2.2 影响半径公式计算

对于一口气井生产后,井底压力开始下降,压力降从井底逐渐向周围扩展,形成一个压降漏斗,随着时间的推移,压降漏斗不断扩大。根据影响半径的定义,在 t 时刻,压降漏斗的边界扩展到半径 r 的位置,半径 r 以内的这些地层受到生产井的扰动,压力发生了变化,半径 r 以外的地层压力未发生变化[12]。

影响半径的公式:

$$r = 0.12\sqrt{\frac{Kt}{\mu \phi C_t}}$$

依据气田的储层物性,$K=0.07\text{mD}$,$\mu=0.023\text{mPa}\cdot\text{s}$,$\phi=0.09$,$C_t=0.02\text{MPa}^{-1}$ 时,代入公式,得到影响半径与时间的关系(表 2)。气田有效砂体的大小主要为 600m×800m,根据表 2,压力传导到边界的时间大致在 85~273 天之间。

表 2 应用影响半径公式计算的影响半径与开井时间的关系表

开井时间(d)	影响半径(m)
4.3	50
17.1	100
68.5	200
85.2	300
273.8	400
427.8	500
616.1	600

2.3 产量不稳定分析法

产量不稳定分析是基于试井理论,利用动态资料评价气井的控制储量和泄气面积。该方法考虑裂缝长度、表皮系数、渗流边界等一系列参数建立解析模型,利用井的生产动态历史数据和储层的基本地质参数进行拟合,使模型计算结果与井的实际生产动态和动储量一致,进而可以确定气井的泄气半径和控制储量。同时,依据拟合模型,可预测气井今后的生产动态,另外根据拟合图版,可以间接得到气体流动进入边界控制流的时间。

气井日产量 Blasingame 典型曲线拟合图见图 2。为了分析气体流动到达边界控制流的时间,选了生产较稳定的 126 口典型气井,通过 Blasingame 典型曲线图版拟合气井的生产动态,得到气井泄流半径和控制储量,同时根据拟合图版读出边界控制流的无因次时间 t_{cD},根据无因次物质平衡时间的定义[13]: $t_{cD}=\dfrac{K}{\phi\mu C_t A}t_c$, 可以求出物质平衡时间 t_c。物质平衡时间的定义: $t_c=\dfrac{Q}{q}$, 即累计产量与当前产量的比值, 图 3 标明了其几何意义, 即建立了变产量生产与定产量生产之间的等效关系。根据物质平衡时间的定义及几何意义, 可以求出对应的开始出现边界控制流的时间。评价 126 口气井发生边界控制流的时间主要集中在 150~300 天。

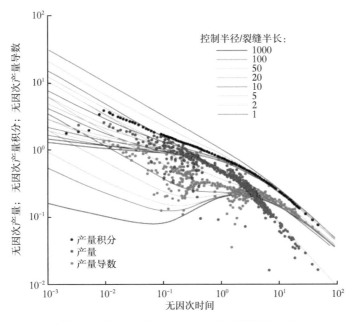

图 2　气井日产量 Blasingame 典型曲线拟合图

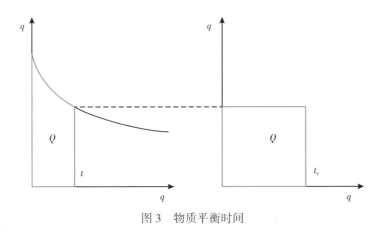

图 3　物质平衡时间

气井开始投产时,压降漏斗的范围不断扩大,当压力前缘到达边界后,流动状态逐渐转化成边界控制流。故压力前缘到达边界的时间会稍早于边界控制流的时间。故 126 口气井的压力前缘到达边界的时间稍早于时间段 150~300 天。

2.4 气井动储量图版

为了分析气井控制储量随时间的变化,选择230口生产较稳定的典型气井,分别应用30天、60天、90天、120天、150天、200天、300天、400天、600天、1000天和1500天的生产动态,选用RTA软件进行拟合分析,拟合不同生产时间的历史动态,得到不同生产时间的控制储量。

气井投产之后,压力逐渐传播到边界,在压力前缘到达边界之前,气井的动储量逐渐增加,当压力传播到边界后,流体进入边界控制流阶段,气井的控制储量保持不变[14]。从图4可以看出,在310天以前,随着生产时间的延长,计算的动储量逐渐增加;超过310天之后,动储量基本保持不变。图4也反映了压力传导过程中气井动储量的变化,所以从图4可以判断,压力前缘到达边界的时间在310天左右。

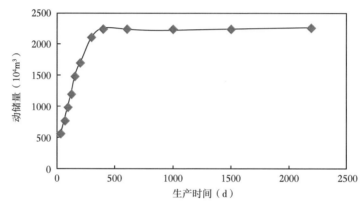

图4 应用不同生产时间的生产动态计算的动储量

3 结论

(1)应用流线模拟技术、影响半径计算公式、产量不稳定分析法和气井动储量图版等四种方法计算,结果表明,某气田气井投产后,压力前缘到达边界的时间大致为200~300天。

(2)气井生产指标评价应选择投产时间超过300天的气井进行,否则生产动态评价的生产指标会偏低。

(3)四种评价方法中流线模拟技术得出的气体到达渗流边界的时间偏小;影响半径计算公式法得出的气体到达渗流边界的时间稍偏小;产量不稳定分析法和气井动储量模板分析法,评价结果更具客观性和科学性。

参 考 文 献

[1] 黄鑫,董秀成,肖春跃,等.非常规油气勘探开发现状及发展前景[J].天然气与石油,2012,30(6):38-41.

[2] 徐博.2020年前中国多气源供应格局展望[J].2012,天然气工业,32(8),53-55.

[3] 李海平,贾爱林,何东博,等.中国石油的天然气开发技术进展及展望[J].天然气工业,2010,30(1),5-7.

[4] 张磊,张文强,李晶,等. 苏6与苏36-11区块水平井井位优选[J]. 天然气与石油,2013,31(3):61-65.
[5] 李建忠,郭斌程,郑民,等. 中国致密砂岩气主要类型、地质特征与资源潜力[J]. 天然气地球科学,2012,23(6):607-615.
[6] 王记俊,廖新武,赵秀娟,等. 注采井网对裂缝形态及压裂规模的影响[J]. 天然气与石油,2014,32(5):49-51.
[7] 王少飞,安文宏,陈鹏,等. 苏里格气田致密气藏特征与开发技术[J]. 天然气地球科学,2013,24(1):138-145.
[8] 余淑明,刘艳侠,武力超,等. 低渗透气藏水平井开发技术难点及攻关建议——以鄂尔多斯盆地为例[J]. 天然气工业,2013,33(1),65-69.
[9] 胡俊坤,李晓平,肖强,等. 利用生产动态资料确定气井产能方程新方法[J]. 天然气地球科学,2013,24(5):1027-1031.
[10] 王军磊,贾爱林,何东博,等. 致密气藏分段压裂水平井产量递减规律及影响因素[J]. 天然气地球科学,2014,25(2):278-285.
[11] 贾成业,姬鹏程,贾爱林,等. 低渗透砂岩气藏开发指标数值模拟预测[J]. 西南石油大学学报:自然科学版,2010,32(5):100-104.
[12] 庄惠农. 气藏动态描述和试井[M]. 北京:石油工业出版社,2004,136-141.
[13] 孙贺东. 油气井现代产量递减分析方法及应用[M]. 北京:石油工业出版社,2013,18-66.
[14] 公言杰,柳少波,姜林,等. 致密砂岩气非达西渗流规律与机制实验研究-以四川盆地须家河组为例[J]. 天然气地球科学,2014,25(6):804-809.

原文刊于《天然气与石油》,2016,34(2):40-43.

低渗透致密砂岩储层充注模拟实验及含气性变化规律
——以鄂尔多斯盆地苏里格气藏为例

徐 轩[1,2]　胡 勇[2]　邵龙义[1]　王继平[3]　陈颖丽[4]　焦春艳[2]

(1. 中国矿业大学,北京 100083;2. 中国石油勘探开发研究院廊坊院区,廊坊 065007;
3. 长庆油田苏里格研究中心,西安 710018;4. 中国石油西南油气田
分公司研究院,成都 610041)

摘要:低渗透致密气藏复杂的储层特征和充注机理是导致其气藏压力与含气性与常规砂岩迥异的关键因素。选取苏里格气藏 42 块岩心,开展气藏充注模拟实验,实验中考虑温度影响,逐级增加充注动力,模拟储层低速缓慢充注过程。系统研究了储层物性、充注动力、充注温度三大关键因素对充注成藏的影响,分析了充注过程中气、水产出机理,提出了含气性与储层物性、充注动力为正相关的对数函数关系。研究表明:(1)低渗透致密储层存在充注门限压力,渗透率 0.10mD 和 0.01mD 的储层,充注门限压力达 2.0MPa 和 10.0MPa。开始进气后,源储压力存在平衡过程,储层越致密,门限压力越高,平衡过程越慢,最终平衡压力越高。(2)门限压力和平衡压力是储层含气性变化的拐点,将充注过程划分为三个不同阶段:临界进气阶段、快速增长阶段和平缓增长阶段。不同阶段含气性上升速度和幅度不同,孔隙出水部位不同。(3)相同储层及充注动力时,提高充注温度,含气性可上升 5% ~ 10%。

关键词:砂岩储层;充注机理;充注物理模拟;源储压力;含气性;苏里格气藏

气藏压力和含气饱和度特征不仅决定了气藏储量丰度,同时对开发过程中储量动用的难易程度具有重要影响[1-4]。大量勘探开发实践及理论研究均表明低渗透致密气藏复杂的储层特征和充注机理是导致其气藏压力与含气性与常规砂岩迥异的关键因素[5-10]。因此,研究此类气藏充注成藏机理,明确充注中储层压力和含气饱和度的主控因素及变化规律,对气藏勘探与开发意义重大。大量学者对此类油气藏充注机理展开了研究工作,获得了较为深入的认识,充注模式研究表明致密储层在成藏过程中对气源压力和饱和度要求较高,存在临界条件;储层非达西渗流对充注存在影响,充注过程中油气饱和度增长过程存在跳跃变化等。然而在实验方法和理论认识上并不统一,一方面,在模拟方法或实验条件上,研究者往往各有侧重,难以完全模拟实际充注条件;另一方面,在相关认识,如储层物性及充注压力范围、充注过程及机理方面还存在争议[11-17]。

本研究以苏里格气藏为研究对象,针对气藏充注过程设计了半封闭充注模拟实验,充分考虑温度影响并最大限度模拟低速注气、缓慢充注过程,同时结合压汞测试和核磁共振分析,系统研究了储层物性、充注动力、充注温度三大关键因素对充注过程的影响,分析了充注过程中气、水移动机理,并绘制了不同物性储层的源储压力、含气性变过程及相应关系图版。利用建

立的相互关系,有助于根据气藏储层物性或压力参数对含气、含水饱和度和含气丰度等进行初步评价,从而为储量预测和有效开发提供理论依据。

1 研究区成藏及储层特征

1.1 研究区成藏特征

苏里格气藏位于鄂尔多斯盆地伊陕斜坡西北侧,主力气层为石盒子组盒八段和山西组山一段,气层埋深为3300~3500m,厚度为5~20m,平均孔隙度在10%左右,各区块平均含气饱和度为45%~65%,气藏异常低压(压力系数0.86),气藏具有低渗、低压、低丰度的"三低"特征。前人针对该气藏成藏过程特征进行了大量研究,认为该气区具有如下特点[18-19]:

(1)储层致密先于成藏,储层孔喉细小,成藏难度加大,充注动力和物性对该区气藏的含气性均具有重要的控制作用。

(2)成藏期源灶的生气强度大,供气量较充分,烃源灶的集中性和规模性都很好,在主要成藏期,烃源岩的流体过剩压力介于13.0~22.0MPa,源储压差可达3.0~10.0MPa。

(3)储集体内部具有很强的非均质性,物性差异较大,具有明显的"甜点"区域,含气性分布受构造影响不明显,主要受砂岩的横向展布和物性变化所控制。

1.2 研究区储层物性特征

研究过程中先后对苏里格气区的六口井进行了取样分析,统计了两个主力产气层近200块岩心的孔渗数据,如图1所示。从累计分布频率图上可以看出,储层总体低渗透致密,各井间物性差异显著,非均质性较强。具体而言,六口井平均中值孔隙度分布在6.7%~9.3%,平均中值渗透率分布在0.09~0.49mD。

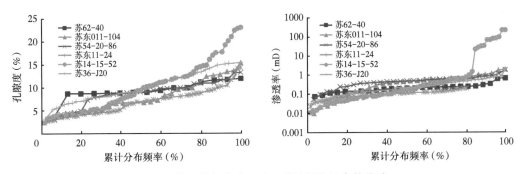

图1 苏里格气藏多口取心井储层物性参数统计

2 实验研究

2.1 岩心参数

取心结果表明,苏里格气藏储层以低渗透致密为主,兼顾苏里格气藏特殊性和研究的普遍性,从六口取心井中主要选取了低渗透致密砂岩样品和少量中高渗透砂岩样品,共计40余块进行了岩心充注实验,部分样品物性及实验参数见表1。

表1 部分样品参数及常温充注过程关键压力及含气饱和度

编号	渗透率（mD）	孔隙度（%）	岩心长度（cm）	充注门限压力（MPa）	门限压力点 S_g（%）	源储平衡压力（MPa）	源储平衡点 S_g（%）	实验结束 S_g（%）
S1	0.034	5.12	3.55	6.92	19.8	19.71	25.2	30.2
S2	0.075	5.87	3.62	5.52	11.3	10.12	24.3	29.1
S3	0.244	6.13	3.47	1.82	18.7	5.02	32.1	37.2
S4	0.505	5.76	3.57	1.01	26.8	3.15	43.3	46.9
S5	1.120	15.30	3.59	0.29	35.7	1.81	50.4	57.9
S6	1.470	10.30	3.37	0.22	24.1	3.03	44.2	48.1
S7	4.770	9.20	3.64	0.13	26.3	1.41	43.1	54.1
S8	10.700	13.80	3.49	0.13	29.2	1.51	44.7	54.8
S9	38.100	17.50	3.53	0.11	35.3	2.00	48.1	59.5
S10	49.100	17.70	3.49	0.10	32.9	1.52	50.6	62.4
S11	99.400	22.40	3.39	—	32.7	1.41	48.9	64.3
S12	126.500	21.70	3.15	—	39.8	1.00	47.6	63.5

2.2 实验方法与步骤

2.2.1 实验思路与特点

针对气藏充注成藏过程和地质条件，设计尽可能符合实际的物理模拟方法和实验条件再现充注成藏过程，是揭示充注机理，获得科学、可靠认识的前提和保证。气藏充注过程是储层在上覆地层压力和温度条件下，天然气逐渐从烃源岩进入储层，缓慢驱替、取代地层水的过程。这个过程中，气源与储层间的压差作为充注的主要动力，是动态变化的，并可能存在多个变化周期。充注过程中，储层并非完全开放或封闭的系统，低渗致密储层可既作为盖层又作为储层，其作用主要取决于充注动力和气水运移阻力的大小，距离烃源岩越远充注动力越小、储层物性越差则储层越可能转变为盖层[18-19]。

针对成藏特点，考虑低速缓慢充注及温度影响，开展充注动力逐级增加的半封闭充注模拟实验，实验装置见图2。实验采用完全饱和水的岩样模拟原始储层，用带有加湿装置的气瓶模拟气源，对储层进行充注实验。实验主要特点详述如下：

（1）逐级增压，低速缓慢充注：从极低充注动力（压差0.1MPa）逐级增加至较强充注动力（30.0MPa），中间设置30个充注充注点，初期每个充注点间压力间隔仅为0.1MPa后期压力间隔逐渐增加到0.2MPa、0.5MPa、1.0MPa、5.0MPa。所有样品采用相同的气源压力序列。

（2）考虑储层半封闭条件：基于充注过程中，储层并非完全开放或封闭的系统。因此，为了模拟这种半封闭条件在夹持器末端设计了具有一定体积的储罐，整个储集罐的体积约为岩心孔隙体积的100~200倍，保证储层中流体充分运移和排出，同时又可模拟由于储层远端封闭，气藏将逐渐聚集能量压力上升的过程；

（3）考虑温度和压力条件：实验充注压力最高可达50MPa，实验分别在20℃、50℃、120℃

开展了充注实验,对比温度对气藏充注的影响。

(4)结合毛细管压力和核磁共振测试:不同压力点充注结束后开展核磁共振测试,在明确不同物性储层孔喉微观结构差异的基础上,研究气水充注中流体微观运聚过程。

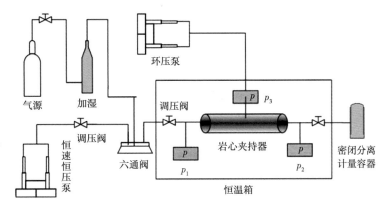

图 2　砂岩气藏充注模拟实验流程

2.2.2　实验方法与步骤

(1)将实验用岩心烘干后称干重,抽真空完全饱和地层水,称湿重,测定核磁共振 T_2 谱;

(2)将饱和水的岩心装入夹持器,对岩心加围压,模拟上覆储层压力,测量样品水相渗透率;

(3)打开恒温箱,根据需要设置不同温度条件。对于120℃高温条件,为保证水为气相,使储层中流体流动过程为两相流动,根据水的PVT属性,夹持器出口可设置回压1.0MPa。

(4)打开气源,通过调压阀设置极低的进气压力开始充注。夹持器两端设置压力传感器,记录充注过程中气源压力 p_1 和储层压力 p_2;压力 p_2 不再变化,并维持24小时以上,认为充注过程达到稳定;两端压力 $p_1=p_2$ 则认为源储压力达到平衡。充注过程中,位于储层末端的储罐储集并分离计量岩心中驱出的水。

(5)每个气源压力充注稳定24小时以上,该压力点充注结束,测试每个气源压力充注结束后岩心的核磁共振 T_2 谱,计量含水饱和度变化情况;

(6)重复步骤(3)(4)(5),从低到高逐级增加气源压力进行充注实验。

2.3　实验结果

逐级增压充注实验过程中,实时记录了不同储层气源、储层压力及含气饱和度变化过程,见图3、图4。

图3源储压力变化曲线显示:渗透率小于1.0mD的低渗透致密储层,其充注存在明显进气门限压力,即只有源储压差达到一定值以后储层才开始进气。而进气后源储压力并不能立刻平衡,而是保持一定的压力差值,表明低充注压力下低渗透致密储层并不能充分充注。随着气源压力逐渐上升,储层进气速度加快,源储压差逐渐减小,当气源压力达到一定值后源储压达到差平衡,称之为源储平衡压力。门限压力和源储平衡压力表现出一定的规律性,即储层渗透率越低,门限压力和源储平衡压力越高,表明储层越致密充注条件越苛刻。

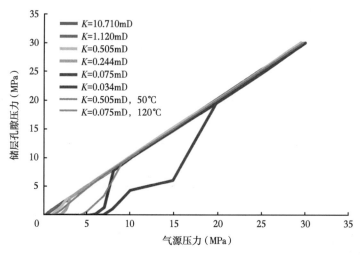

图3 充注实验中源储压力变化过程

图4为不同渗透率储层在不同压力点充注结束后的含气饱和度。含气饱和度变化曲线显示,门限压力下储层开始进气成藏,此时含气饱和度较低,随着气源压力增加,含气饱和度逐渐上升,后期上升幅度减小。

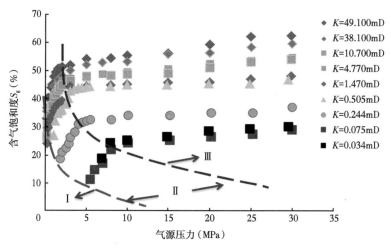

图4 充注实验中含气饱和度随气源压力变化过程

3 储层压力与含气性变化规律

图3、图4为部分储层样品充注实验结果。统计全部40余块样品实验结果进行对比研究,定量分析充注过程中压力与含气性变化规律。

3.1 源储压力变化过程

图3中源储压力变化曲线表明,低渗透、致密储层存在门限压力。统计不同渗透率,不同温度条件下储层的门限压力如图5所示。储层充注门限压力与渗透率关系密切,渗透率越低,

储层充注门限压力越高,渗透率大于1.0mD的储层,进气能力强,源储压差0.3MPa均能进气;渗透率小于1.0mD储层,随渗透率降低,充注门限压力快速升高;渗透率0.5mD储层,充注门限压力约为0.8MPa;而渗透率小于0.1mD的致密储层,充注门限压力约为2.0MPa;渗透率小于0.01mD储层,充注条件更为苛刻,门限压力达到10.0MPa以上。

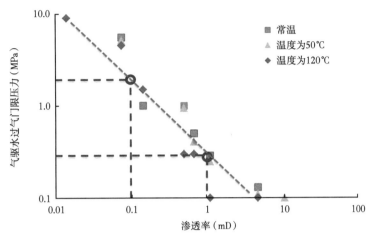

图 5 储层充注门限压力与渗透率关系

图 3 中源储压力变化过程显示,当充注压力高于门限压力以后,在较长一个时期内,储层压力都是低于充注压力的,可以称之为源储压力不平衡。这表明天然气在充注成藏过程中需要损耗一定的能量进行排水,从而充分连通天然气运聚通道。统计不同渗透率储层源储平衡压力见图6,表1。

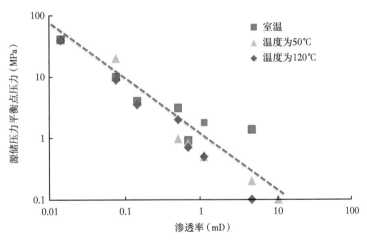

图 6 源储平衡压力与渗透率关系

源储平衡压力随渗透率变化趋势与充注门限压力一致,即储层源储平衡所需的气源压力因储层渗透率差异而不同,高渗层平衡快,平衡压力低,低渗透储层平衡慢,平衡过程长,平衡压力更高。从表1可以看出,渗透率1.12mD的储层,在较低的气源压力(1.81MPa)条件下充注即可达平衡;而渗透率为 0.505mD 和 0.034mD 的储层,分别在 3.15MPa 和 19.71MPa 时达到平衡。这说明对于渗透率小于 0.1mD 的低渗透致密储层,只有当充注动力充分、源储压差足够大时,才能充分充注成藏,并形成高孔隙压力、高储量丰度的气藏,否则当气源压力不足时,只能形成低压、低丰度气藏。

常温条件、50℃ 及 120℃ 条件下平衡压力随渗透率变化规律基本一致,压力值大小接近。

3.2 储层含气性变化过程

图 4 含气性变化过程显示含气饱和度随充注压力增加逐渐升高,但上升速度存在明显转折点。为详细说明饱和度变化过程,绘制渗透率分别为 1.120mD 和 0.075mD 的两块典型储层含气饱和度随气源压力变化过程如图 7,图中分别标识出了两块岩心的源储压力平衡点。从图中可以看出,含气饱和度曲线恰好以充注门限压力点和源储压力平衡点为拐点,分为三个阶段:

(1)临界进气阶段 I:以开始过气的门限压力点为标志,此阶段主要以排水为主,含气饱和度跳跃上升,1.120mD 岩样和 0.075mD 岩样充注门限压力点对应饱和度为 35.7% 和 11.3%。

(2)含气饱和度快速增长阶段 II:从开始进气到源储压力平衡,此阶段,含气饱和度快速上升,排水聚气,是气藏含气性增加的主要阶段,1.120mD 岩样和 0.075mD 岩样含水饱和度分别降低了 14.7% 和 13.0%。

(3)含气饱和度平缓增长阶段 III:从源储压力平衡点到实验结束。相对于前两个阶段,此阶段含气饱和度上升缓慢,幅度有限(仅为 7.5% 和 4.8%)。

图 7 同时表明,源储压力平衡并不代表着充注过程的结束,也不代表储层含气饱和度不在变化,只要气源压力持续增加,储层中的水还能继续被驱替取代,仍然有原生水转化为可动水被驱出,从而使含气饱和度继续上升。多块岩心实验的含气饱和度变化曲线均表明,源储平衡压力点是含气饱和度从快速增长过渡到缓慢增加的转折点。

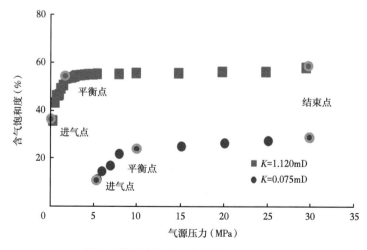

图 7 储层含气饱和度与气源压力关系

源储平衡点作为含气饱和度快速上升的结束,对于气藏充注机理而言具有重要意义。对不同渗透率储层源储平衡点含气饱和度进行统计,如图8、表1所示。渗透率越低,源储平衡点含气饱和度越低。对于渗透率小于1.0mD的低渗透、致密储层而言,这种相关性更强,随着渗透率减小,平衡点含气饱和度显著降低。

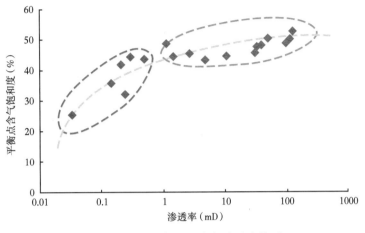

图8 平衡点含气饱和度与渗透率关系

对最大压差下充注实验结束后(压差30MPa)不同储层、不同温度条件下含气饱和度进行统计,统计结果见图9及表1。分析可知:

(1)实验结束后,在储层充注压力相同条件下,渗透率越高,充注越充分,含气饱和度越高。渗透率小于0.1mD的储层,实验结束含气饱和度仅为20%~50%;渗透率大于1.0mD储层,实验结束含气饱和度达到50%~75%。

(2)50℃和120℃温度条件下,含气饱和度随渗透率变化趋势与常温条件基本一致,但是随着温度升高,相同渗透率下储层含气饱和度上升5%~10%。表明,温度是充注成藏的有利条件,对充注具有积极作用,使气藏具有更高的含气饱和度。

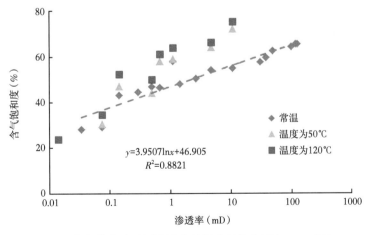

图9 充注结束后不同渗透率储层含气饱和度(30MPa时)

3.3 储层含气性变化规律图版及数学拟合

进行气藏充注实验的重要目的是明确含气饱和度主要影响因素,并获得其影响规律。大量充注实验统计数据表明,储层含气饱和度与渗透率和储层孔隙压力相关性强,可进行数学拟合。

图9显示,不同温度条件,相同储层压力下(30MPa),含气饱和度与渗透率具有较强的相关性,这种相关性可以用对数函数进行拟合,常温条件下拟合关系式为

$$S_g = 3.95 \times \ln K + 46.9 \quad (1)$$

式中　S_g——含气饱和度,%;

　　　K——储层渗透率;mD。

根据不同温度条件,式(1)中系数和常数取值不同。

统计不同渗透率储层的含气饱和度、储层孔隙压力得到 S_g—K—p 变化关系图版,见图10。从图10可以看出,不同渗透率储层其含气性与孔隙压力具有良好的相关性,符合对数关系,可用如下通式拟合:

$$S_g = C_1 \times \ln p + C_2 \quad (2)$$

式中　p——储层压力,MPa;

　　　C_1、C_2——系数,根据具体参数范围拟合得到,一旦渗透率确定则 S_g-p 关系确定。

对于0.034mD岩样,C_1、C_2取值分别为2.77和18.01,C_1、C_2而对于4.77mD岩样,取值为4.50和39.19,大多数储层,拟合相关系数均达到0.95以上。

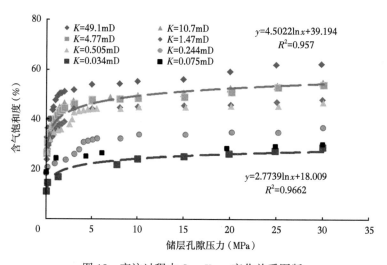

图10　充注过程中 S_g—K—p 变化关系图版

上述式(1)和式(2),揭示了含气性与储层渗透率、储层压力间均符合对数关系。基于单因素研究成果,通过数学推导,不难得出,综合考虑孔隙压力和渗透率影响时,储层含气饱和度应可统一到如下关系式

$$S_g = a \times \ln(K \cdot p) + b \quad (3)$$

式中 a、b——系数,根据具体参数范围拟合得到。

不同温度条件,式(3)中系数取值不同。若储层压力 p 或渗透率已知,则式(3)退化为式(1)或式(2)。实际应用中,一旦气藏或气区确定,成藏温度确定,则 a、b 的具体值可以确定。利用拟合得到的关系式,结合实际气区特点,加以简单校正,可以对苏里格气藏储层含气饱和度和储量进行初步、快速的评价。

4 充注成藏机理分析

在充注过程中储层含气饱和度和储层压力的宏观变化规律,对储层开展了毛细管压力曲线和核磁共振测试,分析其微观孔喉结构特征和微观出水过程。三种实验结合,进一步揭示充注过程中储层内部气水输运过程,研究充注机理。

4.1 储层微观孔喉特征分析

为对比研究不同物性储层孔隙结构差异,分别对苏里格六块代表性岩心(渗透率 0~100.0mD,详细参数见表 1)采用压汞法测试其毛细管压力曲线,统计数据见表 2,毛细管压力曲线和孔隙累计分布曲线如图 11 所示。

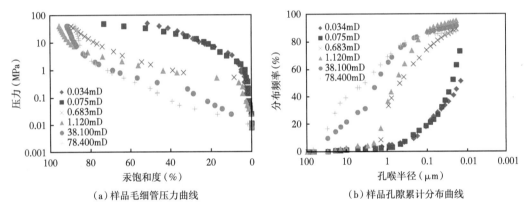

(a)样品毛细管压力曲线   (b)样品孔隙累计分布曲线

图 11 样品毛细管压力测试结果

通过毛细管压力曲线分析可知,不同渗透率级别储层其毛细管压力和孔喉结构差异显著。

渗透率小于 0.1mD 的致密砂岩:流体渗流通道以纳米孔喉(小于 0.01μm)、微毛细管孔喉(0.01~0.10μm)为主,占总孔喉的 80% 以上。这类砂岩孔喉非常细小,渗流阻力极大,排驱压力在 1.0MPa 以上,中值压力大于 10MPa。

渗透率介于 0.1mD 到 5.0mD 的常规低渗透砂岩:流体渗流通道以微毛细管孔喉(0.01~0.10μm)和毛细管孔喉(0.1~1.0μm)为主。这类砂岩渗流阻力较小,其排驱压力小于 0.5MPa,中值压力 1.0MPa 左右。

渗透率大于 5mD 的中、高渗透砂岩:流体渗流通道以毛细管孔喉(0.1~1.0μm)和超毛细管孔喉(大于 1.0μm)为主,占总孔喉的 80% 以上,微纳米孔喉占比仅为 10% 左右。这类砂岩渗流阻力非常小,无排驱压力,中值压力小于 0.5MPa。

储层微观孔喉结构的差异决定了储层中流体运移阻力的大小和流体饱和的难易程度。储层渗透率越高,孔喉尺寸越大,则驱替流体越容易进入储层,最终储层流体饱和度也就越高。在压汞实验中这种差异通过进汞排驱压力、中值压力和进汞饱和度体现出来;而在气藏充注模拟实验中,这种差异则体现为进气门限压力、源储平衡压力和含气饱和度。

表2 砂岩储层毛细管压力曲线特征统计

样品编号	渗透率 (mD)	孔隙度 (%)	排驱压力 (MPa)	最大孔喉半径 (μm)	中值压力 (MPa)	中值半径 (μm)
1	0.034	5.9	2.039	0.361	48.641	0.015
2	0.075	4.6	1.023	0.719	31.830	0.023
3	0.683	11.1	0.372	1.977	1.601	0.462
4	1.120	15.3	0.347	2.119	1.241	0.593
5	38.100	16.4	—	—	0.412	1.785
6	78.400	17.7	—	—	0.212	3.469

4.2 充注过程中核磁共振测试

利用充注过程中核磁共振测试可反映充注过程中动态的气水分布以及残余水的赋存状态,进一步研究充注成藏机理。传统核磁共振曲线横坐标为横向弛豫时间,李海波等[20]研究表明,测得横向弛豫时间 T_2 与孔隙半径 r,然后通过转化系数 C 可获得以孔隙半径为横坐标的核磁共振曲线。基于这一方法,对两块苏里格气藏代表性的样品进行测试分析,获得了1.120mD 和 0.075mD 两块样品充注实验中不同压力点的核磁共振曲线。

结合含气饱和度上升曲线(图7),选取了关键4个关键状态下的核磁曲线进行重点分析,图12a 中 0、0.29MPa、1.81MPa 和 30.00MPa 和图12b 中 0、5.52MPa、10.12MPa 和 30.00MPa 的 T_2 谱分别是样品在原始饱和水,初始过气,源储压力平衡及实验结束四个状态下的核磁共振曲线。

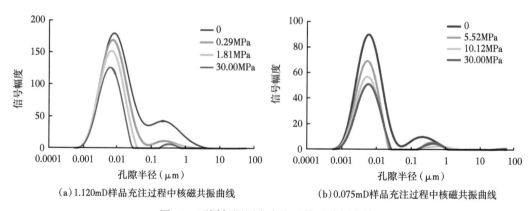

(a)1.120mD样品充注过程中核磁共振曲线　　(b)0.075mD样品充注过程中核磁共振曲线

图12 不同气源压力充注后核磁共振曲线

图中，横轴为样品孔道半径，纵轴为流体信号幅度，代表了流体量的大小。T_2 谱曲线可以反映不同驱替压力下，不同大小孔道中流体的含量。从图中可以看出，随着驱替压力的增加，样品孔道中的水被逐渐驱替出来，T_2 曲线的形态也随之发生变化。从原始 T_2 谱整体形态上看，相对于右侧致密样品，左侧低渗透率样品的 T_2 谱整体偏右且信号幅度更高，表明其微观孔喉中大喉道比例更高，平均孔道半径更大且饱和的水更多。两个样品，在不同驱替压力下，出水的部位也明显不同。如 1.120mD 样品，在 0.29MPa 初始过气压力下，主要为 0.1~1.0 μm 孔道出水，而 0.075mD 样品，初始过气压力则达到 5.52MPa，出水孔道更小，主要出水孔道主要集中在 0.01~0.10μm。

4.3 综合分析

结合储层含气饱和度与气源压力关系(图7、表1)、毛细管力测试结果(图11、表2)及 T_2 谱曲线(图12)，综合分析充注机理。分析可知，气藏充注成藏过程可划分为明显的三个阶段，在不同阶段储层可动水流动产出的孔径范围不同，含气饱和度上升幅度不同。分别为：

(1)临界进气阶段Ⅰ：以开始过气的门限压力点为标志，此阶段主要以排水为主，含气饱和度跳跃上升。从 T_2 谱可以看出，此阶段主要是岩心中相对较大的孔道先连通、出水。对于致密储层，由于其孔道以微纳米孔喉为主(大于80%)，所以压汞排驱压力更大(1.02MPa)，进气门限压力更大(5.5MPa)，充注含气饱和度更低(11.2%)。

(2)含气饱和度快速增长阶段Ⅱ：从开始进气到源储压力平衡，此阶段含气饱和度快速上升，排水聚气，是气藏含气性增加的主要阶段。从 T_2 谱可以看出，此阶段主要为次一级孔道同时也是主流喉道产水。1.120mD 岩样和 0.075mD 岩样对应喉道中值半径为 0.023μm 和 0.593μm，体现在充注过程中平衡压力分别为 1.8MPa 和 10.1MPa。

(3)含气饱和度平缓增长阶段Ⅲ：从源储压力平衡点到实验结束。相对于前两个阶段，此阶段含气饱和度上升缓慢，幅度有限，但储层压力大幅上升，对于气藏储量丰度提升显著。此阶段绝大部分孔隙内水几乎被全部驱出，仅有少量微孔道及孔道壁面薄膜水随着充注压力升高变为可动水，含气饱和度最终将趋于定值，即储层中始终有部分束缚水存在。分析原因在于：一方面，大的孔道壁面会始终吸附一定量的薄膜水，提高充注压力只会使水膜变薄，但水膜始终存在；另一方面，砂岩储层都存在一定比例的半径极小的微纳米孔喉，以及被其封堵的大孔道、盲端，而这部分空间水无法产出，也极难被充注。

5 结论

通过储层气驱水动力充注模拟实验，结合微观测试分析，系统研究了储层物性、充注动力、充注温度三大关键因素对充注成藏的影响，明确了源储压力和储层含气性变化规律，分析了充注过程中气、水产出机理，取得了以下结论与认识：

(1)针对砂岩气藏充注成藏过程，设计了考虑低速缓慢充注的半封闭充注模拟实验。实验系统研究了储层物性、充注动力、充注温度三大关键因素对充注成藏的影响。

(2)低渗透致密储层存在充注门限压力，开始进气后源储压力需要一个平衡过程。储层越致密，渗透率越低，门限压力越高，平衡过程越慢，相同压力下含气性越低。门限压力和平衡压力是储层含气性变化的拐点，将充注成藏过程划分为三个不同阶段。0、50℃、120℃ 这三种温度下实验结果表明,温度是气藏充注成藏的有利条件,随着实验温度的升高,储层含气饱和度增加 5%～10%。

(3)毛细管压力曲线和核磁共振分析研究表明,储层微观孔喉结构的差异决定了其中流体运移阻力的大小,这种运移阻力在压汞实验中表现为进汞的难易程度,而在气藏充注模拟实验中,则体现为储层进气充注的难易程度。不同充注阶段连通的孔喉级别不同,产水孔道不同。驱替压差越大、充注动力越充足,能够连通的孔喉越细小,孔道壁面附着的水膜越薄,充注最终达到的含气性和气藏压力也越高。

参 考 文 献

[1] 朱华银,徐轩,安来志,等. 致密气藏孔隙水赋存状态与流动性实验[J]. 石油学报,37(2):230-236.

[2] 夏鲁,刘震,刘静静,等. 低渗及致密砂岩含油程度动态评价方法研究:以鄂尔多斯盆地南部地区延长组为例[J]. 中国矿业大学学报,2016,45(6):1193-1203.

[3] 胡勇,李熙喆,卢祥国,等. 砂岩气藏衰竭开采过程中含水饱和度变化规律[J]. 石油勘探与开发,2014,41(6):723-726.

[4] 徐轩,胡勇,田姗姗,等. 低渗致密气藏气相启动压力梯度表征及测量[J]. 特种油气藏,2015(04):78-81.

[5] 杨华,付金华,刘新社,等. 苏里格大型致密砂岩气藏形成条件及勘探技术[J]. 石油学报,2012,33(s1):27-36.

[6] 贾承造,邹才能,李建忠,等. 中国致密油评价标准、主要类型、基本特征及资源前景[J]. 石油学报,2012,33(3):343-350.

[7] 庞正炼,陶士振,张琴,等. 致密油二次运移动力和阻力实验研究:以四川盆地中部侏罗系为例[J]. 中国矿业大学学报,2016,45(4):754-764.

[8] 徐轩,刘学伟,杨正明,等. 特低渗透砂岩大型露头模型单相渗流特征实验[J]. 石油学报,2012,33(3):453-458.

[9] 杨正明,于荣泽,苏致新,等. 特低渗透油藏非线性渗流数值模拟[J]. 石油勘探与开发,2010,29(1):94-98.

[10] 胡勇,李熙喆,万玉金,等. 致密砂岩气渗流特征物理模拟[J]. 石油勘探与开发,2013,40(5):580-584.

[11] Shanley K W, Cluff R M, Robinson J W. Factors controlling prolific gas production from low-permeability sandstone reservoirs[J]. AAPG. Bulletin,2004,88(8):1083-1121.

[12] 郑民,李建忠,吴晓智,等. 致密储集层原油充注物理模拟:以准噶尔盆地吉木萨尔凹陷二叠系芦草沟组为例[J]. 石油勘探与开发,2016,43(2):219-227.

[13] 刘震,刘静静,王伟,等. 低孔渗砂岩石油充注临界条件实验:以西峰油田为例[J]. 石油学报,2012,33(6):996-1002.

[14] Cant D. "Unconventional" hydrocarbon accumulations occur in conventional traps[C]// Canadian Society of Petroleum Geologists. Calgary, AB,2011:1-2.

[15] Zhao J Z, Zhang W Z, LI J,et al. Genesis of tight sand gas in the Ordos basin,China[J]. Organic Geochemistry,2014,74(6):76-84.

[16] 赵子龙,赵靖舟,曹磊,等. 基于充注模拟实验的致密砂岩气成藏过程分析:以鄂尔多斯盆地为例[J]. 新疆石油地质,2015,36(5):583-587.

[17] 庞雄奇,金之钧,姜振学,等. 深盆气成藏门限及其物理模拟实验[J]. 天然气地球科学,2003,14(3):207-214.

[18] 赵文智,卞从胜,徐兆辉. 苏里格气田与川中须家河组气田成藏共性与差异[J]. 石油勘探与开发, 2013,40(4):400-408.
[19] 陈占军,任战利,万单夫,等. 鄂尔多斯盆地苏里格气田上古生界气藏充注动力计算方法[J]. 石油勘探与开发,2016(5):39-44.
[20] 李海波,郭和坤,杨正明,等. 鄂尔多斯盆地陕北地区三叠系长7致密油赋存空间[J]. 石油勘探与开发,2015,42(3):396-400.

原文刊于《中国矿业大学学报》,2017,46(6):1323-1331+1339.